Plant Gene Research

Basic Knowledge and Application

Edited by

E.S. Dennis, *Canberra*

B. Hohn, *Basel*

Th. Hohn, *Basel* (Managing Editor)

P.J. King, *Basel*

F. Meins, Jr., *Basel*

J. Schell, *Köln*

D.P.S. Verma, *Columbus*

Springer-Verlag Wien New York

Genes Involved
in Plant Defense

Edited by
T. Boller and F. Meins

Springer-Verlag Wien New York

Dr. Thomas Boller
Botanisches Institut der Universität Basel
Basel

Dr. Frederick Meins, Jr.
Friedrich Miescher Institut
Basel

Printed in Germany by Konrad Triltsch, D-W-8700 Würzburg
Typeset by Macmillan India Ltd., Bangalore 25

With 34 Figures

Library of Congress Cataloging-in-Publication Data

Genes involved in plant defense / edited by T. Boller and F. Meins.
    p.   cm.—(Plant gene research, ISSN 0175-2073)
    Includes bibliographical references and index.
    ISBN 3-211-82312-3 (Springer-Verlag Wien—New York).
    ISBN 0-387-82312-3 (Springer-Verlag New York—Wien)
    1. Plants—Disease and pest resistance—Genetic aspects.
    2. Plant diseases—Genetic aspects.
    I. Boller, T. (Thomas), 1949– . II. Meins, F. (Frederick), 1942– . III. Series.
SB750.G44 1992   92-20768
632'.3—dc20   CIP

ISSN 0175-2073
ISBN 3-211-82312-3 Springer-Verlag Wien New York
ISBN 0-387-82312-3 Springer-Verlag New York Wien

Preface

Many fungi and bacteria that associate with plants are potentially harmful and can cause disease, while others enter into mutually beneficial symbioses. Co-evolution of plants with pathogenic and symbiotic microbes has lead to refined mechanisms of reciprocal recognition, defense and counter-defense. Genes in both partners determine and regulate these mechanisms. A detailed understanding of these genes provides basic biological insights as well as a starting point for developing novel methods of crop protection against pathogens. This volume deals with defense-related genes of plants and their regulation as well as with the genes of microbes involved in their interaction with plants.

Our discussion begins at the level of populations and addresses the complex interaction of plant and microbial genes in multigenic disease resistance and its significance for crop protection as compared to monogenic resistance (Chap. 1). Although monogenic disease resistance may have its problems in the practice of crop protection, it is appealing to the experimentalist: in the so-called gene-for-gene systems, single genes in the plant and in the pathogen specify the compatibility or incompatibility of an interaction providing an ideal experimental system for studying events at the molecular level (Chaps. 2 and 4). Good progress has been made in identifying viral, bacterial, and fungal genes important in virulence and host range (Chaps. 3–6). An important aspect of plant–microbe interactions is the exchange of chemical signals. Microbes can respond to chemical signals of plant origin. This is exemplified by *Agrobacterium* which recognizes plant wound substances and responds by expressing a set of virulence genes (Chap. 7). Plants, on the other hand, recognize substances of microbial origin, so-called elicitors, and these induce an array of defense genes. Chapter 8 discusses the type of elicitor molecules involved, the perception of elicitors by the plant cells, and the transduction of the elicitor signals. The identification of genes induced in plants in the response to potential pathogens and elicitors is a rapidly developing field, and techniques for gene transfer into plants are being used to establish the function of these genes and their regulation. Included in this groups are genes encoding "pathogenesis-related" proteins (Chap. 9), antifungal hydrolases (Chap. 10), thionins (Chap. 11) and key enzymes in the phenylpropanoid pathway leading to the synthesis of several types of phytoalexins (Chap. 12) and lignin (Chap. 13).

The plant's defense against microbes is a much-studied and rapidly evolving topic. We hope that this volume, with its combination of current views of epidemiologists, biochemists and molecular biologists and with its emphasis on both plant and microbial genes, will provide a useful framework for future studies. We thank the authors for their excellent contributions and stimulating treatment of this interesting field. We also give special thanks to Sue Thomas who helped us in preparing the volume.

Basel, January 1992   Thomas Boller and Frederick Meins, Jr.

Contents

Chapter 5 Pathogenicity Determinants in the Smut Fungi of Cereals
F. Banuett and I. Herskowitz, San Francisco, California, U.S.A.

Chapter 6 Identification of Fungal Genes Involved in Plant Pathogenesis and Host Range
W. Schäfer, D. Stahl, and E. Mönke, Berlin, Federal Republic of Germany

Section III *Perception of Pathogens and Signal Transduction*

Chapter 7 Interactions Between *Agrobacterium tumefaciens* and Its Host Plant Cells
S.C. Winans, Ithaca, New York, U.S.A.

Chapter 8 Elicitor Recognition and Signal Transduction
J. Ebel and D. Scheel, Freiburg and Köln, Federal Republic of Germany

Section IV *Plant Genes Induced in the Defense Reaction*

Chapter 9   **Pathogenesis-related Proteins**
J.R. Cutt and D.F. Klessig, Piscataway, New Jersey, U.S.A.

Chapter 10   **The Primary Structure of Plant Pathogenesis-related Glucanohydrolases and Their Genes**
F. Meins, Jr., J.-M. Neuhaus, C. Sperisen, and J. Ryals, Basel, Switzerland, and Research Triangle Park, North Carolina, U.S.A.

Chapter 11   **Characterization and Analysis of Thionin Genes**
F. Garcia-Olmedo, M.J. Carmona, J.J. Lopez-Fando, J.A. Fernandez, A. Castagnaro, A. Molina, C. Hernandez-Lucas, and P. Carbonero, Madrid, Spain

Chapter 12   **Regulatory Elements Controlling Developmental and Stress-induced Expression of Phenylpropanoid Genes**
J.L. Dangl, Köln, Federal Republic of Germany

Section I

Resistance and Susceptibility Genes
of Plants

Chapter 1

The Use of Resistance Genes in Breeding
Epidemiological Considerations

Martin S. Wolfe and Cesare Gessler

Institute for Plant Sciences, Swiss Federal Institute of Technology, CH-8092 Zurich, Switzerland

Contents

I. Introduction

Breeding for resistance should be the best single form of disease control, both from the ecological and economic points-of-view. However, the application of breeding for disease resistance depends on whether or not the crop in question is considered sufficiently valuable to support a breeding or selection programme, whether it is short- or long-cycled and whether the end product is consumed indirectly or directly. At one extreme are the annual cereals, the most valuable crops in the world, which have short cycles allowing several breeding generations per year and which are used indirectly, so that some degree of damage to the crop can be tolerated. With these crops, breeding for resistance often occupies more than half of the resources of breeding programmes.

At the other extreme are perennial quality crops such as coffee or apple, where the length of the crop cycle delays breeding progress and where quality is an overriding factor that implies conservatism (difficulty of introducing a new variety) and stringent market requirements (no tolerance of infection or infestation; fungicides often used intensively). Between these limits are crops such as potato or strawberry where breeding can be accelerated through 'capture' of appropriate hybrids by vegetative propagation, but quality for market is still paramount. Conversely, plantains are of low commercial value and perennial, but cultivated more for quantity than for quality.

The differences among these categories of crops are reflected in the immense success of breeding in annual crops for yield, appropriate quality and disease and pest control. Disease and pest control in perennial crops, on the other hand, has had to be achieved largely by other means, such as quarantine, sanitation and frequent application of pesticides.

In general, the cost/benefit ratio for fungicide use has been positive for the producer. In the last two decades, however, the success of chemical control has come under increasing scrutiny.

There is concern about possible side-effects of pesticides in the agricultural ecosystem and about residues in the products for consumers (for example, the use of phenyl mercury compounds and arsenate of lead against apple scab). This was partly countered by the development and use of less harmful pesticides, but, as these became more specific towards target organisms, selection of resistance in the targets became an increasing problem. This led to the development of integrated control, to try to optimise use of all available measures. Despite considerable achievements in this direction, there is still further pressure for the reduction of pesticide use on all crops, whether or not they are consumed directly. Consequently, in recent years, selection and breeding for disease resistance has been elevated to the same high priority in the majority of our more valuable crops.

To breed successfully for disease resistance requires attention to three basic questions (see also Buddenhagen, 1983) which are essentially epidemiological: (1) Against how many, and which, diseases should a new variety be resistant? (2) What is the level of resistance needed for each disease? (3) How much durability is needed for each of the resistances used, bearing in mind the likely length of life of the variety?

Following a consideration of these questions, we need then to consider the techniques available to implement the answers and the degree to which these techniques constrain the answers to the questions. Finally, a major consideration is the way in which resistant varieties are subsequently used in agriculture in relation to the maintenance of resistance.

A. Against How Many, and Which, Diseases Should a New Variety Be Resistant?

To answer this question, it is necessary to know the area that may be occupied by the variety in terms of the potential range of diseases. Moreover, if the area is large, it is necessary to know whether the objective is to produce a broadly adapted variety that may be grown across many different ecological zones, or to produce varieties that are highly adapted to more restricted localities.

The most satisfactory method of determining the disease potential of each area involved is to have survey data, or at least, extensive experience, of the range and importance of the more common diseases. Such knowledge needs to come from as long a time scale as possible, because the current or recent range of important diseases may reflect only the recent range of varieties grown and their major susceptibilities. It is clearly more difficult to breed a single variety to grow over a large area than one to be grown only locally since more different resistances will have to be packaged into a single genotype. On the other hand, if broadly adapted varieties are successful, then only a few will be needed.

For perennial quality crops, there is a need for sharp definition of priorities, including market considerations, since the task of introducing resistance even to a single disease is very considerable (Aldwinkle, 1989).

Whatever the kind of crop, it is highly likely that there will be a need to combine resistances to different diseases. A brief survey of the plant breeding literature clearly shows the importance, and difficulty, of this task and, indeed, of the difficulty of combining different resistances to the same disease. Nevertheless, in annual crops, combining resistances is regarded as a routine. In perennials, however, this is much less common (Kellerhals, 1989; Lespinasse, 1989; Alston, 1989) and is usually regarded as a need to ensure that the resistance of a new variety to secondary diseases is no less than in currently popular varieties.

Experience has shown repeatedly how, in a new variety, omission of resistance to a potentially important pathogen can lead to the demise of the variety in question. This point needs to be emphasised because of the tendency to reductionism encouraged by current emphasis on the potential for transgenic plants to control a particular disease. Sophisticated manipulation of a single gene for disease resistance may not give long-term control of the disease in question and may have no influence on other important diseases.

As far as possible, breeders obtain parental material which already has combined resistance to various diseases, from gene banks, international nurseries, other breeding programmes and existing varieties. However, as soon as these sources are hybridised with a parent chosen for other reasons, the resistance gene complex is broken up and multiple resistance screening is needed. This can be avoided in vegetatively reproduced crops by 'freezing' appropriate combinations in the F1.

For sexually reproducing crops, disturbance of a desirable gene complex can be reduced if linked genes are available. For example, in wheat and barley, there are many examples of linked genes for mildew and rust resistance. Moreover, there are several examples of alien chromosomes or chromosome segments introduced into wheat that carry resistance to more than one disease. Rye chromosome 1B substituted into the wheat genome carries four: Sr31 resistance to *Puccinia graminis*, Lr26 resistance to *P. recondita*, Yr9 resistance to *P. striiformis* and Pm8 resistance to *Erysiphe graminis* f. sp. *tritici*; Sr9g is similarly linked with Yr7, and Sr24 with Lr24 (McIntosh, 1988; Roelfs, 1988; Johnson, 1988). In barley, resistance to *Drechslera graminea* is linked with M1-La resistance to *E. graminis* f. sp. *hordei* (Haahr et al., 1989). Such linkages are relatively easy to maintain and to follow through a breeding programme.

In addition, there may be pleiotropic effects that can be exploited. For example, Pink et al. (1983) found that the short arms of the homoeologous group 5 chromosomes in wheat carry genes that appear to promote, simultaneously, resistance to yellow rust and to powdery mildew. Mathis et al. (1986) found that resistance to powdery mildew in wheat severely constrained the ability of *Fusarium culmorum* to infect the host. It is also likely that, in the presence of the diseases, mildew or rust resistant plants are less affected by soil-borne pathogens that attack the roots of cereal hosts than are plants that are susceptible to these diseases. Against these positive pleiotropic effects, the recent trend to reduced height among European wheat varieties has been partly responsible for the increase in importance of the diseases caused by *Leptosphaeria nodorum* and *Septoria tritici*, with splash-dispersed spores.

Recently, the possibility for maintaining gene complexes in sexually reproduced crops has been improved by techniques for producing haploids from F1 or later generation plants since the diploid chromosome number can then be restored easily without genetic segregation. In practice, there may be difficulties, because of accompanying male sterility and outcrossing. Rapid progress in the same direction is also being made for some perennial crops through clonal selection and the development of various regeneration techniques. An alternative approach is to take lines which already have desirable gene complexes for disease resistance, built up over many generations of field selection, and then to attempt to improve those lines agronomically with as little disturbance as possible, for example, by selecting for dwarf mutants.

Although there is a critical need to combine resistance to the major pathogens, resistance to all abiotic and biotic stresses cannot be packaged into a single genotype. Indeed, it may be undesirable to do so because a number of the characters may work against each other. For example, different resistance mechanisms are likely to be more or less energy-dependent and may thus affect yield potential; short straw may improve lodging resistance but may also increase the potential for infection by splash-dispersed spores, and so on.

B. What Is the Level of Resistance Needed for Each Disease?

Theoretically, from epidemiological considerations, it should be possible to calculate the minimum resistance level needed to prevent serious epidemic development of a particular disease. However, little information or theory is available on the relative value of different levels of resistance. In practice, moreover, because of practical constraints, the breeder is forced to take a more limited view of the possible options.

1. The Tendency to Select for Qualitative Resistance

Among all crops, the level of resistance needed to prevent epidemic development of many diseases caused by less well-adapted pathogens is often low and may be achieved by " . . . a strategy of watchful neglect" (Simmonds, 1979), for example, by selecting against excessive susceptibility. Nevertheless, on the occasions when a normally sporadic disease does occur severely, the breeder will tend to select, for safety, the highest levels of resistance available in his populations.

At the other extreme, demands for perfect quality in fruit and vegetable marketing require a level of cosmetic control of all pathogens often achievable only by integration of resistance breeding with fungicide application. The reasons for this non-tolerance of disease symptoms or their effects are partly due to consumer preference (largely led by supermarket competition) and partly due to the problem of storage and transport of products that are already infected. Consequently, breeders need to select for high levels of resistance using qualitative selection procedures.

For diseases that do not degrade the product but reduce the productivity of a tree (for example, canker of apple caused by *Nectria galligena*), a quantitative decrease in susceptibility would be acceptable. However, the level of resistance selected is determined by the parental sources available and the screening technology rather than by epidemiological and strategic considerations. Apple cultivars with a high level of resistance against *N. galligena* are needed in Europe but such resistance is unknown. Moreover, suitable screening methods to detect differences in seedling susceptibility have been developed only recently (van der Weg, 1989). The correlation between seedling reaction and adult plant reaction is not yet proven, but there are indications that it does exist.

Apples are screened for scab resistance on seedling leaves under the assumption that resistance expressed at this stage will also be evident in the fruit; this assumption has proved to be correct up to now for high levels of resistance. On the other hand, for quantitative resistance, such correlations may not hold. For example, the extent of fireblight damage caused by *Erwinia amylovora* on pome crops is strongly affected by the age and vigour of the host plants, by environmental factors especially during flowering, soil type and cultural practices. Clearly, artificial screening for resistance at the seedling stage could not take all of these factors into account. Consequently, screening for fireblight resistance can only be used for the high levels of resistance that are available. The susceptibility of commercial apple cultivars towards powdery mildew (*Podosphaera leucotricha*) ranges from unacceptable to high levels of resistance. Moderate levels of resistance are acceptable in practice, but the available screening methods differentiate these levels on seedlings only with difficulty and at great cost (Lespinasse, 1989). Consequently, high levels of resistance introduced from wild species appear to be more useful. These examples demonstrate that for perennial crops, the level of resistance to be used in selection depends principally on the availability of high levels of resistance.

For similar practical reasons with crops that are consumed indirectly, breeders tend to select for high resistance since it is distinct and easily recognised. Furthermore, even though tolerance of disease should be greater with these crops, there is a tendency to a high level of risk aversion in intensive production, so that any amount of disease is regarded as dangerous and must therefore be controlled.

Related to the tendency to select for high levels of disease control is the necessity to select for resistance that is highly heritable, which usually means that it is simply inherited. Such resistance may or may not be specific and this may be difficult to predict. Where the resistance turns out to be non-specific, it should, by definition, be durable. Where it turns out to be specific, it may lead to selection of a corresponding pathogen race adapted to the new variety. However, the rate at which this may occur is usually unpredictable; there is considerable variation in the degree of durability among different resistances, which may also vary among environments. Some programmes link breeding for specific resistance with monitoring of the pathogen population to avoid potential calamities (see Sect. III); this can be highly effective.

2. Developing Approaches to Quantitative Resistance

To try to broaden the range of resistances used in annual crops, and perhaps avoid some of the problems associated with qualitative specific resistance, there is now considerable interest in exploiting quantitative resistance (see Sect. II.C), which does not give complete disease control. Components of quantitative resistance can be selected in the glasshouse (Parlevliet, 1988), but the full effect of quantitative resistance in combination with other desirable agronomic characters can only be selected in the field. Here, a number of problems arise (Parlevliet, 1988; Buddenhagen, 1983).

(*i*) First is the question of inter-plant and inter-plot interference, particularly in the early generations of selection of pure lines of self-pollinated crops. At this stage, the crop is in the form of a mixed population so that mechanisms operate that tend to reduce infection of susceptible plants and to increase infection of more resistant plants (Chin and Wolfe, 1984). Control varieties whose field performance is known, are therefore needed to provide appropriate comparisons. Also at these early stages of selection, there will be errors in selection due to heterozygosity.

(*ii*) Timing of evaluation of the field material is critical, not only from the point of view of maximising the differences in disease response among lines, but also in relation to the problem of variation in earliness and maturity influencing the range of lines selected.

(*iii*) A third group of potential errors in selection arises from the choice of pathogen material used for testing. This can never be fully representative of the range of pathogen variation currently in existence or that might develop in the future.

Despite these difficulties, it is evident that levels of quantitative resistance that appear to be relatively low in breeders trials because of the continuous impact of incoming inoculum from nearby susceptible lines, may be perfectly adequate in large-scale field use. For example, the wheat variety Cappelle-Desprez was almost eliminated from official consideration in Britain because of its susceptibility to *Puccinia striiformis* in early field trials. Later, as the variety gained popularity and occupied more than 60% of the wheat area during the 1960's, infection with yellow rust was rarely of concern.

In practice, attempts to introduce quantitative resistance into breeding programmes are no doubt widening the range of resistance genes that are being exploited. Nevertheless, the breeders' concern to select for high levels of qualitative resistance that are highly heritable, means that the genes that are used may not differ greatly from the qualitative resistances outlined above. Moreover, there is no improvement in the prediction of the durability of the resistance, unless it can be shown that the selected resistance is identical with that of a durably resistant parent.

C. How Much Durability Is Needed for Each of the Resistances Used, Bearing in Mind the Likely Length of Life of the Variety?

Essential to this question is the definition of durability and a consideration of the best approach to handle the problem. To define durability, we follow Johnson (1981, 1988), in the sense that durable resistance is that resistance which is not overcome despite large-scale exposure over a long period to a pathogen, under conditions favourable for pathogen development. This problem is most acute for biotrophic pathogens that grow on the exposed parts of plants and that are easily dispersed.

Thus, durability of resistance cannot be quantified and varies greatly depending on the nature of the resistance, the environment and the current status of the pathogen population; whether or not a particular resistance is accepted as durable depends on experience. Furthermore, it is not possible to define durable resistance in terms of form or mechanism. For example, some researchers regard quantitative resistance as durable and qualitative resistance as ephemeral. Others say that rate-reducing resistance is durable, whereas resistance preventing infection is not and that polygenically controlled resistance is durable, whereas monogenic resistance is not. However, there are numerous exceptions to each of these propositions (see Johnson, 1988) and durable resistance, though clearly evident, remains as probably the most important unexplained phenomenon in plant pathology.

All that we can say at present is that resistance that is invoked by some form of signal interaction between host and pathogen may be ephemeral because of the possible selection of genes that interfere with the signal. On the other hand, constitutive resistance due to the inability of the host to provide some essential need for the pathogen, as, for example, in age resistance, is likely to be durable.

One proposition to improve the durability of specific resistance is to combine, or pyramid, resistances in a single genotype. Apart from the technical difficulty involved, the method depends on the assumption that none of the genes involved have been previously exposed to the pathogen in the area involved. It also depends on the assumption that a pathogen race able to overcome the first acting gene in the pyramid is not able to establish a population on the host sufficiently large to allow selection of mutants able to overcome the second gene, and so on. If these assumptions are not fulfilled, the pyramid is unlikely to be significantly more durable than the components.

One consequence of our inability to recognise characters involved in durability is that sources of durable resistance to important pathogens have emerged or have been recognised by accident; they have not been selected by design. This is well-illustrated in the cereal rusts where durable resistance is now recognised against stem, leaf and yellow rust of wheat. In all three examples, the original donor cross was made many years ago and it is only in the long-term performance of these sources compared with other and more recently introduced resistances that they have filtered through as consistent and durable. Against stem rust, the single genes Sr2, Sr26, Sr31, and Sr36, can all be described as providing durable resistance. However, Sr2 is probably the most widely used since it is now in many of the widely grown CIMMYT wheats. It was first introduced from Yaroslav emmer wheat into the hexaploid varieties Hope and H44 in the 1920s (Roelfs, 1988).

Against leaf rust, durable resistance is due to a combination of genes associated with Lr13, Lr12, and Lr34. This complex came from South American land races around 1920.

Against yellow rust, a number of durably resistant varieties are recognised (Johnson, 1988), including the French wheat, Cappelle-Desprez. Genetical and pathological analyses of Cappelle revealed that the variety has race-specific and apparently non-race-specific resistance genes, combined with genes on the short arms of the group 5 chromosomes that enhance resistance and genes on the long arms of the same chromosomes that promote susceptibility. Although there is considerable knowledge of the many genes on different chromosomes with positive and negative effects on the rust phenotype of the variety, it is not known whether durability is due to the whole complex of resistance-affecting genes or to specific parts of the complex. It is important to point out, incidentally, that the long-term success of Cappelle was due not only to the durability of resistance against yellow rust but also to durable resistance against eyespot (*Pseudocercosporella herpotrichoides*) and only moderate susceptibility to other potentially important diseases.

More recently, durable host resistance has been recognised for the first time against powdery mildew of barley, due to the resistance gene, *mlo*. Previously, probably all of the resistances exploited consciously by breeders against this disease have been inherited as single genes. They were ephemeral and typified by some form of hypersensitive reaction to the pathogen. The resistance due to *mlo* is also monogenic but it has a different mechanism, affecting the rate of papilla development in the host and so restricting entry of the pathogen into the epidermis (Stolzenburg et al., 1984a, b).

For perennials, it is essential that disease resistance should be durable, because of the size of the investment involved in breeding and introducing new cultivars. The use of polygenic resistance due to the additive effects of independent genes or groups of genes that are easily separated by crossing-over, is severely hampered by erosion during backcrosses with susceptible parents. Consequently, for this type of resistance, only recurrent selection among generally acceptable parents is possible. Resistance from inferior cultivars or wild species is valuable only if it does not erode and it can be followed easily in small populations of segregating generations. Dayton (1977) stated the general opinion: "For greatest breeding efficiency the character should be dominant, simply inherited, and stable."

Resistance to powdery mildew among commercial apple cultivars is insufficient so breeders have had to consider wild species as sources of resistance (Knight and Alston, 1969). In several examples, the resistance disappeared under field conditions or when tested in similar crosses at different places. For example, P11 resistance from *Malus robusta* was transmitted to half of the progeny of a cross at Angers, France, but, in similar crosses, none of the progeny were resistant in Geneva, New York (Lespinasse, 1989). To further confuse the situation, selections carrying P11 or P12 (from *M. zumi*) from crosses with susceptible parents gave an acceptable proportion of resistant progeny only if they carried additional major and minor modifying factors (Watkins, 1974).

The commercial apple cultivars, Cox and Golden Delicious, are considered mildew susceptible in Europe, but are classified as moderately resistant in Argentina (Alston, 1969). Siebs (1959) inoculated different commercial cultivars generally regarded as susceptible with pathogen cultures originating from those hosts and was able to detect several physiological races after cross-infecting the cultivars. Thus, cultivars generally regarded as susceptible had specific resistances which were ineffective against the current field inoculum. Similarly, Gravensteiner, generally considered as moderately susceptible (Alston, 1969; Aldwinckle, 1974) was susceptible or resistant to different strains of the mildew pathogen from Jonathan. It seems that two forms of resistance to mildew in apple can be recognised. On the one hand, there is immunity from *M. robusta* and *M. zumi*, both of which may be specific and ephemeral. On the other, there are differences in quantitative resistance which leads to similar ranking of a range of cultivars worldwide and which may therefore be horizontal and perhaps polygenic (Alston, 1989; Lespinasse, 1989).

Resistance of apple to scab (*Venturia inaequalis*) is also complex. Vertical resistance introduced from *Malus micromalus* and *M. atrosanguinea* is overcome by race 5 of the pathogen and is probably inherited as a single dominant gene. Some genes expressed as quantitative resistance in many commercial cultivars are overcome by commonly occurring races of the scab pathogen. Others that cause minor differences in susceptibility among cultivars are insufficient to restrict scab epidemics and their effects have to be supplemented by fungicides. Resistance introduced from *M. floribunda* 821 was first thought to be due to a single dominant gene but is now recognised as being more complex involving modifiers that may change the expression of resistance from almost undetectable to complete. This resistance appears to be durable and similar in action to the ontogenetic durable resistance present in all commercial cultivars.

Until more precise knowledge emerges, the only safe approach to breeding for durable resistance is to further exploit known sources (Johnson, 1988). Even then, the breeder must prepare for possible shocks. For example, so-called Type A resistance to cabbage wilt (*Fusarium oxysporum* f. sp. *conglutinans*) has been effective for more than 50 years in the cabbage-growing areas of the mid-West of the U.S.A. Recently, however, Ramirez-Villupadua et al., (1985) found a race that overcomes Type A resistance. Curiously, but fortunately, the race was found in California in an area where cabbage production is unimportant.

II. Current Techniques for Selecting for Disease Resistance

In attempting to follow resistance genes through a breeding programme, the breeder is faced with the problem of deciding among the methods available, their long-term effectiveness, and the trade-off involved in terms of the time and facilities needed which might be used for selecting other characters. Indeed, traditionally, selection for resistance has been empirical based simply on the lack, or a reduction, of disease. The continuous introduction of new and improved techniques should reduce the work and time needed and gradually allow an escape from empiricism to a more rational approach. For convenience, we divide our comments on methods into laboratory, glasshouse (or growth chamber), and field, to briefly consider some of the epidemiological implications.

A. Laboratory

A potentially powerful approach for recognising the presence of disease resistance genes without the use of disease testing is already underway through the use of isozyme and RFLP (restriction fragment length polymorphism) markers. Very recently, an exciting additional method (RAPD markers: random amplified polymorphic DNA) has become available that utilises the polymerase chain reaction (PCR) to amplify short pieces of DNA that occur between marker sequences of, say, 10 base pairs recognised by randomly synthesised oligonucleotides (Williams et al., 1990). It is likely that this new technique will provide a powerful supplement to the RFLP approach. Both methods are also beginning to be of great value in studies of the population genetics of plant pathogens (Michelmore and Hulbert, 1987; O'Dell et al., 1989; Brown et al., 1990).

Genomes of the more valuable crops (wheat, rice, maize) are being 'saturated' with markers such that, theoretically at least, different disease resistance genes could be assembled in a single genotype by checking for the presence of the markers without recourse to disease screening. However, this depends on knowing which genes are important and where they are, and also on having tight linkage of the resistance gene with an appropriate marker, or preferably two flanking markers (Stuber, 1989). The ideal example is that of resistance to *Pseudocercosporella herpotrichoides* in wheat derived from *Aegilops ventricosa*, in which the single gene for resistance on wheat chromosome 7D (Worland et al., 1988) probably covers an endopeptidase locus (McMillin et al., 1986; Koebner et al., 1988), such that disease resistance can be selected reliably by testing for the isozyme product, using only part of a single grain.

Thus there are considerable advantages for molecular markers in allowing combinations of alleles for resistance to the same and to different diseases to be combined in a single host genotype with only one laboratory screen, avoiding the uncertainties associated particularly with field tests. Moreover, molecular markers are plentiful, probably lacking any effect on the host, and co-dominant, which means that homozygotes and heterozygotes can be easily separated. They can also be identified in any form of plant tissue, whether in cell culture or from whole plants. Consequently, there is an even greater gain to be considered by combining this technique with the use of doubled haploids to establish a desired homozygous genotype from the earliest possible stage.

Abundantly available RFLP markers could also be used for tracking quantitative resistance characters, assuming that the latter exist in the form of definable QTLs (quantitative trait loci).

A further advantage of the marker techniques lies in the possibility for minimising the amount of unwanted DNA transferred with an alien resistance. For example, since all sources of resistance against scab and mildew in apple come from wild *Malus* species, it is necessary to backcross to a recurrent susceptible parent to restore a genome with desired commercial qualities. Minimum generation time for apple is four years and it needs five backcross generations to reduce the wild parent genome to less than 5%. DNA from the wild parent that is independent from the target resistance gene(s) diminishes on average by a factor of two in each recurrent backcross generation with an individual variation of unknown magnitude. On the other hand, removal of linked segments occurs only through random and unpredictable cross-overs so that little reduction of unwanted DNA may occur (Stamm and Zeven, 1981; Young and Tanksley, 1989). Linkage drag in which undesired traits are carried along with the resistance is a major problem in the breeding process. To reduce the time needed to eliminate the unlinked wild genome and to ensure that the resistance-linked drag of the wild genome is minimal would be a considerable advantage in breeding disease-resistant perennials. An example of how to estimate the size of linked DNA around resistance genes is given by Young and Tanksley (1989). Using RFLPs as markers, it was possible to estimate the size of chromosomal segments retained around the Tm-2 locus of tomato during backcross breeding. Thus with RFLP mapping of genomes a technique is available to identify chromosome segments and their origins and so to rapidly and effectively overcome linkage drag by selecting individuals with cross-overs near the target gene(s). Chromosome separation by pulsed field electrophoresis further increases the potential for this technique.

The main disadvantages of molecular marker techniques are the need to obtain very close linkages to the desired characters and, following from this difficulty, the likelihood that, if these new techniques are used extensively, resistance breeding may be concentrated on a limited number of resistances for which appropriate markers are available. The problem here is that effective response of a host genotype to a complex of different pathogens may be dependent on the interaction of many genes in the host background and selection for appropriate combinations may not be possible except under field conditions with whole plants.

Moreover, where the marker methods are combined with the use of doubled haploids, a resistance gene may be introduced into a new genetic background without any possibility for adjustment of the functioning of the whole genotype to the effects of the new gene. This concern applies also to the engineering of new resistant varieties (see below). Whether or not this factor is of any importance is unknown, but this may turn out to underline an advantage of conventional breeding methods: sufficient generations of selection are available to allow selection of combinations of genes whose overall interactions are positive for the field performance of the line selected.

In the longer term, as pointed out by Beckmann and Soller (1989), extensive use of molecular marker techniques could lead to more rapid turnover of varieties (which may have advantages in terms of maintenance of disease resistance) and the further concentration of breeding in the hands of large companies.

Other laboratory methods involve selection of, for example, host protoplasts by screening with fungal toxins or culture filtrates of pathogenic fungi (Buiatti, 1989; Roebbelen et al., 1989). Such methods again have advantages in terms of the amount of material that can be tested and the speed of doing so, given the initial investment in the laboratory procedures. On the other hand, however, it is most unlikely that all features of either host resistance or of pathogenicity are represented in such tests. It may thus be optimistic to expect that such methods will provide more than a useful indication of host resistance to some isolates of the pathogen at some stage of host growth in the field. Clearly, experience will help to show how far such methods could be developed for their applicability to long-term practical breeding.

B. Glasshouse

Glasshouse, or growth chamber, testing is also expensive relative to field screening. It is particularly useful, therefore, for diseases that are important but sporadic or that are difficult to generate in the field. The greatest value of such controlled testing, however, is probably in screening progeny with different races of a specialised pathogen, for example, to confirm inheritance of a combination of resistance genes. Because of the expense and handling difficulties involved in such tests, they are usually carried out only on seedling plants; useful features of ontogenetic resistance may thus be missed.

The main disadvantage of these test methods, apart from the expense, is that the environmental variation that is available is different from and much more limited than that in the field. This often leads to differences in the expression of resistance between glasshouse and field that may be undesirable.

C. Field

For less important diseases, the assumption is usually made that, during the course of several generations of selection, such diseases will occur naturally at least once, so that selection against excessive susceptibility can be made. These opportunities may be lost if, for example, shuttle breeding (making use of a distant site to allow an increased number of generations per year in the breeding cycle) involves a site whose environment is not relevant to the production area of the new variety.

For more important diseases, methods of artificial inoculation have been widely perfected and are in common use. Compared with the methods indicated above, these mimic most closely the infection conditions that the selected host line will face in agriculture so that the phenotypes selected in this way should be the most appropriate. However, there are still various sources of error in field selection of resistance, some of which were referred to above (quantitative resistance). They can be summarised:

(*i*) The inoculum is applied in different quantity (usually greater) and at lower frequency (usually only once) than natural inoculum. This can often lead to undesired variation in the expression of resistance.

(*ii*) It may be difficult to recognise the difference in resistance between desired homozygotes and heterozygotes.

(*iii*) In early generations there will be mixture effects in the breeding populations which may give an impression of greater resistance than is available for progeny lines when grown in pure stands.

(*iv*) In later generations there may be inter-plot interference among selected lines which may reduce the resistance levels of some lines relative to their likely performance in large pure stands.

(*v*) Among progeny lines with effective qualitative resistance, it is not possible to determine directly the underlying quantitative resistance; indirect methods are needed (see below) to ensure selection of both forms of resistance. Without such an approach, there is the possibility (but not inevitability) of a "Vertifolia" effect (Vanderplank, 1968) in which a variety with vertical resistance is selected without the support of quantitative resistance so that it can become excessively susceptible to a race able to overcome the vertical resistance.

(*vi*) With respect to specialised pathogens, the manifestation of resistance will be related directly to the current status of the pathogen population at the test site(s). The resistance selected cannot represent likely performance against the whole range of pathogen variability, present or future.

Methods have been developed to try to overcome some of these potential problems, particularly in relation to the selection of quantitative resistance (see Sect. I B 2). For example, concerning the performance of wheat against different races of the pathogen causing yellow rust, Johnson (1979), developed a field nursery system in which selected lines can be tested in hill plots against different races and compared with control varieties whose field performance against the same or similar races is well-known. Material can thus be selected which is quantitatively resistant to races that are capable of overcoming the effects of the resistance genes of major effect that they contain; this resistance should be at least equal to that of the quantitative resistant parent inoculated with the race which shows the greatest degree of virulence against the parent. The success of this method depends on the fact that artificial field infections can be established successfully when the natural occurrence of the disease is low.

This system was developed for existing varieties or lines. Breeders, however, need a method to combine qualitative resistance that has not been overcome with quantitative resistance so that if the qualitative resistance is overcome, the overall plant resistance is largely retained, i.e., to avoid the

"Vertifolia" effect. A simple and workable system involving selection in the F3 generation of a cross between a qualitative and a quantitative resistant parent was first proposed by Dr. W. Q. Loegering. In the F3, all homozygous resistant and homozygous susceptible families are discarded. Among the segregating families, resistant plants are selected from those families in which the "susceptible" segregants show the least degree of susceptibility. This system maximises the probability that the selected resisters combine both qualitative and quantitative resistance.

Because of the uncertainty of recognising durable resistance, attempts have been made to rationalise methods of selection and exploitation based on the theory of vertical and horizontal resistance (Vanderplank, 1963). Robinson (1976) argued that if host plants were inter-crossed and the progeny tested with a single pathogen race that overcame all specific resistances in the parents, then it should be possible to accumulate genes for horizontal or non-specific resistance in selected lines. A programme of recurrent selection in wheat using this concept was attempted by Beek (1983). The populations that were produced were selected rapidly for improved resistance to rusts and powdery mildew, but it could not be proved that the resistance was either horizontal or durable. Even if horizontal resistance can be selected, the possibility remains that selection for increased non-specific pathogenicity could reduce the effectiveness of the resistance.

Proceeding from laboratory to field methods, there is a loss of direct control or observation of the fate of individual resistance genes, but a gain in the potential for combining genes that affect resistance to a particular pathogen in the environment in which the crop is likely to be grown. For some crops and diseases, there is no real choice; only one or the other of these approaches is possible in practice. Where a choice is possible, and particularly with respect to specialised pathogens, then, in the absence of a comprehensive understanding of the durability of disease resistance, it seems more likely that combinations of genes for resistance, rather than "naked single" genes, will be more durable in the field.

It hardly needs stressing again, that for practical breeding, resistance to a range of diseases is always necessary. This will usually require exploitation of a range of the methods indicated above. However, a further important consideration then arises: the interaction of different diseases. This may have negative aspects, for example, resistance to one disease, which implies removal of competition by one pathogen, may allow access to a second, unexpected pathogen. The opposite may also be true. For example, Mathis et al. (1986) found that leaf infection of wheat by *Fusarium* spp. was much reduced if there was no prior infection by powdery mildew or rusts; thus selection for resistance for the latter may remove the problem of *Fusarium* resistance from the breeders' list of objectives.

III. Strategies for Using Resistant Varieties

The system of breeding and exploiting varieties of field crops that has developed over the last hundred years or so has tended to minimise genetic diversity over large areas. Farmers now prefer to grow only one variety, the one that provides the highest return for the lowest risk. However, this system maximises selection on the pathogen population for phenotypes that overcome the most popular varieties. There is a great need to increase host diversity, relevant to the important pathogens, by using different resistances and durable resistances regionally or at the farm level.

Currently, the most common strategy is that of replacement (Duvick, 1977). It is expected that existing varieties will be replaced by new ones: this is the basis of the breeding business. It is not our objective to discuss the politics of this approach, but from the farmer's point of view, the rapid turnover of varieties that are used extensively, with the catastrophic "breakdown" against some disease that often occurs, is highly undesirable.

From an epidemiological point of view, much could be done in the direction of improved strategies of use of resistant varieties to reduce such problems (Wolfe, 1983). For example, the "anticipatory" breeding approach to resistance in wheat against the major rust diseases in Australia, the United States and Canada (McIntosh, 1988; Roelfs, 1988) is proving to be highly successful. This involves continuous surveillance of the range and dynamics of genetic variation in the pathogen population with glasshouse and field tests of breeding material against the crucial part of the pathogen spectrum. This is being extended to the material being used in breeding programmes in the developing world. A remarkably small investment can help to ensure safety of the wheat crop against stem rust (*Puccinia graminis* f. sp. *tritici*) across many millions of hectares in Asia, South America and Africa.

This particular strategy is adapted to monoculture and is not appropriate to all pathogens. An alternative approach has been to try to encourage variety deployment either regionally or at the farm level so as to reduce the potential for pathogen multiplication and spread. Unfortunately, a number of proposed schemes based on the existing monocultural system have been unsuccessful simply because it is not possible to impose control of the choice of varieties on farmers.

A different approach based on mixed cultivation of lines, varieties or species of cereals, has been advocated for a number of years in different countries (Wolfe, 1985). This has reached a critical stage of large-scale application in eastern Germany where virtually the whole of the spring barley crop (approximately 350,000 hectares) is now grown as mixtures of four or five different resistant varieties. The effect of such mixtures is to slow down the rate of disease spread (in this case, particularly of powdery mildew and leaf rust), thus reducing massively the need for fungicide application. From other evidence (Wolfe, 1988), it is also likely that yields of the barley mixtures are both higher and more stable relative to the component varieties grown as individual pure stands. There is yet no answer available to the key question, whether or not this strategy affects the durability of the resistance of the components relative to their expected durability in monoculture, although preliminary evidence (Schaffner, unpubl.) indicates that it is, at least, unlikely that there would be a rapid "breakdown" of the effectiveness of the resistance of the mixtures.

It is likely that mixed plantings could be used with advantage for a number of other field crops and for plantation and forest crops where quality is not the primary objective. In orchard crops and others where quality is the main consideration, introduction of mixed cropping is regarded as impracticable and the goal of simply inherited durable resistance remains paramount. However, in a recent development in coffee production in Columbia, it has been found possible to release populations of clones that vary in resistance to coffee leaf rust but that maintain an acceptable uniformity of quality (Moreno-Ruiz and Castillo-Zapata, 1990). Sufficient lines are available to allow the seed producers to change the release from time to time in order to impose disruptive selection on the pathogen (Wolfe and Barrett, 1980).

Experiments with mixtures have been carried out either by pathologists or agronomists. For the former, it is clear that analysis of disease restriction and its effects on yield and yield stability were the most important objectives. Agronomists have been more interested in competition and yield, but it is often not clear whether positive results arose from disease restriction or other factors. More data are needed to support the view that, even in the complete absence of disease, mixtures should help to improve yield stability. One example of control of the effect of an abiotic factor is provided by Maillard and Vez (1983) who showed that in a mixture of a spring and winter wheat, the winter wheat could compensate to a large extent where severe winter conditions damaged the spring wheat component.

Increasingly, it is being recognised that the question of maintaining and exploiting varietal resistance is not a question of resistance deployment alone. Rather, it is necessary to consider host resistance as one powerful tool among many that can be combined within the system of farming to reduce the potential for disease development. It is beyond the scope of this chapter to develop such a theme, but we believe that this is a crucially important development for the future.

IV. New Approaches

There has been vast and controversial speculation about the potential for genetic engineering to deliver plant genotypes with novel forms of disease resistance. This has afforded an immense expenditure on appropriate research, but so far with little direct evidence for major advances. It may be argued that these are early days and that some encouraging examples have emerged, for instance in host plants transformed to produce TMV coat protein (Fischoff, 1989) or CaMV satellite (Baulcombe, 1989).

However, there are many scientific and technical problems to be faced, apart from ethical questions. For example, Ellis et al. (1988), in reviewing the prospects for cloning disease resistance genes, concluded with some guarded optimism for the long-term future while highlighting some of the considerable technical difficulties that still need to be overcome.

However, their review did not extend to the most important question which is the expected performance of novel resistances under field conditions. Current evidence is not encouraging. For example, over the last forty years or so, breeders and cytogeneticists have become remarkably adept at crossing species barriers and introducing resistance genes from species that would otherwise be unable to hybridise with the target crop species. Unfortunately, the durability of such "accessions" has often been less than that of immediately available resistances. For example, the wheat variety Compair with resistance derived from *Aegilops comosa* was overcome by a previously unknown race of *Puccinia striiformis* before it could be introduced into a breeding programme. In contrast, the most widely used source of durable resistance to this pathogen comes from a group of French winter wheat varieties which contain DNA only from *Triticum aestivum*.

The problem will continue to be that our knowledge of the mechanisms of resistance and pathogenicity will depend on the variation available for investigation—and we will never know the total available range of variation in the pathogen. Indeed, we still do not know whether there is more than one genetic response available for overcoming a specific resistance gene.

A theoretical problem relating to the direct introduction of novel resistance into an existing genotype is the unknown interaction that may, or not, occur between the new mechanism and the physiological expression of the existing genotype; this could relate to the reason for the evolution of species barriers. Such problems of mechanistic adjustment may not be immediately apparent. It has certainly been observed in earlier breeding programmes, that where resistance from distant species has been introduced, then various other negative characters were also expressed. Clearly, these were often related to the unnecessary DNA dragged along with the resistance, but the question remains open as to whether or not some of those effects were due to the resistance gene itself.

These problems suggest to us that approaches to the engineering of resistance should be targeted at crops for which other methods are unsuitable or unavailable (perennial, quality crops), and at pathogens for which resistance is unknown (take-all in wheat). Alternatively, greater consideration should be given to enhancing known mechanisms, for example, as suggested by the approach to improving constitutive resistance which may also have a higher probability of durability.

In the meantime, constant refining of breeding methods and of pathogen screening will continue to improve the kinds of resistance available in all crop species, and to increase the rate at which they can be introduced.

V. Conclusions

The fundamental problems of breeding for disease resistance still exist: how to recognise and combine durable resistance to the major diseases with least possible effort and in the shortest possible time. Different disciplines are needed for the solution of these problems, among which epidemiology has a key role to play.

Nevertheless, we feel that breeding for disease resistance stands to gain much from the wide array of new approaches that can be considered under the umbrellas of biotechnology and recombinant DNA technology. However, the methods cannot be expected to revolutionise this aspect of breeding simply because the concern of breeders is not only with manipulation of the evolution of the crop as a whole, but also with that of many other organisms.

For this reason, epidemiology and studies of plant pathogen populations should be drawn more directly into the development and exploitation of host resistance. The new technologies also have a role to play in this area, for example, in the population biology of major pathogens to allow us to understand better the micro-evolution of the pathogen and thus to be more clear about the meaning of durability and the value of existing and future strategies.

We should remain aware that, in the end, we are dealing with whole organisms, whole populations and farming systems. The balance of research funding and activities should reflect this and not only the areas that are currently fashionable.

VI. References

Aldwinckle H (1974) Field susceptibility of 51 apple cultivars to apple scab and apple powdery mildew. Plant Dis Rep 58: 625–629

Aldwinckle H (1989) Recommendations for future work. In: Gessler C, Koller B, Butt D (eds) Integrated control of pome fruit diseases, vol II. International Union of Biological Sciences, Paris, pp 344–346 (IOBC/WPRS Bulletin XII/6)

Alston FH (1969) Response of apple cultivars to mildew, *Podosphaera leucotricha*. Ann Rep East Malling Res Stat 1968: 133–135

Alston FH (1989) Breeding pome fruits with stable resistance to diseases. In: Gessler C, Koller B, Butt D (eds) Integrated control of pome fruit diseases, vol II. International Union of Biological Sciences, Paris, pp 90–99 (IOBC/WPRS Bulletin XII/6)

Baulcombe DC (1989) Genetic engineering of virus resistance in plants. In: XII Eucarpia Congress Göttingen. Vorträge Pflanzenzücht 16: 243–252

Beckmann JS, Soller M (1989) Genomic genetics in plant breeding. In: XII Eucarpia Congress Göttingen. Vorträge Pflanzenzücht 16: 91–106

Beek MA (1983) Breeding for disease resistance in wheat: the Brazilian experience. In: Lamberti F, Waller JM, van der Graaff NA (eds) Durable resistance in crops. Plenum, New York, pp 377–384

Brown JKM, O'Dell M, Simpson CG, Wolfe MS (1990) The use of DNA polymorphisms to test hypotheses about a population of *Erysiphe graminis* f. sp. *hordei*. Plant Pathol 39: 391–401

Buddenhagen IW (1983) Breeding strategies for stress and disease resistance in developing countries. Annu Rev Phytopathol 21: 385–409

Buiatti M (1989) Use of cell and tissue cultures for mutation breeding. In: XII Eucarpia Congress Göttingen. Vorträge Pflanzenzücht 16: 179–200

Chin KM, Wolfe MS (1984) The spread of *Erysiphe graminis* f. sp. *hordei* in mixtures of barley varieties. Plant Pathol 33: 89–100

Dayton DF (1977) Genetic immunity to apple mildew incited by *Podosphaera leucotricha*. Horticult Sci 12: 225–226

Duvick DN (1977) Major United States crops in 1976. Ann NY Acad Sci 287: 86–96

Ellis JG, Lawrence JG, Peacock WJ, Pryor AJ (1988) Approaches to cloning plant genes coferring resistance to fungal pathogens. Annu Rev Phytopathol 26: 245–263

Fischhoff DA (1989) Applications of plant genetic engineering to crop protection. Phytopathology 79: 38–40

Haahr V, Skou JP, Jensen HP (1989) Inheritance of resistance to barley leaf stripe (*Drechslera graminea*). Vorträge Pflanzenzücht 15: 3–15

Johnson R (1979) Practical breeding for durable resistance to rust diseases in self-pollinating cereals. Euphytica 27: 529–540

Johnson R (1981) Durable resistance: definition of genetic control, and attainment in plant breeding. Phytopathology 71: 567–568

Johnson R (1988) Durable resistance to yellow (stripe) rust in wheat and its implications in plant breeding. In: CIMMYT 1988. Breeding strategies for resistance to the rusts of wheat. CIMMYT, Mexico D.F., pp 63–75

Kellerhals M (1989) Breeding disease resistant apple cultivars in Switzerland. In: Gessler C, Koller B, Butt D (eds) Integrated control of pome fruit diseases, vol II. International Union of Biological Sciences, Paris, pp 130–136 (IOBC/WPRS Bulletin XII/6)

Knight RL, Alston FH (1969) Developments in apple breeding. Ann Rep East Malling Res Stat 1968: 125–132

Koebner RMD, Miller TE, Snape JW, Law CN (1988) Wheat endopeptidase: genetic control, polymorphism, interchromosomal gene location and alien variation. Genome 30: 186–192

Lespinasse Y (1989) Breeding pome fruits with stable resistance to diseases: genes, resistance mechanisms, present works and prospects. In: Gessler C, Koller B, Butt D (eds) Integrated control of pome fruit diseases, vol II. International Union of Biological Sciences, Paris, pp 100–116 (IOBC/WPRS Bulletin XII/6)

Maillard A, Vez A (1983) La culture de mélanges de variétés de blé. Rev Suisse Agricult 15: 195–198

Mathis A, Forrer HR, Gessler C (1986) Powdery mildew pustules supporting *Fusarium culmorum* infection of wheat leaves. Plant Dis 70: 53–54

McIntosh RA (1988) The role of specific genes in breeding for durable stem rust resistance in wheat and triticale. In: CIMMYT 1988. Breeding strategies for resistance to the rusts of wheat. CIMMYT, Mexico D.F., pp 1–9

McMillin DE, Allan RE, Roberts DE (1986) Association of an isozyme locus and straw-breaker foot rot resistance derived from *Aegilops ventricosa* in wheat. Theor Appl Genet 72: 743–747

Michelmore RW, Hulbert SH (1987) Molecular markers for genetic analysis of phyto-pathogenic fungi. Annu Rev Phytopathol 25: 383–404

Moreno-Ruiz G, Castillo-Zapata J (1990) The variety Colombia: a variety of coffee with resistance to rust (*Hemileia vastatrix* Berk. & Br.). Cenicafé Tech Bull 9

O'Dell M, Wolfe MS, Flavell RB, Simpson CG, Summers RW (1989) Molecular variation in populations of *Erysiphe graminis* on barley, wheat, oats and rye. Plant Pathol 38: 340–351

Parlevliet JE 1988: Strategies for the utilization of partial resistance for the control of cereal rusts. In: CIMMYT 1988. Breeding strategies for resistance to the rusts of wheat. CIMMYT, Mexico D.F., pp 48–62

Pink DAC, Bennett FGA, Caten CE, Law CN (1983) The effects of homoeologous group 5 chromosomes on disease resistance. Z Pflanzenzücht 91: 278–294

Ramirez-Villupadua J, Endo RM, Bosland P, Williams PH (1985) A new race of *Fusarium oxysporum* f. sp. *conglutinans* the attacks cabbage with Type A resistance. Plant Dis 69: 612–613

Roebbelen G, Spanier A, Spanier K (1989) Genetic manipulation in oil crop improvement. Vorträge Pflanzenzücht 16: 411–422

Robinson RA (1976) Plant pathosystems. Springer, Berlin Heidelberg New York

Roelfs AP (1988) Resistance to leaf and stem rusts in wheat. In: CIMMYT 1988. Breeding strategies for resistance to the rusts of wheat. CIMMYT, Mexico D.F., pp 10–22

Siebs E (1959) Ergebnisse zu Problemen des Mehltaus und der Mehltauresistenz des Apfels. J Phytopathol 34: 86–102

Simmonds NW (1979) Principles of crop improvement. Longmans, London

Stamm P, Zeven C (1981) The theoretical proportion of the donor genome in near-isogenic lines of self-fertilizers bred by backcross breeding. Euphytica 30: 227–238

Stolzenburg MC, Aist JR, Israel HW (1984a) The role of papillae in resistance to powdery mildew conditioned by the *mlo* gene in barley. I Correlative evidence. Physiol Plant Pathol 25: 337–346

Stolzenburg MC, Aist JR, Israel HW (1984b) The role of papillae in resistance to powdery mildew conditioned by the *mlo* gene in barley. I Experimental evidence. Physiol Plant Pathol 25: 347–361

Stuber CW (1989) Marker-based selection for quantitative traits. In: XII Eucarpia Congress Göttingen. Vorträge Pflanzenzücht 16: 31–49

Vanderplank JE (1963) Plant diseases: epidemics and control. Academic Press, New York

Vanderplank JE (1968) Disease resistance in plants. Academic Press, New York

van der Weg WE (1989) Breeding for resistance to *Nectria galligena* in apple: differences in resistance between seedling populations. In: Gessler C, Koller B, Butt D (eds) Integrated control of pome fruit diseases, vol II. International Union of Biological Sciences, Paris, pp 137–145 (IOBC/WPRS Bulletin XII/6)

Watkins R (1974) Fruit breeding. Ann Rep East Malling Res Stat 1973: 121–129

Williams JGK, Kubelik AR, Livak KJ, Rafalski JA, Scott VT (1990) DNA polymorphisms amplified by arbitrary primers are useful as genetic markers. Nucleic Acids Res 18: 6531–6535

Wolfe MS (1983) Genetic strategies and their value in disease control. In: Kommedahl T, Williams PH (eds) Challenging problems in plant health. The American Phytopathological Society, St Paul, MN, pp 461–473

Wolfe MS (1985) The current status and prospects of multiline cultivars and variety mixtures for disease resistance. Annu Rev Phytopathol 23: 251–273

Wolfe MS (1988) The use of variety mixtures to control diseases and stabilize yield. In: CIMMYT 1988. Breeding strategies for resistance to the rusts of wheat. CIMMYT, Mexico D.F., 91–100

Wolfe MS, Barrett JA (1980) Can we lead the pathogen astray? Plant Dis 64: 148–155

Worland AJ, Law CN, Hollins TW, Koebner RMD, Giura A (1988) Location of a gene for resistance to eyespot (*Pseudocercosporella herpotrichoides*) on chromosome 7D of bread wheat. Plant Breed 101: 43–51

Young ND, Tanksley SD (1989) RFLP analysis of the size of chromosomal segments retained around the Tm-2 locus of tomato during backcross breeding. Theor Appl Genet 77: 353–359

Chapter 2

Functional Models to Explain Gene-for-Gene Relationships in Plant–Pathogen Interactions

Pierre J.G.M. de Wit

Department of Phytopathology, Wageningen Agricultural University,
NL-6700EE Wageningen, The Netherlands

Contents

I. Introduction

Flor's pioneering genetic studies on flax (*Linum usitatissimum*) and its pathogen, the obligate parasite *Melampsora lini*, proved the genetic interdependence of host and parasite and eventually led to the gene-for-gene hypothesis (Flor, 1946). His work has strongly influenced hypotheses and investigations on the basis of host–parasite interactions during the last five decades.

Since Flor's discovery, gene-for-gene relationships have been reported for many other fungus–plant interactions but also for virus-, bacterium-, nematode-, and insect–plant relationships (Ellingboe, 1984; Sidhu, 1980). Geneticists, epidemiologists, biochemists and molecular plant pathologists have studied numerous host–parasite interactions for which a gene-for-gene relationship has been proven or assumed. Although much research has been devoted to the identification of genes and gene products involved, it is only recently that we are beginning to understand some of the molecular basis of gene-for-gene relationships. This understanding is most advanced for interactions involving bacteria and their host plants.

In this chapter a brief review is given on the genetic basis of gene-for-gene systems involving plant pathogenic fungi and bacteria and their genetic, biochemical and molecular implications. A few model systems which have been studied rather extensively at the biochemical and molecular level in recent years will be discussed in more detail.

II. Physiologic Races and Differentials

Race is a term commonly used in biology to denote a group of individuals possessing common features which distinguish them from other groups of individuals within formally recognized species or subspecies (Caten, 1987). Intraspecific variation in plant pathogenic fungi became apparent from studies of pathogenic specialization towards different host species and genotypes within one host species (Stakman, 1917). Stakman (1917) introduced the term physiologic race which is often abbreviated as race. Although the term race has widely been accepted many authors have criticized the concept of race (Caten, 1987). The potential number of races that can be recognized depends on the number of resistance factors present in a set of differentials. When each resistance factor–race interaction gives two possible outcomes (compatible or incompatible) one can theoretically differentiate 2^n races and 2^n cultivars. It is clear that identical races may be identical only for a small part of their genomes, i.e., the part involved in the virulence/avirulence towards differentials. In practice this means that the classification and the grouping of races is completely dependent on the differentials used. The name of a certain race can never be absolute as new virulence or avirulence factors can be detected when the race is tested on a larger set of differentials; adding a new differential can separate previously described identical isolates into two different races.

The race nomenclatures routinely used for plant pathogenic fungi and bacteria are completely different. Pathogenic fungi are named after their avirulence gene(s) c.q. the resistance gene(s) they can 'overcome', while pathogenic bacteria are named after their avirulence gene(s) c.q. the resistance gene(s) they cannot 'overcome' (Table 1) (Collinge and Slusarenko, 1987; Gabriel et al., 1986; Swanson et al., 1989; Keen and Staskawicz, 1988; Yoder et al., 1986).

Table 1. Nomenclature of fungal and a bacterial races fitting into a hypothetical
gene-for-gene relationship with their respective host cultivars

	Fungal races		Bacterial races	
Host	race 1	race 2	race 1	race 2
cultivar	a1A2	A1a2	A1a2	a1A2
R1R1 r2r2	C	I	I	C
r1r1 R2R2	I	C	C	I

C, compatible, plant is susceptible; I, incompatible, plant is resistant

III. Genetic Analysis of Gene-for-Gene Relationships

The gene-for-gene relationship as discovered by Flor (1946) was proposed as the simplest explanation of the observations on the inheritance of virulence in the flax rust fungus *M. lini.* On varieties of flax, *L. usitatissimum* with one gene for resistance to the avirulent parent race, F2 cultures of the fungus segregated in monofactorial ratios while on varieties with 2, 3, or 4 genes for resistance, the F2 cultures segregated in bi-, tri-, or tetrafactoral ratios. This suggests that for each gene that conditions a response in the host, there is a corresponding gene in the pathogen conditioning virulence. Each gene in each member of the host pathogen system can only be identified by its counterpart in the other member of the system (Flor, 1971). Characterization of a gene-for-gene relationship for a pathogen–host interaction is easy when the virulence spectrum of a given race can be established by a selfing study of the pathogen on differentials of the host carrying single resistance genes. However, these criteria are not often met in pathogens. Many plant pathogenic bacteria and some plant pathogenic fungi have no sexual stage, which means that the conventional genetic studies cannot be carried out.

When genetic studies with the pathogen are impossible or have not yet been carried out the so-called Person analysis (Person, 1959) can be very helpful. Person (1959) formally redefined the gene-for-gene hypothesis. He illustrated a new method to analyse a theoretical gene-for-gene system involving 5 loci in the host interacting with 5 loci in the pathogen (Table 2). One premise which simplifies the system is that the outcome of the interaction is either compatible (host susceptible) or incompatible (host resistant). When all the genotypes can be fitted in such a theoretical model, a gene-for-gene relationship can be established without the necessity of genetic studies of host and pathogen. However, in practice host-pathogen systems are never so ideal. Often some of the putative host genotypes and pathogen genotypes are unknown or have not yet been discovered, which causes gaps in the theoretical interaction scheme. If too many genotypes of host and/or pathogen are missing, the definite genetic constitution of the

Table 2. Host–pathogen interactions with gene-for-gene relationships at five pairs of loci

```
              H    HHHHH  HHHHHHHHH  HHHHHHHHHH  HHHHH  H
              0    00000  000111111  1112222222  22233  3
              1    23456  789012345 6 7890123456  27801  2

                   A      AAAA       AAAAAA       AAAA   A
                    B      B    BBB  BBB    BBB   BBB  B B
                     C     C C   CC  C  CC  CC C  CC  CC  C
                      D        D  D D D  D  DD  DD  D  DDD  D    2^n
                       E      E   E EE      E  EE  EEE   EEEE  E   ·
                                                                 N  △  n
 P01          s                                                 1  1  0

 P02  a       s    s                                            2     1
 P03    b     s    s                                            2     1
 P04      c   s       s                                         2  5  1
 P05       d  s        s                                        2     1
 P06        e s         s                                       2     1

 P07  a b     s    s s        s                                 4     2
 P08  a   c   s    s  s     s                                   4     2
 P09  a     d s    s   s       s                                4     2
 P10  a      e s    s    s       s                              4     2
 P11    b c   s      s s          s                             4 10  2
 P12    b   d s      s s         s                              4     2
 P13    b    e s      s  s         s                            4     2
 P14      c d s       s s           s                           4     2
 P15      c  e s       s s            s                         4     2
 P16        d e s        s s            s                       4     2

 P17  a b c   s    s s s    s s   s         s                   8     3
 P18  a b   d s    s s  s   s s   s          s                  8     3
 P19  a b    e s    s s  s  s s   s          s                  8     3
 P20  a   c d s    s  s s    s s   s           s                8     3
 P21  a   c  e s    s  s  s  s s    s           s               8     3
 P22  a     d e s    s  s s  s s    s            s              8 10  3
 P23    b c d s      s s s     s s  s              s            8     3
 P24    b c  e s      s s s    s s  s               s           8     3
 P25    b   d e s      s s s   s s   s               s          8     3
 P26      c d e s       s s s      s s s              s         8     3

 P27  a b c d s    s s s s   s s s   s s   s           s       16     4
 P28  a b c  e s    s s s  s s s  s s s   s   s            s   16     4
 P29  a b   d e s    s s s  s s s  s s s   s   s            s  16  5  4
 P30  a   c d e s    s  s s s   s s s    s s s     s s s        s 16     4
 P31    b c d e s      s s s s     s s s s s s      s s s s       s 16     4

 P32  a b c d e s    s s s s s  s s s s s s s s s s  s s s s s s s s s s  s 32  1  5
```

N, number of attacking races; m, total number of genes involved; n, number of genes involved; △, partition according to Pascal's triangle

```
        m-n
 N = 2      32  1616161616  8 8 8 8 8 8 8 8 8 8  4 4 4 4 4 4 4 4 4 4  2 2 2 2 2  1
 △          1      5            10                  10                   5        1
 n          0   1 1 1 1 1   2 2 2 2 2 2 2 2 2 2  3 3 3 3 3 3 3 3 3 3  4 4 4 4 4  5
```

phenotypes of the differentials and races cannot be established with certainty.

Characteristics of the gene-for-gene system as illustrated by Person (1959) (Table 2) for interactions of 5 loci in the pathogen with 5 loci in the host are:

(*i*) In a gene-for-gene system resistance in the host may be either dominant or recessive.

(*ii*) In a gene-for-gene system avirulence in the pathogen may be either dominant or recessive.

(*iii*) The differentials carrying single genes for resistance (H2, H3, H4, H5, H6) are capable of differentiating completely all the races of the pathogen; these differentials are attacked by half of the total number of races.

(*iv*) The races carrying 4 genes for virulence (P27, P28, P29, P30, P31) are capable of differentiating completely all the differentials of the host; these races attack half of the total number of differentials.

(*v*) The 32 phenotypes in each population of host and pathogen show frequencies following the integral values of the expansion $(x + y)^n$, where n is one less than the total number of genes (Pascal's triangle). Analysis of published data on the potato–*Phytophthora infestans* interaction (Toxopeus, 1956) proved that the gene-for-gene relationship is operative in field situations. The genetic constitution of *P. infestans* (for which a sexual stage was not yet reported in Europe at that time) could be established with a high degree of certainty.

Person and Mayo (1974) have, however, drawn the attention to the importance of the conditional nature of resistance genes and the existence of epistatic relationships between functional resistance genes and those hidden due to either the matching of virulence alleles or conditioning a less completely resistant phenotype. For this reason it is never possible to exclude that undetected genes in either the host or the pathogen are contributing to the outcome of any host-pathogen combination. This implies that in addition to the Person analysis genetic studies of both partners are necessary to establish their definite genetic constitution.

A. Exceptions to the General Characteristics of the Gene-for-Gene Systems

Christ et al. (1987) reviewed a number of exceptions to the 'rule' of gene-for-gene systems concerning the host as well as the pathogen.

1. The Inheritance of Resistance and Avirulence

The expectation that in gene-for-gene systems resistance and avirulence are dominant has so widely been accepted that recessive resistance and recessive avirulence would be considered as exceptions to this rule. However, dominance relations in plants are based on investigations of a small number of cases where gene-for-gene interactions have been demonstrated (Barret, 1985). The fact that gene-for-gene interactions have mainly been described for cultivated species indicates that the prevalence of dominant resistance could be a consequence of plant breeding. The breeder will detect dominant resistance easier than recessive resistance and consequently overlook or miss potential recessive resistance sources. Heterozygotes which express any level of resistance are lumped together in one category and will be classified as resistant and dominant; the breeding process thus tends to favour the selection of dominant resistance. Indeed, in the literature a number of clear examples of recessive resistance towards the rusts, smuts and *Xanthomonas campestris* pv. *oryzae* are described (see Christ et al., 1987, for a review).

Similarly within the pathogen dominant avirulence has been described more frequently than recessive avirulence. However, under natural conditions the concept is not applicable for plant pathogenic bacteria and Ascomycetes as they are haploid. Therefore no dominance relations can be studied under natural conditions. Studying forced parasexual recombinations and transformation can shed some light on dominance or epistatic relationships within these organisms. The cases where real dominant relationships can be investigated are confined almost exclusively to the Basidiomycetes, and Oomycetes (Christ et al., 1987; Ilott et al., 1989; Spielman et al., 1989, 1990).

However, the fact that resistance genes are often dominant has no bearing on the mechanism by which they bring about their effects. Similarly, the fact that avirulence genes are often dominant is unlikely to have biochemical or functional significance in terms of mechanisms.

2. Control of Resistance and Avirulence by Two Genes

In some host–pathogen interactions, involving rusts, smuts but also Oomycetes such as *Bremia lactucae* and *Phytophthora infestans* there have been indications for the involvement of two genes governing resistance towards a certain race and of two genes matching towards one resistance gene (Christ et al., 1987; Michelmore et al., 1984; Ilott et al., 1989; Spielman et al., 1989, 1990). Often, however, this apparent involvement of two genes can be explained by non-allelic interactions between the locus for resistance in the host (Knott and Anderson, 1956; Martens et al., 1981; Dyck and Samborski, 1982) or the locus for virulence or avirulence in the pathogen (Lawrence et al., 1981; Christ and Groth, 1982; Michelmore et al., 1984; Crute, 1985; Ellingboe pers. comm.), by so-called dominant inhibitor or suppressor genes resulting in non-functioning of resistance genes, virulence or avirulence genes respectively. Sometimes the presence of inhibitor genes is less clear than initially thought because of uncharacterized background in either host or pathogen (Ilott et al., 1989). The establishment of non-allelic modification of virulence and avirulence genes in a number of host–pathogen interactions suggests that there are different levels of control in gene-for-gene interactions. These findings have implications for studies directed towards a further understanding of the molecular basis of gene-for-gene relationships. However, besides all the reported apparent exceptions, the gene-for-gene theory still appears to give an adequate genetic description of most differential host–pathogen interactions. Clearly the existence of inhibitor genes does not weaken the gene-for-gene hypothesis.

3. Polygenically Determined Resistance

Christ et al. (1987) in their review consider resistance controlled by several genes that have small additive effects to be completely separate from major gene resistance. It is, however, questionable whether the facts concerning polygenically determined resistance substantiate the existence of two fundamentally different types of resistance. Parlevliet and Zadoks (1977) examined the characteristics of a polygenic model with 5 loci both in the host and the pathogen, where each locus exerted an equal and additive effect. The interaction between loci of host and pathogen was considered to follow the rules of gene-for-gene relations in one case. In another case loci in the host and the pathogen did not interact in a gene-for-gene fashion but were purely additive. Analysis of the model demonstrated that the statistical interaction in the gene-for-gene model would be too small to be discerned. This means that such a relationship would appear to be horizontal despite being the result of a gene-for-gene relationship. It is therefore very well possible that vertical as well as horizontal resistance have a gene-for-gene basis.

B. Models Explaining Molecular Aspects of Gene-for-Gene Interactions

The development of methods for the genetic manipulation and transformation of bacteria, fungi, and plants has positively influenced progress in research concerning molecular aspects of gene-for-gene interactions. Unfortunately some genetically well defined gene-for-gene systems are difficult to study at the molecular level as a number of molecular techniques cannot yet be applied to these systems easily. Table 3 gives a number of desired criteria which make plant-pathogen systems amenable for studies at the molecular level.

It is clear that none of the host-pathogen systems presently studied meets all these criteria. Some are more important than others, but lack of an efficient transformation system for one or both partners will greatly hamper

Table 3. Characteristics which make a plant-pathogen system amenable for studies at the molecular level

1. Host and pathogen are genetically well defined.
2. Host and pathogen can easily be cultured and have a short generation cycle.
3. Sexual cycle of both host and pathogen can easily be studied.
4. Host and pathogen can be transformed in order to introduce new genes or inactivate existent genes.
5. Host and pathogen have a relative small genome.
6. The existence of putative products of either resistance or avirulence genes.

progress in the long term. Eventually, one has to prove the role of a putative resistance or avirulence gene by transforming it back into the host or pathogen, respectively. Table 4 gives the state of the art concerning transformation of pathogens and their host plants involved in gene-for-gene interactions. In practice absence of the desired characteristics has limited molecular research to only a few host-pathogen systems. Despite the fact that the genetics of plant–bacterium interactions have not been studied extensively (Cook and Stall, 1963; Brinkerhoff, 1970; Chatterjee and Vidaver, 1986; Gabriel et al., 1988), molecular studies on plant pathogenic bacteria have become very fashionable recently, mainly because bacteria have small genomes and can be transformed efficiently. The availability of molecular techniques made it possible to dissect the bacterial genome of certain pathovars or species and express their genomes part by part in another background with the result of discovering new avirulence genes (Keen and Dawson 1992; Keen and Staskawicz, 1988). Similarly dissection of plants with small genomes such as *Arabidopsis thaliana* will probably generate a number of yet unknown resistance genes to pathogens of the plant and to pathogens in other crop plants (Somerville, 1989).

Table 4. The presence or absence of a transformation system for host and pathogen for which a gene-for-gene relationship has been reported which have been studied at the biochemical and molecular level (data obtained from Leong and Holden, 1989; Chatterjee and Vidaver, 1986, pers. comm.).

Host–pathogen interaction	Transformation of the host	Transformation of the pathogen
Fungi		
Flax–*M. lini*	yes	no
Wheat–*P. graminis*	no	no
Barley–*E. graminis*	no	no
Corn–*P. sorghi*	yes	no
Rice–*M. grisea*	yes	yes
Bean–*C. lindemuthianum*	no	yes
Potato–*P. infestans*	yes	no
Soybean–*P. megasperma*	yes	no
Lettuce–*B. lactucae*	yes	no
Tomato–*C. fulvum*	yes	yes
Bacteria		
Soybean–*P. s.* pv. *glycinea*	yes	yes
Tomato–*P. s.* pv. *tomato*	yes	yes
Bean–*P. s.* pv. *phaseolicola*	no	yes
Pea–*P. s.* pv. *pisi*	no	yes
Pepper–*X. c.* pv. *vesicatoria*	yes	yes
Cotton–*X. c.* pv. *malvacearum*	yes	yes
Rice–*X. c.* pv. *oryzae*	yes	yes

It is impossible here to discuss all of the different host–pathogen interactions which have been studied successfully at the biochemical and molecular level in recent years (Crute et al., 1985). I will limit myself to briefly mentioning a small number of them and to discussing only a few in greater detail. These include the interactions between species of the bacteria *Pseudomonas* and *Xanthomonas* and their host plants, and the fungal pathogen–plant interactions *Bremia lactucae*–lettuce, *Magnaporthe grisea*–rice and *Cladosporium fulvum*–tomato.

IV. Gene-for-Gene Interactions Involving Plant Pathogenic Bacteria

A. Pseudomonas syringae Species

One of the great successes of molecular genetic analysis of plant pathogenic bacteria has been the cloning of avirulence genes. Staskawicz et al. (1984) cloned the first avirulence gene from race 6 of *P. syringae* pv. *glycinea* by a shotgun approach. A genomic library of race 6 was constructed in the wide host range cosmid-cloning vehicle pLAFR1 and individual clones were conjugated into the universally virulent race 5. Individual transconjugants were tested on the appropriate cultivars for induction of the hypersensitive response (HR). Such a transconjugant would have obtained a piece of DNA expressing an avirulence gene from race 6 (Napoli and Staskawicz, 1987). Indeed in this way the first avirulence gene called *avr A* was cloned from race 6. From race 0 of the same pathogen two additional avirulence genes *avr Bo* and *avr C* were cloned (Staskawicz et al., 1987). They specify the induction of HR on differentials of soybean (Tamaki et al., 1988). Race 1 carried the *avr B1* gene which appeared to be identical to the *avr Bo* gene which prompted the authors to rename *avr Bo* and *avr B1* into *avr B*. A complication with the *P. syringae* pv. *glycinea*-soybean system is that only the resistance gene *Rpg 1* present in the soybean cultivars Norchief and Harosoy, and conferring resistance towards race 1 (carrying *avr B*) has been genetically well characterized. Resistance factors present in the other cultivars are not characterized yet. The identification and cloning of the *avr B* gene interacting with the resistance gene *Rpg 1* thus provides formal genetic evidence for a gene-for-gene relationship. For the other avirulence genes a gene-for-gene relationship is very suggestive but not yet conclusive.

One of the obvious things to do with the first cloned avirulence genes was to check whether their products could act as race-specific elicitors, the putative products of avirulence genes as put forward in the specific elicitor-receptor model (Ellingboe, 1984; Keen, 1982; De Wit, 1987). The avirulence genes *avr A*, *avr B*, and *avr C* encode proteins with molecular weights of 100 kDa, 36 kDa, and 39 kDa, respectively (Table 5). The proteins themselves, however, do not induce HR on the appropriate host cultivars indicating that in addition to the avirulence gene protein another factor is needed. Evidence for such a factor came from experiments where the

Table 5. Bacterial avirulence genes involved in putative gene-for-gene relationships

Bacterial pathogen	Avir. gene	Protein M.W. (kDa)	Homology within			HR activity in nonhost plants	Reference(s)
			races		other pathovars		
			vir.	avir.			
P. s. pv. *glycinea*	*avrA* (chrom.)	100	no	no	yes	yes	Staskawicz et al., 1984
	avrB (chrom.)	36	no	yes	?	yes	Staskawicz et al., 1987
	avrC (plasm.)	39	no	yes	?	yes	Tamaki et al., 1988
P. s. pv. *tomato*	*avrA* (chrom.)	100	no	no	yes	yes	Kobayashi et al., 1989
	avrD (plasm.)	34	yes	?	yes	yes	Kobayashi et al., 1989
	avrE (chrom.)	?	?	?	yes	yes	Keen et al., 1990
P. s. pv. *phaseolicola*	*avrTend. Gr.*	?	?	?	?	?	Hitchin et al., 1989
	avrRed Mex.	?	?	?	?	?	Shintaku et al., 1989
P. s. pv. *pisi*	*avrAspi*	?	?	?	?	?	Vivian et al., 1989
X. c. pv. *vesicatoria* (pepper strains)	*avrBs1* (plasm.)	50	no	no	?	yes	Swanson et al., 1988 Ronald and Staskawicz, 1988
	avrBs2 (chrom.)	69	yes	no	?	yes	Minsavage et al., 1990
	avrBs3 (plasm.)	125	no	no	no	yes	Bonas et al., 1989
	avrRxv (chrom.)	38	no	no	no	yes	Whalen et al., 1988
X. c. pv. *vesicatoria* (tomato strain)	*avrBsT* (plasm.)	?	no	no	no	yes	Whalen et al., 1988 Minsavage et al., 1990
X. c. pv. *malvacearum*	*avrB2* (chrom.)	?	yes	no	?	?	Gabriel et al., 1986
	avrB3 (chrom.)	?	?	no	?	?	Gabriel et al., 1986
	avrB6 (chrom.)	?	?	no	?	?	Gabriel et al., 1986
	*avrB*_N (chrom.)	?	?	no	?	?	Gabriel et al., 1986
	avrln (chrom.)	?	yes	no	?	?	Gabriel et al., 1986
X. c. pv. *oryzae*	*avr10* (chrom.)	35	yes	yes	yes	?	Kelemu and Leach, 1990

avirulence genes were expressed in different bacterial pathogens. Bacterial plant pathogens contain a cluster of genes called hypersensitive response and pathogenicity (*Hrp*) genes which seem to be needed for the induction of HR by the avirulence gene products. The biological function of this gene cluster in the pathogen is not understood yet. The *Hrp* products could either interact directly with the *avr* gene products or could facilitate the transport of the *avr* proteins to the correct cellular location in order to induce HR. The *Hrp* cluster occurs in many different plant pathogenic bacteria, is interchangeable between different bacteria and is highly conserved (Lindgren et al., 1988).

Another bacterial pathogen which has been intensively studied is *P. syringae* pv. *tomato* (Kobayashi et al., 1989, 1990a, b; Huynh et al., 1989; Keen et al., 1990). Surprisingly from this pathogen one avirulence gene, *avrA*, has been isolated which was fully identical to *avrA* of *P. syringae* pv. *glycinea*. It was rather expected that avirulence genes would only act at the race-cultivar level. From *P. syringae* pv. *tomato* other surprising results were obtained. A cosmid library of *P. syringae* pv. *tomato* transconjugated into the universally virulent race 4 of *P. syringae* pv. *glycinea* identified three new avirulence genes with completely different HR-inducing spectra on the available soybean differentials (Table 5). One of these avirulence genes *avrD* codes for a protein of 34 kDa. The protein itself does not induce HR on the appropriate cultivars but it does not need a functioning *Hrp* gene cluster. When expressed in *Escherichia coli* the protein does not induce HR on its own but the homogenate of the bacterium expressing *avrD* does induce HR indicating that the protein is an enzyme that catabolizes the conversion of a normal bacterial metabolite present in *E. coli* and different *Pseudomonas* species. The molecule inducing HR has been purified but not yet fully characterized (see Keen and Dawson, 1992). Interestingly races of *P. syringae* pv. *glycinea* contain a gene which is homologous to *avrD* of *P. syringae* pv. *tomato* encoding a protein with 86% identical amino acids but lacking HR-inducing activity. This gene could be considered as a recessive allele of *avrD*. However, as has been pointed out before the term recessive allele is rather misleading in relation to bacterial pathogens. In Table 5 a few other avirulence genes from *P. syringae* pv. *phaseolicola* (Shintaku et al., 1989; Hitchen et al., 1989) and *P. syringae* pv. *pisi* (Vivian et al., 1989) are shown but as their structure has not been characterized yet they will not be discussed here in detail.

B. *Xanthomonas campestris* Species

As mentioned above a disadvantage of the *P. syringae* pv. *glycinea*–soybean interaction is that only the resistance gene *Rpg 1* has been genetically well characterized. More genetic studies have been carried out concerning resistance towards species of *Xanthomonas* such as *X. campestris* pv. *vesicatoria* (tomato and pepper strains), *X. campestris* pv. *malvacearum* and

X. campestris pv. *oryzae* (Mew, 1987; Brinkerhoff, 1970; Cook and Stall, 1963; Hibberd et al., 1987).

Gabriel et al. (1986) have cloned five avirulence genes from *X. campestris* pv. *malvacearum*, the causal agent of angular leaf spot of cotton (Table 5). The products of these genes have not yet been identified. They found that virulent races contained nonfunctional DNA sequences homologous to the avirulence DNA present in avirulent races. These results are similar to those for the *avr D* gene of *P. syringae* pv. *tomato* of which homologous nonfunctioning DNA sequences are present in *P. syringae* pv. *glycinea* (Kobayashi et al., 1990a).

From *X. campestris* pv. *vesicatoria* three avirulence genes called *avr Bs1, avr Bs2,* and *avr Bs3* have been cloned (Ronald and Staskawicz, 1988; Swanson et al., 1988; Bonas et al., 1989; Minsavage et al., 1990). The genes encode proteins with molecular weights of 50, 69, and 125 kDa, respectively (Table 5). Interestingly the different avirulence genes can lose their activity in different ways. The *avr Bs1* gene is located on a plasmid and has been reported to be inactivated by a transposable element IS476 (Kearney et al., 1988) rendering the avirulent race into a virulent one on the cultivar carrying the resistance gene *Bs 1*. Alternatively, an avirulence gene located on a plasmid can frequently be lost under natural conditions as has been the case with the *avr Bs3*. The avirulence gene *avr Bs2* seems to be present in all races of *X. campestris* pv. *vesicatoria* indicating that this gene possibly encodes a factor important for the fitness of the pathogen. Loss could create survival problems to the pathogen. This led Minshavage et al. (1990) to speculate that the resistance gene *Bs 2* in pepper would probably be more durable than the other resistance genes, *Bs 1* and *Bs 3*. The corresponding avirulence genes *avr Bs1* and *avr Bs3* can easily be lost from the bacterium under natural conditions without causing visible loss of fitness. Gene replacement studies on the *avr Bs2* gene could indicate the relevance of this gene for fitness of the bacterium. Until recently none of the avirulence genes sequenced show any homology with sequences of other genes known in different data banks. The sequence data obtained thus far cannot yet shed light on how bacterial avirulence gene products are recognized by the host and induce HR, nor what the functions of the genes in the pathogen are. One exception to this seems to be the avirulence gene *avr 10* cloned from *X. campestris* pv. *oryzae* (Kelemu and Leach, 1990; Leach, pers. comm.). The *avr 10* shows homology with the *phoS* gene in *E. coli*, a gene encoding a protein in the outer membrane of the bacterium involved in phosphate uptake and/or transport. However, future experiments are necessary to determine whether this homology has any significance.

C. Possible Functions of Bacterial Avirulence Genes

It has been demonstrated that bacterial avirulence genes are able to induce HR not only on resistant cultivars of host plants but also on nonhost plants.

These results were obtained by introducing different avirulence genes of *P. syringae* pv. *glycinea* in different other *P. syringae* pathovars and inoculating them on their normal host plants. Kobayashi et al. (1989) introduced the *avrA* gene into *P. syringae* pv. *lachrymans*, *P. syringae* pv. *phaseolicola*, *P. syringae* pv. *pisi* and *P. syringae* pv. *tabaci* and inoculated the bacteria on their normal host plants. None of the transformed bacteria became avirulent on their host plants except *P. syringae* pv. *tabaci* which became avirulent on a number of tobacco cultivars by invoking HR. Similarly, Whalen et al. (1988) cloned avirulence gene *avrRxv* from the tomato strain of *X. campestris* pv. *vesicatoria* and introduced it into other species of *X. campestris* such as pv. *glycines* pathogenic on soybean, pv. *vignicola* pathogenic on cowpea, pv. *alfalfae* pathogenic on alfalfa, pv. *holcicola* pathogenic on corn, pv. *malvacearum* pathogenic on cotton and pv. *phaseoli* pathogenic on bean. *AvrRxv* introduced in *X. campestris* pv. *phaseoli* induced HR on cultivar Sprite but not on Bush Blue Lake. These results indicate that specific bacterial avirulence genes can be recognized by different plant species suggesting that these species may contain similar resistance genes. It will be very interesting to find out whether these different plant species indeed carry identical resistance genes. Unfortunately no resistance genes have been cloned yet but this may occur in the near future (Bennetzen et al., 1988, pers. comm.).

Alternatively, as has been discussed before, similar avirulence genes have been found in different bacterial species indicating that these avirulence genes possibly have similar functions in these species. It is not known yet how these avirulence genes have spread through the bacterial population during the evolutionary process.

V. Gene-for-Gene Interactions Involving Fungal Plant Pathogens

A. Bremia lactucae–Lettuce

Bremia lactucae is an Oomycete which is genetically very well characterized (Ilott et al., 1987, 1989; Hulbert et al., 1988). The major determinants of specificity are involved in a gene-for-gene interaction (Judelson and Michelmore, 1989). One of the goals of the research group of Michelmore is to identify and clone avirulence genes of *B. lactucae*. The major disadvantage of this obligate parasite is that until now no DNA-mediated transformation has been described. The genetics of both host and pathogen have been analysed simultaneously (Michelmore et al., 1984; Ilott et al., 1987; Hulbert et al., 1988; Farrara et al., 1988). Thirteen single dominant genes for avirulence have been characterized in *B. lactucae*. These loci are matched by 13 dominant loci for resistance to downy mildew (*Dm*) in lettuce. There is clear evidence for the existence of one dominant inhibitor of avirulence epistatic to *avr5*. Other inhibitor genes or genes of minor effect may exist but they do not alter the specificity determined by the major

avirulence and resistance genes. Very little is known about putative products of avirulence genes conferring an incompatible response in the host
cultivars carrying the appropriate resistance genes. The shotgun approach
to clone avirulence genes from bacteria as used by Staskawicz et al. (1984)
and others cannot be used for *B. lactucae*. Restriction fragment length
polymorphism (RFLP) markers that are linked to avirulence genes will be
identified (Judelson and Michelmore, 1989; Hulbert et al., 1988). These
probes will be used as starting points of chromosome walks to the
avirulence genes using a cosmid library. Clones of the fungus containing
putative avirulence genes will be transformed into virulent races as soon as
a transformation system becomes available. By this approach only a limited
number of clones will need to be introduced into *B. lactucae* rather than the
large numbers required for the shotgun approach. By using RFLP markers
one linkage of 6.5 cM between the RFLP probe (*G 538*) and avirulence gene
avr 6 has been identified. By assuming a constant relationship between
physical and genetic distance 6.5 cM equals on average approximately
160 kb in *B. lactucae*. Each walking step through this region will provide
new polymorphic probes which can be used to estimate the proximity to the
avirulence gene and the direction of the walk. Also pulsed-field gel electro-
phoretic separation of chromosomes of *B. lactucae* could facilitate chromosome walking.

However, independent of the approach used to clone an avirulence gene
there is need to develop an efficient transformation system for *B. lactucae*.
Many of the methods used to transform other fungi cannot be used for
B. lactucae because of its strict obligate nature and possibly because
Oomycetes are quite different from Ascomycetes and Basidiomycetes (Michelmore et al., 1988). To achieve transformation of *B. lactucae*, vectors
need to be constructed that will express markers for selection. The DNA
must be introduced into intact spores or germlings and the selection must
work in planta.

B. Magnaporthe grisea (Pyricularia oryzae)–Rice

Magnaporthe grisea is a fungal pathogen which has received much
attention in recent years as a model pathogen in molecular plant pathology
by several research groups. Cultivars of rice that carry single dominant
resistance genes effective against certain races of the pathogen have been
developed by Yamada et al. (1976). Genetic studies with the pathogen were
previously impossible due to lack of fertile isolates of *M. grisea* that infect
rice. Presently genetic studies are performed by crossing a sterile field
isolate pathogenic on rice and on weeping lovegrass with a highly fertile
strain that is only pathogenic on weeping lovegrass. The progeny that were
still pathogenic on rice were backcrossed several times with the sterile
rice/weeping lovegrass isolate as recurrent parent (Valent and Chumley,
1989; Leung et al., 1988). By checking the progeny on differentials of rice

several avirulence genes were identified (Valent and Chumley, 1989). The identified avirulence genes *avr C039*, *avr M 201* and *avr 1YM* appeared to be inherited from the parent that is nonpathogenic on rice, as the rice pathogen parent is pathogenic on the three rice cultivars C039, M201, and Yashiro-mochi, respectively. A similar result was obtained by Yaegashi and Asaga (1981) who reported that a finger millet pathogen carries an avirulence gene corresponding to the *M. grisea* resistance gene *Pi-a*. These results are also similar to those obtained with *P. syringae* pv. *tomato* where the *avrD* gene was isolated that specifies differential responses on a set of soybean cultivars (Kobayashi et al., 1989), and also to those obtained with *X. campestris* pv. *vesicatoria* (tomato strain) which contains the avirulence gene *avr Rxv* effective on pepper (Whalen et al., 1988). Additional crosses have identified other avirulence genes: *avr 2YM* and *Pw12* which appear to be unstable. From strains carrying those avirulence genes spontaneous virulent mutants often appear in standard differential assays.

Whether the identified avirulence genes are dominant or not remains to be seen. *M. grisea* is a haploid organism and presently it is impossible to obtain stable vegetative diploids or heterokaryotic conidia (Crawford et al., 1986). Linkage of RFLPs to avirulence genes will be one of the first steps towards cloning of avirulence genes. In this respect *M. grisea* is easier to handle than *B. lactucae* as the former can be transformed efficiently (Parsons et al., 1987). On the other hand transformation of the host plant rice is more difficult than transformation of lettuce. Transformation of rice is a prerequisite to eventual isolation and characterization of the corresponding resistance genes. There is yet much to be done before the cloning of avirulence genes and the isolation of the corresponding resistance genes in the interactions of *B. lactucae*–lettuce and *M. grisea*–rice become a reality.

C. Cladosporium fulvum–Tomato

Cladosporium fulvum is a biotrophic fungal parasite which enters tomato leaves through stomata, colonizes the intercellular spaces between meso-phyll cells and is confined to the apoplast during the main part of its life cycle (De Wit, 1977; Lazarovits and Higgins, 1976). The fungus does not form haustoria and causes no visible damage to the mesophyll cells and their walls. The interaction between *C. fulvum* and tomato is supposed to fit into a gene-for-gene relationship (De Wit and Oliver, 1989). Many genes for resistance have been identified in tomato cultivars. Of these resistance genes, Cf2, Cf4, Cf5, and Cf9 are available in near-isogenic lines of the cultivar Moneymaker. These lines give a clear differential response to the presently known races, resulting either in a compatible or an incompatible interaction. The *C. fulvum*–tomato interaction is an ideal model system to study communication between plant and pathogen, as the interface exchange of molecules is confined to the apoplast, from which washing fluids can be easily obtained (De Wit and Oliver, 1989; De Wit and Spikman,

1982; De Wit et al., 1984, 1985). Apoplastic fluids of infected leaves can be obtained by in vacuo infiltration with water or buffer followed by low speed centrifugation. In the apoplastic fluids of *C. fulvum*-infected leaves, in addition to many other compounds, fungal proteins occur which are constitutively produced or specifically induced by the plant. The fungal proteins can be divided in two categories. The first are those proteins important for obtaining or establishing basic compatibility (De Wit et al., 1986; Joosten and De Wit, 1988). The second category contains proteins which play a role as inducers of HR in resistant cultivars of tomato (De Wit and Spikman, 1982; De Wit et al., 1984, 1985; Scholtens-Toma et al., 1988, 1989). Apoplastic fluids obtained from different compatible race–cultivar interactions have been shown to contain race-specific elicitors which specifically induce HR in tomato cultivars carrying the corresponding genes for resistance (De Wit and Spikman, 1982; De Wit et al., 1984, 1985; Scholtens-Toma et al., 1989). Here the state of the art concerning the biochemical and molecular characterization of one of the race-specific elicitors and the cloning of its encoding avirulence gene *avr 9* will be discussed in detail.

1. Cloning and Characterization of the Avirulence Gene *avr 9*

The intercellular fluids isolated from different compatible *C. fulvum*–tomato interactions contain race-specific proteinaceous elicitors (De Wit and Spikman, 1982; De Wit et al., 1984, 1985; Higgins and De Wit, 1985). One of these elicitors, the putative product of avirulence gene *avr 9* which specifically interacts with a product of resistance gene Cf9 has been purified and its amino acid sequence has been determined (Scholtens-Toma and De Wit, 1988). It was isolated from apoplastic fluids of compatible interactions with races carrying the *avr 9* gene. The elicitor was not detectable in compatible interactions involving race 2.4.5.9, which does not carry *avr 9*. Recently we have found that new Dutch, French, and Polish isolates of *C. fulvum* carrying virulence gene a9 (race 2.5.9, 2.4.9.11, and 2.4.5.9.11, respectively) also do not produce the necrosis-inducing peptide indicating that absence of this peptide and virulence towards Cf9 genotypes is strongly associated (Scholtens-Toma et al., 1989). The production of the peptide was strongly induced in the host plant and could not be detected in culture filtrates of the fungus grown in vitro. The peptide could also not be detected in incompatible interactions, presumably because it is only present in very low amounts and it may bind to receptor sites as soon as it is produced.

To clone the *avr 9* gene of *C. fulvum* an approach significantly different from that used to clone bacterial avirulence genes has been followed. As the amino acid sequence of the race-specific elicitor, the putative product of avirulence gene *avr 9*, is known (Scholtens-Toma and De Wit, 1988) the approach from protein to gene was followed rather than the shotgun approach. A degenerated oligonucleotide probe derived from the amino

acid sequence was used to screen a cDNA library of a compatible
C. fulvum–tomato interaction involving a race carrying the *avr 9* gene. The
cDNA library was prepared from the interaction at the time that there was a
high production of the race-specific elicitor. From this cDNA library one
clone was obtained with an insert that contained the entire amino acid
sequence of the race-specific *avr 9* elicitor as determined previously
(Scholtens-Toma and De Wit, 1988; Van Kan et al., 1991). Analysis of the
cDNA clone revealed that the *avr 9* gene encodes a protein of 63 amino acids,
containing the sequence of the mature elicitor at the C-terminal end (Van
Kan et al., 1991). Southern blot analysis of DNA isolated from races of *C.
fulvum* with and without *avr 9* revealed that the cDNA clone hybridized to
single bands in various restriction enzyme digests of the races with *avr 9*, i.e.,
races 0, 2, 4, 5, 2.4, 2.5, 2.4.11, 2.4.5, and 2.4.5.11, respectively, but not to DNA
from races which lack *avr 9*, i.e., races 2.5.9, 2.4.5.9, 2.4.9.11, and 2.4.5.9.11,
respectively. Apparently the races that are virulent on tomato genotypes
that carry resistance gene Cf9 have no DNA homologous to the coding
sequence of *avr 9*. This indicates that there is no evidence for a recessive
allele of *avr 9* present in races which are virulent on Cf9 genotypes. From
these results it may be concluded that *avr 9* is located on an episomal factor,
on a B-chromosome or on an unstable part of a chromosome. Avirulence
gene *avr 9* is the first fungal avirulence gene ever cloned. It will be interesting
to know whether the other avirulence genes inducing HR on tomato
genotypes carrying resistance genes Cf2, Cf4, and Cf5 can be lost from the
C. fulvum genome in a similar way as avirulence gene *avr 9*. If this would be
the case then the term recessive allele is misleading and may not be
applicable to *C. fulvum*.

2. Transformation of Virulent Races of *C. fulvum* with the Avirulence Gene *avr 9*

A further proof for having cloned the avirulence gene *avr 9* will be obtained
by transforming virulent races from *C. fulvum* such as race 2.5.9, 2.4.9.11,
and 2.4.5.9.11 with the clone carrying the *avr 9* gene. Transformants should
become avirulent on tomato genotypes carrying resistance gene Cf9. A
transformation system for *C. fulvum* has been established (Oliver et al.,
1987; Roberts et al., 1989). As mentioned above the *avr 9* gene is completely
lacking in the virulent races 2.5.9, 2.4.9.11, and 2.4.5.9.11. There are some
preliminary indications that the *avr 9* gene is located on part of a chromosome which can be lost quite easily. A race lacking *avr 9* carries a chromosome which seems to be 500 kb shorter than the corresponding chromosome in the races carrying *avr 9* (Oliver et al., unpubl. results). It will be
interesting to know whether position effects play a role for the functioning
of the *avr 9* gene once introduced in the transformants. A disadvantage of *C.
fulvum* is the lack of a sexual stage which makes it impossible to perform
conventional genetic studies. However, it seems possible to perform para-

sexual studies with this fungus as reported by Talbot et al. (1988a), who constructed diploids by protoplast fusion. To this end a small collection of selectable mutants has been made using UV and transformation (Talbot et al., 1988a). Regeneration frequencies of the double-selected phenotypes were in the range of 10^{-5} to 10^{-3}. Ploidy was determined using a novel system based on 4′,6-diamidino-2-phenylindole (DAPI) fluorescence of nuclei and a computer controlled system (Talbot et al., 1988b). Some stable diploids were obtained, but in most cases spontaneous haploidisation occurred. Preliminary evidence suggests that haploids are recombinant but heterokaryosis has not been ruled out.

Combining genetic and biochemical approaches has led to the discovery of retrotransposon-like elements in the genome of *C. fulvum* (McHale et al., 1989). If this element proves to be a classical retrotransposon, it may help explain the ability of fungal races to mutate to overcome plant resistance genes. However, this conclusion is very preliminary as transposition of this element still has to be proven.

VI. Conclusion

Many gene-for-gene relationships have been reported of which some are genetically well characterized. However, only a few of them have become amenable to study at the molecular level. Data of molecular genetic studies on model pathogen–plant interactions reported here support the gene-for-gene hypothesis as has been put forward by Flor nearly fifty years ago. Many bacterial avirulence genes and one fungal avirulence gene have been cloned until now, but the first resistance gene has yet to be cloned. Based on the data obtained the specific elicitor-receptor model is valid to explain gene-for-gene interactions at the molecular level. However, resistance genes and their products need to be identified and their interaction with specific elicitors needs to be studied before the specific elicitor-receptor model can be fully accepted.

Acknowledgements

I wish to thank Dr. Jan A. L. van Kan and Prof. William E. Fry for helpful comments on the manuscript. The authors' research is supported by grants from the EEC (in the framework of the Biotechnology Action Programme; contract no. BAP-0074-NL) and the Dutch Science Foundation (NWO).

VII. References*

Barret JA (1985) The gene-for-gene hypothesis: parable or paradigm. In: Rollinson D, Anderson RM (eds) Ecology and genetics of host–parasite interactions. Academic Press, London, pp 215–225 (Linnean Society of London)

*The literature was updated until May 1990.

Bennetzen JL, Qin MM, Ingels S, Ellingboe AH (1988) Allele specific and mutator-associated instability at the Rpl disease-resistance locus of maize. Nature 332: 369–370

Bonas U, Stall RE, Staskawicz BJ (1989) Genetic and structural characterization of the avirulence gene *avrBs3* from *Xanthomonas campestris* pv. *vesicatoria*. Mol Gen Genet 218: 127–136

Brinkerhoff LA (1970) Variation in *Xanthomonas malvacearum* and its relation to control. Annu Rev Phytopathol 8: 85–110

Caten CE (1987) The concept of race in plant pathology. In: Wolfe MS, Caten CE (eds) Populations of plant pathogens, their dynamics and genetics. Blackwell, Oxford, pp 21–37 (British Society of Plant Pathology)

Chatterjee AK, Vidaver AK (1986) Genetics of pathogenicity factors: application to phytopathogenic bacteria. Adv Plant Pathol 4: 1–224

Christ BJ, Groth JV (1982) Inheritance of virulence to three bean cultivars in three isolates from the bean rust pathogen. Phytopathology 72: 767–770

Christ BJ, Person CO, Pope DD (1987) The genetic determination of variation in pathogenicity. In: Wolfe MS, Caten CE (eds) Populations of plant pathogens, their dynamics and genetics. Blackwell, Oxford, pp 21–37 (British Society of Plant Pathology)

Collinge DB, Slusarenko AJ (1987) Plant gene expression in response to pathogens. Plant Mol Biol 9: 389–410

Cook AA, Stall RE (1963) Inheritance of resistance in pepper to bacterial spot. Phytopathology 53: 1060–1062

Crawford MS, Chumley FG, Weaver CG, Valent B (1986) Characterization of the heterokaryotic and vegetative diploid phases of *Magnaporthe grisea*. Genetics 114: 1111–1129

Crute IAM, De Wit PJGM, Wade M (1985) Mechanisms by which genetically controlled resistance and virulence influence host colonization by fungal and bacterial parasites. In: Fraser RSS (ed) Mechanisms of resistance to plant diseases. Martinus Nijhoff/Dr W Junk, Dordrecht, pp 197–309

Crute IR (1985) The genetic bases of relationships between microbial parasites and their hosts. In: Fraser RSS (ed) Mechanisms of resistance to plant diseases. Martinus Nijhoff/Dr W Junk, Dordrecht, pp 80–142

De Wit PJGM (1977) A light and scanning-electron microscopic study of infection of tomato plants by virulent and avirulent races of *Cladosporium fulvum*. Neth J Plant Pathol 83: 109–122

De Wit PJGM (1987) Specificity of active resistance mechanisms in plant-fungus interactions. In: Pegg GF, Ayres PG (eds) Fungal infection of plants. Cambridge University Press, Cambridge, pp 1–24

De Wit PJGM, Oliver RP (1989) The interaction between *Cladosporium fulvum* (syn. *Fulvia fulva*) and tomato: a model system in molecular plant pathology. In: Nevalainen H, Penttilä M (eds) Molecular biology of filamentous fungi. Found Biotechn Industr Ferment Res 6: 227–236

De Wit PJGM, Spikman G (1982) Evidence for the occurrence of race and cultivar-specific elicitors of necrosis in intercellular fluids of compatible interactions of *Cladosporium fulvum* and tomato. Physiol Plant Pathol 21: 1–11

De Wit PJGM, Hofman JE, Aarts JMMJG (1984) Origin of specific elicitors of chlorosis and necrosis occurring in intercellular fluids of compatible interactions of *Cladosporium fulvum* (syn. *Fulvia fulva*) and tomato. Physiol Plant Pathol 24: 17–23

De Wit PJGM, Hofman JE, Velthuis GCM, Kuc JA (1985) Isolation and characterization of an elicitor of necrosis isolated from intercellular fluids of compatible interactions of *Cladosporium fulvum* (syn. *Fulvia fulva*) and tomato. Plant Physiol 77: 642–647

De Wit PJGM, Buurlage MB, Hammond KE (1986) The occurrence of host, pathogen and interaction-specific proteins in the apoplast of *Cladosporium fulvum* (syn. *Fulvia fulva*) infected tomato-leaves. Physiol Mol Plant Pathol 29: 159–172

Dyck PL, Samborski DJ (1982) The inheritance of resistance to *Puccinia recondita* in a group of common wheat cultivars. Can J Gen Cytol 24: 273–283

Ellingboe AH (1984) Genetics of host–parasite relations: an essay. Adv Plant Pathol 2: 131–151

Farrara BT, Illot TW, Michelmore RW (1988) Genetic analysis of factors for resistance to downy mildew (*Bremia lactucae*) in lettuce (*Lactuca sativa*). Plant Pathol 36: 499–514.

Flor HH (1946) Genetics of pathogenicity in *Melampsora lini*. J Agricult Res 73: 335–357

Flor HH (1971) Current status of the gene-for-gene concept. Annu Rev Phytopathol 9: 275–296

Gabriel DW, Burges A, Lazo GR (1986) Gene-for-gene interactions of five cloned avirulence genes from *Xanthomonas campestris* pv. *malvacearum* with specific resistance genes in cotton. Proc Natl Acad Sci USA 83: 6415–6419

Gabriel DW, Loschke DC, Rolfe BG (1988) Gene-for-gene recognition: the ion channel defense model. In: Palacios R, Verma DPS (eds) Molecular genetics of plant–microbe interactions. American Phytopathological Society, St. Paul, MN, pp 3–14

Hibberd AM, Basset MJ, Stall RE (1987) Allelism tests of three dominant genes for hypersensitive resistance to bacterial spot of pepper. Phytopathology 77: 1304–1307

Higgins VJ, De Wit PJGM (1985) Use of race and cultivar specific elicitors from intercellular fluids for characterizing races of *Cladosporium fulvum* and resistant tomato cultivars. Phytopathology 75: 695–699

Hitchen FE, Jenner CE, Harper S, Mansfield JW, Barber CE, Daniels MJ (1989) Determinant of cultivar specific avirulence cloned from *Pseudomonas syringae* pv. *phaseolicola* race 3. Physiol Mol Plant Pathol 34: 309–322

Hulbert SH, Ilott TW, Legg EJ, Lincoln SE, Lander ES, Michelmore RW (1988) Genetic analysis of the fungus *Bremia lactucae*, using restriction fragment length polymorfisms. Genetics 120: 947–958

Huynh TV, Dahlbeck D, Staskawicz BJ (1989) Bacterial blight of soybean: regulation of a pathogen gene determining host cultivar specificity. Science 245: 1374–1377

Ilott TW, Durgan ME, Michelmore RW (1987) Genetics of virulence in Californian populations of *Bremia lactucae* (lettuce downy mildew). Phytopathology 77: 1381–1386

Ilott TW, Hulbert SH, Michelmore RW (1989) Genetic analysis for the gene-for-gene interaction between lettuce (*Lactucae sativa*) and *Bremia lactucae*. Phytopathology 79: 888–897

Joosten MHAJ, De Wit PJGM (1988) Isolation, purification and preliminary characterization of a protein specific for compatible *Cladosporium fulvum* (syn. *Fulvia fulva*)–tomato interactions. Physiol Mol Plant Pathol 33: 142–253

Judelson HS, Michelmore RW (1989) Strategies for cloning avirulence genes from *Bremia lactucae*. In: Staskawicz B, Ahlquist P, Yoder O (eds) Molecular biology of plant–pathogen interactions. AR Liss, New York, pp 71–85 (UCLA Symposia on Molecular and Cellular Biology, vol 101)

Kearney B, Ronald PC, Dahlbeck D, Staskawicz BJ (1988) Molecular basis of evasion of plant host defense in bacterial spot disease of pepper. Nature 332: 541–543

Keen NT (1982) Specific recognition in gene-for-gene host parasite systems. Adv Plant Pathol 1: 35–82

Keen NT, Dawson WO (1992) Pathogen avirulence genes and elicitors of plant defense. In: Boller T, Meins F (eds) Genes involved in plant defense. Springer, Wien New York, pp 85–114 [Dennis ES et al (eds) Plant gene research. Basic knowledge and application]

Keen NT, Staskawicz BJ (1988) Host range determinants in plant pathogens and symbionts. Annu Rev Microbiol 42: 421–440

Keen NT, Tamaki S, Kobayashi D, Gerhold D, Stayton M, Shen H, Gold S, Lorang J, Thordal-Christensen H, Dahlbeck D, Staskawicz B (1990) Bacteria expressing avirulence gene D produce a specific elicitor of the soybean hypersensitive reaction. Mol Plant Microbe Interact 3: 112–121

Kelemu S, Leach JE, (1990) Cloning and characterization of an avirulence gene from Xanthomonas campestris pv. oryzae. Mol Plant Microbe Interact 3: 59–65

Knott DR, Anderson RG (1956) The inheritance of rust resistance. I. The inheritance of stem rust resistance in ten varieties of common wheat. Can J Agricult Sci 36: 174–195

Kobayashi DY, Tamaki SJ, Keen NT (1989) Cloned avirulence genes from the tomato pathogen Pseudomonas syringae pv. tomato confer cultivar specificity on the non-host soybean. Proc Natl Acad Sci USA 86: 157–161

Kobayashi DY, Tamaki SJ Trollinger DJ, Gold S, Keen NT (1990a) A gene from Pseudomonas syringae pv. glycinea with homology to avirulence gene D from P. s. pv. tomato but devoid of the avirulence phenotype. Mol Plant Microbe Interact 3: 103–111

Kobayashi DY, Tamaki SJ, Keen NT (1990b) Molecular characterization of avirulence gene D from Pseudomonas syringae pv. tomato. Mol Plant Microbe Interact 3: 94–102

Lawrence GJ, Mayo GME, Shepherd KW (1981) Interaction between genes controlling pathogenicity in the flax rust fungus. Phytopathology 71: 12–19

Lazarovits G, Higgins VJ (1976) Ultrastructure of susceptible, resistant and immune reactions of tomato to races of Cladosporium fulvum. Can J Bot 54: 255–249

Leong SA, Holden DW (1989) Molecular genetic approaches to the study of fungal pathogenesis. Annu Rev Plant Pathol 27: 463–481

Leung H, Borromeo ES, Bernardo MA, Notteghem JJ (1988) Genetic analysis of virulence in the rice blast fungus Magnaporthe grisea. Phytopathology 78: 1227–1233

Lindgren PB, Panopoulos NJ, Staskawicz BJ, Dahlbeck D (1988) Genes required for pathogenicity and hypersensitivity are conserved and interchangeable among pathovars of Pseudomonas syringae. Mol Gen Genet 211: 499–506

Martens JW, Rothmana PG, Mckenzie RIH, Brown PD (1981) Evidence of complementary gene action conferring resistance to Puccinia graminis avena in Avena sativa. Can J Gen Cytol 23: 581–595

McHale MT, Roberts IN, Talbot NJ, Oliver RP (1989) Expression of reverse transcriptase genes in Fulvia fulva. Mol Plant Microbe Interact 2: 165–168

Mew TW (1987) Current status and future prospects of research on bacterial blight of rice. Annu Rev Plant Pathol 25: 359–382

Michelmore RW, Norwood JM, Ingram DS, Crute IR, Nicholson P (1984) The inheritance of virulence in Bremia lactucae to match resistance factors 3, 5, 6, 8, 9, 10 and 11 in lettuce (Lactuca sativa). Plant Pathol 13: 301–315

Michelmore RW, Ilott TW, Hulbert SH, Farrara B (1988) The downey mildews. Adv Plant Pathol 6: 53–79

Minsavage GV, Dahlbeck D, Whalen MC, Kearney B, Bonas U, Staskawicz BJ, Stall RE (1990) Gene-for-gene relationships specifying disease resistance in Xanthomonas campestris pv. vesicatoria-interactions. Mol Plant Microbe Interact 3: 41–47

Napoli C, Staskawicz BJ (1987) Molecular characterization and nucleic acid sequence of an avirulence gene from race 6 of Pseudomonas syringae pv. glycinea. J Bacteriol 169: 572–578

Oliver RP, Roberts IN, Harling R, Kenyon L, Punt PJ, Dingemanse MA, Van den Hondel CAMJJ (1987) Transformation of Fulvia fulva, a fungal pathogen of tomato, to hygromycin B resistance. Curr Genet 12: 231–233

Parlevliet JE, Zadoks JC (1977) The integrated concept of disease resistance; a new view including horizontal and vertical resistance in plants. Euphytica 26: 5–21

Parsons KA, Chumley FG, Valent B (1987) Genetic transformation of the fungal pathogen responsible of rice blast disease. Proc Natl Acad Sci USA 84: 4161–4165

Person CO (1959) Gene-for-gene relationships in host : parasite systems. Can J Bot 37: 1101–1130

Person CO, Mayo GME (1974) Genetic limitations of specific interactions between a host and its parasite. Can J Bot 52: 1339–1347

Roberts IN, Oliver RP, Punt PJ, van den Hondel CAMJJ (1989) Expression of the *E. coli* β-glucuronidase gene in filamentous fungi. Curr Genet 15: 177–180

Ronald PC, Staskawicz BJ (1988) The avirulence gene *avrBs 1* from *Xanthomonas campestris* pv. *vesicatoria* encodes a 50-kD protein. Mol Plant Microbe Interact 1: 191–198

Scholtens-Toma IMJ, De Wit PJGM (1988) Purification and primary structure of a necrosis inducing peptide from apoplastic fluids of tomato infected with *Cladosporium fulvum* (syn. *Fulvia fulva*). Physiol Mol Plant Pathol 33: 59–67

Scholtens-Toma IMJ, De Wit GJM, De Wit PJGM (1989) Characterization of apoplastic fluids isolated from tomato lines inoculated with new races of *Cladosporium fulvum*. Neth J Plant Pathol 95: 161–168

Shintaku MH, Klueppel DA, Yacoub A, Patil SS (1989) Cloning and partial characterization of an avirulence determinant from race 1 of *Pseudomonas syringae* pv. *phaseolicola*. Physiol Mol Plant Pathol 35: 313–322

Sidhu GS (1980) Genetic analysis of plant parasitic systems. In: Proceedings XIV International Congress of Genetics 1: 391–408

Somerville C (1989) *Arabidopsis* blooms. Plant Cell 1: 1131–1135

Spielman LJ, McMaster BJ, Fry WE (1989) Dominance and recessiveness at loci for virulence against potato and tomato in *Phytophthora infestans*. Theor Appl Genet 77: 832–838

Spielman LJ, Sweigard JA, Shattock RC, Fry WE (1990) The genetics of *Phytophthora infestans*: segregation of allozyme markers in F2 and backcross progeney and the inheritance of virulence against potato resistance genes R2 and R4 in F1 progeny. Exp Mycol 14: 57–69

Stakman EC (1917) Biologic forms of *Puccinia graminis* on cereals and grasses. J Agricult Res 10: 429–495

Staskawicz BJ, Dahlbeck D, Keen NT (1984) Cloned avirulence gene of *Pseudomonas syringae* pv. *glycinea* determines race-specific incompatibility on *Glycine max* (L.) Merr. Proc Natl Acad Sci USA 81: 6024–6028

Staskawicz B, Dahlbeck D, Keen N, Napoli C (1987) Molecular characterization of cloned avirulence genes from race 0 and race 1 of *Pseudomonas syringae* pv. *glycinea*. J Bacteriol 169: 5789–5794

Swanson J, Kearney B, Dahlbeck D, Staskawicz BB (1988) Cloned avirulence gene of *Xanthomonas campestris* pv. *vesicatoria* complements spontaneous race-change mutants. Mol Plant Microbe Interact 1: 5–9

Talbot NJ, Coddington A, Roberts IN, Oliver RP (1988a) Diploid construction by protoplast fusion in *Fulvia fulva* (syn. *Cladosporium fulvum*): genetic analysis of an imperfect fungal plant pathogen. Curr Genet 14: 567–572

Talbot NJ, Rawlins D, Coddington A (1988b) A rapid method for ploidy determination in fungal cells. Curr Genet 14: 51–52

Tamaki S, Dahlbeck D, Staskawicz BJ, Keen NT (1988) Characterization and expression of two avirulence genes cloned from *Pseudomonas syringae* pv. *glycinea*. J Bacteriol 170: 4846–4854

Toxopeus HJ (1956) Reflections on the origin of new physiologic races in *Phytophthora infestans* and the breeding for resistance in potatoes. Euphytica 5: 221–237

Valent B, Chumley F (1989) Genes for cultivar specificity in the rice blast fungus, *Magnaporthe grisea*. In: Lugtenberg BJJ (ed) Signal molecules in plants and plant–microbe interactions. Springer, Berlin Heidelberg New York Tokyo, pp 415–422

Van Kan JAL, Van Den Ackerveken AFJM, De Wit PJGM (1991) Cloning and characterization of the avirulence gene *avr 9* of the fungal pathogen *Cladosporium fulvum*. Mol Plant Microbe Interact 4: 52–59

Vivian A, Atherton GT, Bevan JR, Crute IR, Mur LAJ, Taylor JD (1989) Isolation and characterization of cloned DNA conferring specific avirulence in *Pseudomonas syringae* pv. *pisi* to pea (*Pisum sativum*) cultivars, which possess the resistance allele, R2. Physiol Mol Plant Pathol 34: 335–344

Whalen M, Stall RE, Staskawicz BJ (1988) Characterization of a gene from a tomato pathogen determining hypersensitive resistance in non host species and genetic analysis of this resistance in bean. Proc Natl Acad Sci USA 85: 6743–6747

Yamada M, Kiyosawa S, Yamamuchi T, Hirano T, Kobayashi T, Kushibuchi K, Watanabe S (1976) Proposal of a new method for differentiating races of *Pyricularia oryzae* Cavara in Japan. Ann Phytopathol Soc Jpn 42: 216–219

Yaegashi H, Asaga K (1981) Further studies on inheritance of pathogenicity in crosses of *Pyricularia oryzae* with *Pyricularia* sp. from finger millet. Ann Phytopathol Soc Jpn 47: 677–679

Yoder OC, Valent B, Chumley F (1986) Genetic nomenclature and practice for plant pathogenic fungi. Phytopathology 76: 383–385

Section II
Virulence and Avirulence Genes of Pathogens

Chapter 3

An Analysis of Host Range Specificity Genes of *Rhizobium* as a Model System for Virulence Genes in Phytobacteria

Michael A. Djordjevic, Barry G. Rolfe, and Wendy Lewis-Henderson

Plant–Microbe Interaction Group, Research School of Biological Sciences, Australian National University, Canberra, ACT 2601, Australia

With 2 Figures

Contents

I. Introduction

Microbes have long exploited plants as favourable niches for colonisation. The surfaces and intercellular spaces of leaves, vascular tissue (phloem and xylem), and the ecto- and endorhizospheres (root surface and intercellular spaces between root cortical cells, respectively) are colonised by microbes to varying extents (Agrios, 1988). About 80 species of bacteria are known to interact with plants either beneficially or detrimentally. The consequence of colonisation or penetration of the plant tissue varies from asymptomatic, to disease, to symbiosis and is influenced greatly by the environment. Products of photosynthesis, such as organic acids and sucrose, can provide energy sources for microbial growth. The roots are not only surrounded by a plethora of soil microbes but up to 40% of fixed carbon can be exported from root cells as organic chemicals or as cellular debris sloughed from the root cap (Lynch and Whipps, 1989). The soil, by comparison, offers a relatively nutrient-poor environment (Suslow, 1982). The cellular energy reserves of microbes bound to soil particles will influence the rapidity of their responsiveness to changes in their environment, for example, the passing of a growing root in their vicinity.

Plants have evolved elaborate defense systems (Dickinson and Lucas, 1982; Dixon, 1986; Lamb et al., 1989). Most microbes are successfully excluded from colonising plant tissue simply through passive resistance. Those microbes that do penetrate and colonise plants, have evolved an equally elaborate system for circumventing or subverting plant defenses (Daniels et al., 1988; Djordjevic et al., 1987a; Keen and Staskawicz, 1988; Long, 1989). Bacteria of the family Rhizobiaceae (including rhizobia, bradyrhizobia, azorhizobia and agrobacteria) provide good examples of this type of complex interaction which leads to minimal cell death (Vance, 1983; Billing, 1987). Biotrophic microbes, which require living tissue, may be thought of as the "confidence tricksters" of the parasite world (Keen, 1986). In contrast, the necrotrophs kill invaded host cells often through the action of toxins and then metabolize the remains. They are commonly referred to as parasitic "thugs" (Keen, 1986). This contrast in strategy for parasitism between biotrophs and necrotrophs is generally reflected in the level of genetic sophistication of parasitism genes in these organisms. Because of the requirement to maintain the cellular integrity of the plant cells, biotrophs tend to be highly specific for host species (Gabriel, 1986). The fine balance of biotrophic interactions often can be influenced by single gene traits in the host and the biotroph which lead to variations in infectivity between cultivars of a single species (Gabriel, 1986).

In this review, a comparison of the genetic and molecular requirements for infection by nitrogen-fixing plant symbionts belonging to the Rhizobiaceae, collectively called "rhizobia", will be addressed to determine if general principles can be identified that apply to biotrophic plant pathogens such as pseudomonads, xanthomonads and agrobacteria.

Rhizobia are soil bacteria capable of infecting and colonising the roots of select legumes and inducing root out-growths called nodules where nitrogen-fixation can occur. Inside nodules, rhizobia may undergo a specialized alteration in their physiological and biochemical status and become (usually) terminally differentiated bacteroids (Zhou et al., 1985), which can fix nitrogen and make a major contribution to the plant's nitrogen requirements (Vance et al., 1988). The complexity of this interaction is such that bacteroids are thought of as being sub-cellular plant organelles (Morrison et al., 1988).

A focus of this review will be a comparison of *Rhizobium* host range genes to the virulence genes in biotrophic plant pathogens. Bacteria should be ideal plant invaders because of their rapid environmental responsiveness and multiplication rates, and their ability to synthesize and secrete low and high molecular weight molecules which can be potential mediators of communication with the plant host.

Several terms are used in plant pathology to describe different interactions that occur in plant–pathogen interactions which are generally not applied to *Rhizobium*–plant interactions (Fraser, 1985). The "jargon" used can sometimes create unnecessary barriers that restrict useful comparisons. Definitions of the terms which will be used in this review are covered here.

A plant that supports the multiplication of a particular microbe is *compatible* to that microbe (Gabriel, 1989). Although the term compatible appears somewhat counter-intuitive it implies that there is a sophisticated interaction of host and bacterial genes which leads to a complex interplay— even if the interaction is unfavourable to the plant. If, for example, this compatible interaction results in a disturbance of normal physiological growth or development, that is usually manifested in signs of *macroscopic dysfunction*, the plant is *diseased*, the organism is termed a *pathogen* and a *virulent* response results (Agrios, 1988; Billing, 1987; Dickinson and Lucas, 1982). The plant host is also compatible if proliferation of the bacterium inside plant tissues occurs without any recognisable response (for example, in the growth of certain xanthomonads in apple plants; Mass et al., 1985) or if symbiosis is established. *Incompatibility* occurs where a host is induced to restrict the colonisation of the invader resulting in a reduction of the reproductive potential of the bacterium. This can refer to variations in colonisation on different cultivars of the same host or situations where reactions to invaders are seen on non-host plants. An incompatible interaction is often, but not always, associated with the induction of rapid plant cell death (called a hypersensitive response: HR) at the site of penetration of the host, which leads to an unfavourable niche for colonisation by the biotroph and hence reduced colonisation potential. Compatibility and incompatibility are best assayed by the determination of the reproductive fitness and extent of colonisation by the bacteria in planta (Gabriel, 1989). Cellular recognition is an integral component of plant–pathogen interactions as well as normal plant developmental processes. Compatibility at different stages of infection is required for *symbiosis*. In the early stages of plant infection by rhizobia a parasitic interaction takes place. The plant host does not derive a benefit until nitrogen fixation occurs (Djordjevic et al., 1987a).

II. Genes Required for Plant–Microbe Interactions

The evolution of plant–microbial interactions has led to the generation of several types of genes which condition these interactions. Other reviews have examined *Rhizobium*–plant interactions and showed that this plant symbiont could provide good examples of the stages of selective evolution which reflect the different types of genes present in plant pathogens (Djordjevic et al., 1987a; Halverson and Stacey, 1986; Long, 1989; Long and Cooper, 1988; Vance, 1983; Downie and Johnston, 1986; Kondorosi, 1989; Long, 1984; Martinez et al., 1990; Rolfe and Gresshoff, 1988; Rolfe et al., 1989).

Gabriel (1986) has provided a useful framework for categorising genes of bacteria involved in interactions with plants. Superimposed onto a basic set of maintenance genes, are additional genes which confer parasitism pathogenicity, host range, and cultivar specificity. This classification of the genetic determinants of phytobacteria into those genes required for primary metabolism and essential maintenance and those required for plant interactions has been adopted here. Genes in the plant host, some of which will be addressed here, will also influence the outcomes of plant–bacterial interactions (Collinge and Slusarenko, 1987; Dixon, 1986; Franssen et al., 1987; Gloudemans et al., 1989; Holl 1975; Lie, 1984; Sequeira, 1983; Vance et al., 1988).

Phytobacteria induce different plant growth and developmental responses ranging from plant cell death, to uncontrolled growth of infected plant cells (galls or cankers), to highly organised root nodules, to asymptomatic responses (Agrios, 1988; Al-Mousawai et al., 1982; Djordjevic et al., 1987a; Gabriel, 1986; Maas et al., 1985; Sequeira, 1983). Signal exchange between the participating organisms mediate these responses (Halverson and Stacey, 1986). The potential array of signal molecules produced by phytobacteria include: extracellular enzymes and other secreted proteins, nucleic acids, polysaccharides, phytohormones, and secondary metabolites such as antibiotics, plant toxins and bacteriocins (Collmer et al., 1982; Daniels et al., 1988; Martin and Demain, 1980; Morris, 1986; Vidaver, 1983; Klement and Goodman, 1967; Ryan, 1988).

Since saprophytes do not multiply in or on living plants an argument can be made for the existence of microbial genes that condition parasitism (Gabriel, 1986, 1989; Keen and Staskawicz, 1988). One might also expect that parasitism genes are not required for saprophytic growth and, therefore, that they might be evolved and transmitted through soil bacterial populations on plasmids. Genetic studies of the Rhizobiaceae and several xanthomonads and pseudomonads has confirmed that many of the genes required for parasitism are plasmid located (Djordjevic et al., 1987a; Keen and Staskawicz, 1988; Panopoulos and Peet, 1985). Mutation of a parasitism gene does not affect saprophytic growth. Moreover, curing of entire plasmids from *Rhizobium* and *Agrobacterium* strains results in mutants which no longer visibly interact with plants although these strains appear to multiply on the root surface and remain as commensal organisms. This commensal ability is best demonstrated in mixed inoculation experiments where *Rhizobium* mutants, unable to nodulate either through mutation or deletion of plasmid genes, can be helped into nodules by wild-type strains where they can form a substantial proportion of the nodule population (Rolfe et al., 1980). Agrobacteria deleted for genes required for crown gall formation on plants can affect the ability of wild-type strains to colonise and cause disease through competition for infection sites and antibiotic synthesis (Kerr, 1987).

Plasmid located genes required for compatibility in rhizobia include the nodulation (*nod*) genes, some of which condition host specific nodulation (*hsn*), the nitrogen fixation (*nif* and *fix*) genes which confer the ability to establish symbiosis and the genes coding for exopolysaccharide (*exo*) production which, in many cases appear to be important for colonisation ability in planta. In some cases all of these genes are located on the chromosome (e.g., in bradyrhizobia), however, individual loss of any of these genes through mutation or deletion does not affect saprophytic growth ability.

III. *Rhizobium* as a Useful Model System

The *Rhizobium*-legume interaction is a useful model system for the study of virulence genes for the following reasons: (1) rhizobia and their plant hosts are easily grown in laboratory conditions; (2) the extracellular molecules secreted from rhizobia and plants have been studied extensively; (3) a large range of defined bacterial and plant mutant strains are available; (4) many bacterial and plant genes which condition compatibility during nodule establishment have been isolated, cloned, sequenced and their potential functions determined; (5) the host cells initially affected (root hairs) can be observed using microscopy techniques, are amenable to physiological measurements and can be isolated using simple techniques (Gerhold et al., 1985); (6) the anatomical steps that occur during symbiosis have been extensively studied microscopically (Callaham and Torrey, 1981; Calvert et al., 1984; Dudley et al., 1987; Ridge and Rolfe, 1986). The *Rhizobium*-legume symbiosis represents the best studied system with a great number of tools available to permit further advances.

A. The Infection Process

The infection of legume roots by *Rhizobium* is an intricate process and has been extensively reviewed elsewhere (Dart, 1974; Rolfe et al., 1989; Vincent, 1974, 1980). The unique aspects are summarised here. Rhizobia are attracted towards plant roots by secreted compounds (Caetano-Anolles et al., 1988). One set of signal molecules secreted by plant roots, the flavonoids, are important for nodule initiation (Djordjevic et al., 1987b; Firmin et al., 1986; Peters et al., 1986; Peters and Long, 1988; Peters and Verma, 1990; Redmond et al., 1986; Rolfe, 1988; Rolfe et al.,1988; Zaat et al., 1989; Innes et al., 1985; Mulligan and Long, 1985; Spaink et al., 1987b). A specific flavone recognition occurs which facilitates the switch in rhizobia from a saprophytic mode to an infectious (parasitic) mode and thus contributes to host species specificity. The root cells which occur just behind the root tip, secrete the highest concentrations of preferred flavonoids and cells in other areas of the root secrete anti-inducer flavonoids in higher concentrations than inducers (Rolfe et al., 1988). These anti-inducers appear to compete with inducer flavonoids to antagonise *nod* gene induction (Djordjevic et al., 1987b; Firmin et al., 1986; Peters and Long, 1988). The relative concentrations of inducers and anti-inducers at any one location on the root would, therefore, affect the ability of bacteria to rapidly colonise the root hair cells and this may result in the majority of rhizobia remaining in a commensal association instead of becoming true parasites or symbionts. Recognition of flavonoid molecules represents an adaptation by *Rhizobium* to the exudates of the host plant, as many flavonoids adversely affect microbial growth (Firmin et al., 1986). Similarly, the nitrogen levels in the rhizosphere also affect the expression of nodulation genes (Dusha et al., 1989).

Root hair cells located immediately behind the root tip are most susceptible to infection (Bhuvaneswari et al., 1980; Ridge and Rolfe, 1986; Sargent et al., 1987). As the root grows, the zone of the root which is highest in susceptibility also moves. A dynamic interaction occurs between the rapidity of bacterial responsiveness and (*i*) adaptation to the new host environment and (*ii*) the short time period to initiate infection events. The commensal population of rhizobia may remain on the root surface and be in a favourable location for subsequent infections at a later point if, say, a lateral root is initiated in the proximity of these bacteria.

Rhizobia responding rapidly to the presence of a passing root tip, may colonise the surface of the root and attach to the sites where root hair cells will emerge. As root hairs emerge from differentiating epidermal cells, flavonoid inducers have already induced *nod* gene expression and as a result rhizobia secrete low molecular weight molecules which (*i*) directly cause root hair cells to grow in an aberrant manner (root hair curling) and (*ii*), directly or indirectly, induce normally non-dividing cortical cells to undergo cell division. Cortical cell division may be induced directly by *Rhizobium* signals or via a secondary signal released from an affected root hair (Bauer et al., 1985; Dudley et al., 1987; Long and Cooper, 1988). The identity of the compounds inducing cortical cell division is not known but their activity can be partially mimicked by certain chemicals. Recently, polar auxin transport inhibitors have been shown to induce nodule formation on certain legume roots without the addition of *Rhizobium* cells (Hirsch et al., 1989). Moreover, a genetic variant of alfalfa has been isolated which is capable of forming nodules in the absence of rhizobia (Truchet et al., 1989). Collectively, these observations suggest that if *Rhizobium* signals cause cortical cell division directly, these signals act through a perturbation of normal hormonal gradients in a manner akin to polar auxin transport inhibitors. However, since nodule formation can occur in alfalfa variants in the absence of rhizobia, it becomes more likely that nodule initiation occurs through the generation of a secondary signal from the root hair cell as proposed by Long and Cooper (1988). Repair of the cell wall caused by *Rhizobium* penetration or root hair curling activity may generate a secondary signal which is transmitted to the root cortical cells by the symplast between the root hair cells and the stele.

Root hair curling occurs 15 h after exposure to rhizobia. Rhizobia trapped in the folds of the newly curled root hair cell appear to degrade plant cell wall material in the immediate vicinity of the attachment site of the bacterial cell (Callaham and Torrey, 1981; Ridge and Rolfe, 1985). Despite this, there exists no conclusive molecular evidence for the existence of *Rhizobium* genes coding for enzymes that degrade the pectin or cellulose. Rhizobia which have degraded the cell wall have the potential to kill the root hair cell. In some *Rhizobium* mutants, this is known to occur (see below) and a response similar to a HR is induced. However in the vast majority of cases, instead of initiating a plant-defense response which would prevent *Rhizobium* growth, the challenged plant root hair cell responds by elaborating newly synthesised cell wall material in the vicinity of the degraded area and an "infection thread" is initiated (Dart, 1974). This can be considered a "mild" defense or wounding response akin to lignification in sites of microbial colonisation in planta (Rolfe et al., 1989). However, while *Rhizobium* access to the plant cytoplasm is denied, there seems little impedence to *Rhizobium* colonisation at the site of infection thread initiation. The infection thread wall has a different chemical composition to the original cell wall as indicated by differential staining observed in electron micrographs and using cytological dyes (Ridge and Rolfe, 1986; Vasse and Truchet, 1984).

Infection threads grow just ahead of the multiplying rhizobia. A clear life style change, from saprophyte to parasite, is established in the new micro-niche. The infection thread grows to the base of the root hair, fuses there with the root hair cell wall and infection thread formation is re-initiated in adjacent plant cells (Libbenga and Bogers, 1974). Within 48 h, the infection thread has grown into the area of cortical cell division initiated earlier in the process (Calvert et al., 1984; Ridge and Rolfe, 1986). A continual release of signal molecules from rhizobia contained within infection threads is probably needed to sustain the number of cortical cell divisions in some plants (e.g., clover) while in other plants (e.g., soybean and alfalfa), once the initial signal is sent to the cortical cells, a number of sustained rounds of cortical cell divisions ensue as if an irreversible developmental sequence is triggered. In some plants, infection threads do not continue to grow further then the root hair or sub-epidermal cells (Djordjevic et al., 1986). This phenomenon occurs not only in infections initiated by certain mutant strains (Debelle et al., 1986; Djordjevic et al.,

1985b; Huang et al., 1988), but is also quite common with wild-type strains (Djordjevic et al., 1986).

Compounds regulating nodulation are synthesised in the plant shoot as a result of signals released from either the nodule meristem or from the colonising rhizobia which control nodule numbers (Delves et al., 1986). Plant mutants affected in this autoregulation of nodule number are hyper-nodulated (Carroll et al., 1985). The transmissibility of the signals between root and shoot tissue can be demonstrated using "split root" plants where two equally robust lateral root systems have been initiated by destroying apical dominance (Sargent et al., 1987). Nodulation by one strain on a lateral root can reduce nodulation ability by another strain on the sepa-rated lateral root system of the same plant.

Infection threads enter the cortical tissue and bacteria are packaged into plant-derived "peribacteroid" membranes within the infected cell (Van Den Bosch and Newcomb, 1986). Rhizobia undergo a differentiation process which results in physiological and metabolic changes: the bacteria become "bacteroids" which may then specifically express genes involved in nitrogen fixation so that a symbiosis is established (Gussin et al., 1986; Triplett et al., 1989; Vincent, 1974, 1980).

B. Plant Genes Involved in Nodule Formation

The differential expression of a series of nodule-specific plant genes is involved in the formation of nodules. The proteins encoded by these genes are called "nodulins" (Van Kammen, 1984). Early nodulins are expressed well before the onset of nitrogen fixation and are involved in the formation of infection thread wall and cell wall components of specific nodule tissues. Late nodulins such as leghemoglobin, uricase and peribacteroid membrane proteins are expressed around the onset of nitrogen fixation after a complete nodule has developed and are involved in establishing the physiological conditions necessary for nitrogen fixation. Early nodulins such as ENOD2 and ENOD12, appear to be structural proteins with high proline contents and resemble hydroxy-proline rich glycoproteins (which are thought to be involved in host-pathogen interactions) (Franssen et al., 1987; Scheres et al., 1990). The genes encoding nodulins are highly con-served amongst legumes and do not appear to be involved in host specificity.

Lectins have been shown recently to be plant-derived determinants of host specificity. Lectins are plant proteins capable of specifically binding to polysaccharides and recognise the three-dimensional shape of the target molecules in a manner analogous to antibody-antigen coupling (Dazzo and Truchet, 1983). The precise targets are ill-defined because little is known about the tertiary folding of polysaccharides. For example, despite identical carbohydrate constituents being present in the exopolysaccharides (EPSS) secreted by two *Rhizobium* strains, a clover root lectin, called trifoliinA, appears to be capable of preferentially binding to EPS from the rhizobia strain that nodulates clover but not pea (Dazzo and Truchet, 1983). Thus, differences in non-carbohydrate substitutions between the EPSs secreted by these two highly similar *Rhizobium* strains may determine their differential binding ability to the clover lectin (Philip-Hollingsworth et al., 1989).

Recently, clover roots were transformed with a pea lectin gene using *Agrobacterium rhizogenes* as a vector. These transformed clover roots were infected by rhizobia that normally nodulate pea (Diaz et al., 1989). Rhizobia which normally infect clover retained the ability to infect the transformed roots. Therefore root lectins contribute to the specificity seen in *Rhizobium*–plant interactions possibly by mediating bacterial agglutination and immobilization to plant cell walls which facilitates the binding of rhizobia to the most susceptible point of the root hair cell for penetration to occur.

C. Rhizobium Genes Required to Establish a Compatible Response on Legume Roots

Speciation in *Rhizobium* is partly determined by the range of host plants that can be nodulated by a particular nodule isolate. However, rhizobia have been classified into biovars because host range ability is determined by minor genetic differences.

Genes in *Rhizobium* required to establish infection on the roots of specific legumes are located in clusters usually on large indigenous plasmids (Banfalvi et al., 1981; Djordjevic et al., 1982; Downie et al., 1983; Johnston et al., 1978; Schofield et al., 1983, 1984). Allelic variants of many *nod* genes occur in a range of rhizobia. Several *nod* genes essential for nodulation are functionally conserved across different genera and species of rhizobia (Canter-Cremers et al., 1989; Djordjevic et al., 1985a; Fisher et al., 1985; Horvath et al., 1986; Spaink et al., 1989b; Surin and Downie, 1988). Mutations in other *nod* genes do not lead to a Nod⁻ phenotype, but rather, to a diminution in the ability to colonise one particular host. Some *nod* genes, which have no counterparts in other strains, appear to condition host range and, in some cases, cultivar specificity (see below).

1. Rhizobium Genes Required for Intercellular Communication

The *nod* D gene, which is expressed constitutively at low levels, is important in intercellular communication and is essential for nodulation. Nod D acts both as a membrane bound environmental sensor (Schlaman et al., 1989) and a positive regulatory transcriptional factor (Burn et al., 1989; Fisher et al., 1988; Hong et al., 1987; Honma and Ausubel, 1987; Kondorosi et al., 1989; McIver et al., 1989; Spaink et al., 1989a, 1987b). The interaction of Nod D with inducers (flavones) effects a conformational change resulting in the ability to mediate transcription from conserved promoter sequences

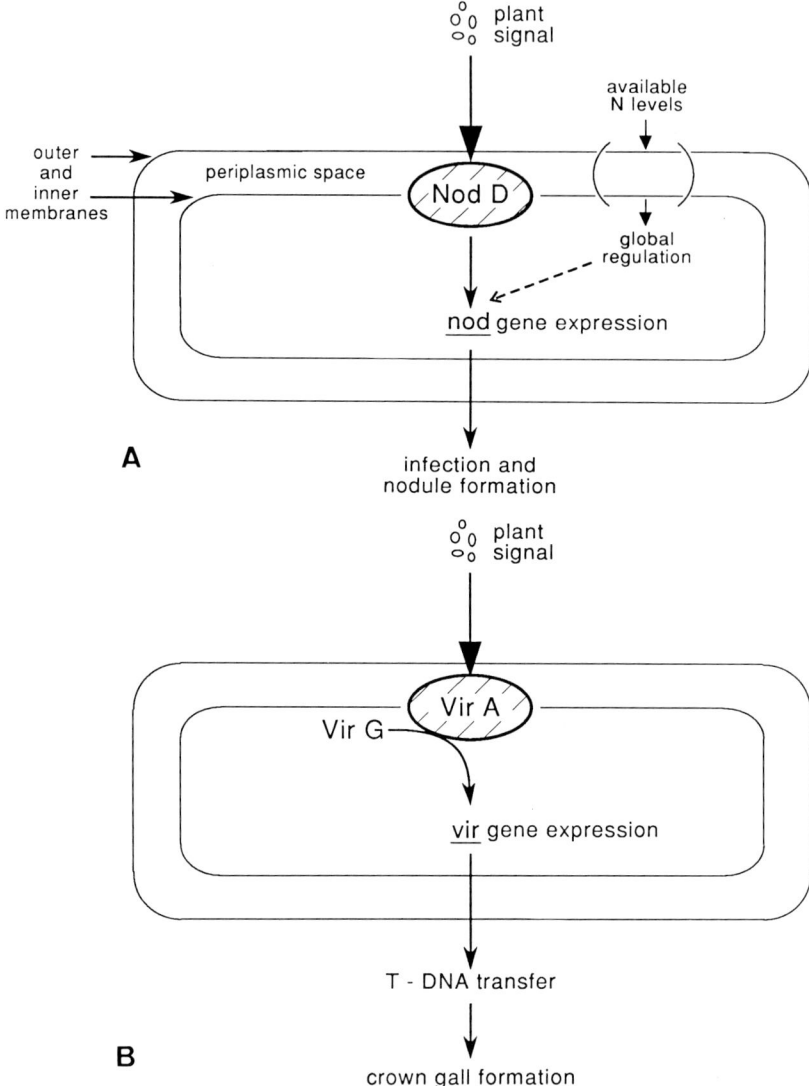

Fig. 1. Bacterial gene products required for intercellular communication. A proposed interaction occurs at the cytoplasmic membrane between the bacterial gene product and plant signals which results in an activation of these "environmental sensor" molecules. A, in *Rhizobium* the *nod D* gene product forms a single component system that interacts with plant flavonoid compounds. The activated Nod D, along with the RNA polymerase, induces the expression of a series of *Rhizobium nod* genes. The presence of combined nitrogen can influence this *nod* gene expression via the general (global) nitrogen regulation (*ntr*) system (Dusha et al., 1989). B, in *Agrobacterium* a two-component regulatory system is involved in intercellular communication. The membrane bound *vir A* gene product acts as the environmental sensor interacting with a plant signal and subsequently becomes activated, exhibiting autokinase activity. Activated Vir A is then thought to phosphorylate Vir G which is a DNA-binding protein. Enhanced transcription of several *vir* genes results

(called "*nod*-boxes") occurring 5′ to several *nod* gene operons (Fisher et al., 1988; Hong et al., 1987; Kondorosi et al., 1989; Spaink et al., 1987a). Thus co-ordinated regulation of the *nod* genes occurs, and this is coupled to the presence of inducing flavonoids from the plant root through Nod D-mediated transcriptional activation (Fig. 1 A). Anti-inducer flavonoids antagonise this interaction (Djordjevic et al., 1987b; Peters and Long, 1988; Spaink et al., 1989c). Point mutation analysis and the generation of *nod D* fusions have confirmed that Nod D can exist in two "states" (Burn et al., 1989; McIver et al., 1989; Spaink et al., 1989c). Inducer and anti-inducer-independent variants of Nod D "locked" into the positive regulatory form, can confer the extension of the host range of the *Rhizobium* strain to other genera of legumes (McIver et al., 1989; Spaink et al., 1989c). This demonstrates that *nod D* is a host specificity determinant (Spaink et al., 1987b). Reiteration of *nod D* in many *Rhizobium* species occurs and particular *nod D*s appear to be required to sense different types of plant inducer signals (Gyorgypal et al., 1988; Honma and Ausubel, 1987). Mutations in *nod D* can also affect chemotaxis to roots mediated by the presence of flavonoid molecules at the appropriate concentration (Caetano-Anolles et al., 1988).

2. Regulation of Pathogenicity Genes in Agrobacteria and Pseudomonads

In agrobacteria and certain pseudomonads, two-protein component regulatory systems sense the external environment for the appropriate conditions favouring the expression of genes required for plant colonisation and pathogenesis (Jin et al., 1990a, b) (Fig. 1 B). The *Agrobacterium* regulatory protein Vir A, like Nod D, is membrane bound and senses the presence of simple phenolic molecules such as vanillin and acetosyringone, released from wounded plant tissue (Stachel et al., 1985). Inducers mediate Vir A-dependent phosphorylation of Vir G. Vir A is also capable of autophosphorylation in vitro (Huang et al., 1990; Jin et al., 1990a; Leroux et al., 1987). The activated Vir G has a positive regulatory activity and conditions transcription from *vir*-box promoters. *Vir*-box promoters occur 5′ to several genes required to establish bacterial colonisation and transfer of the T-DNA to the plant tissue (Jin et al., 1990b).

In certain pseudomonads, two component regulatory systems control the expression of genes required for pathogenicity. In *Ps. syringae* pv. *phaseolicola* the *hrp S* gene is involved in the regulation of pathogenicity genes (Grimm and Panopoulus, 1989). In these bacteria, the nutritional and physiological conditions that are encountered in the apoplast during infection are necessary to stimulate the transcription of genes (e.g., *hrp A, B, C, D*) required for plant colonisation and pathogenicity (Rahme et al., 1989). In *X. campestris*, a cluster of regulatory genes occurs. Some genes in this cluster have homology to genes coding for two component regulatory systems (Daniels and Dow, 1989). In addition, a component of

xylem fluid is capable of inducing the expression of a pathogenicity gene in
X. c. pv. *campestris* (Kamoun and Kado, 1989), suggesting that pathogenicity genes are specifically regulated by plant substances.

It is not surprising that phytobacteria have adapted to the presence of
host environments by coupling the expression of genes required for pathogenicity and infection to compounds or physiological conditions encountered in these special environments. While the signal molecules recognised
are often unknown, the presence of two component regulatory systems to
sense the presence of the host are ubiquitous in phytobacteria (Nixon et al.,
1986). As with rhizobia, if the type of molecule recognised is specific,
variations in host range could occur due to differential signal recognition.

3. Genes in *Rhizobium* Essential for Plant Colonisation

The *nod A, B, C, I, J* genes (Table 1) occur universally in rhizobia and are
structurally and functionally conserved. The *nod A–C* genes are essential for
root hair curling, cortical cell division and, hence, nodulation (Djordjevic
et al., 1985a, b; Dudley et al., 1987; Rossen et al., 1984). In most rhizobia,
mutation of *nod I* or *nod J* have little discernible effect upon nodulation
ability. However, in clover-nodulating rhizobia, mutations in *nod I* or *nod J*
lead to poor nodulation ability, typified by poor infection thread synthesis
and underdeveloped nodules containing few bacteria (Huang et al., 1988).

4. Genes in Pathogens Essential for Plant Colonisation
and Virulence

Aside from regulatory genes, essential pathogenicity genes in phytobacteria
include those involved in the production and export of degradation
enzymes, phytohormones or other extracellular factors. The production of
extracellular polysaccharides and toxins can play an important role in
pathogenicity. In addition, many other genes of unknown function (including *hrp* genes) are essential for pathogenicity and mutation of these genes
lead to a Path⁻ phenotype. These genes often occur in clusters in pseudomonads and xanthomonads. Path⁻ mutants exhibit a different range of
phenotypes from: (*i*) asymptomatic, to (*ii*) unable to colonise the tissues of
plants, to (*iii*) able to colonise but unable to elicit disease symptoms.
Mutations in those genes which result in unreactive phenotypes or which
affect colonisation ability on all plant hosts show that these genes are
essential for pathogenicity and are more akin to the essential *nod* genes of
Rhizobium.

D. Genes Required for Host-Species Specificity

Several types of genes condition host specificity in phytobacteria. These
include genes which alter the physiological or developmental state of the

plant: some cause disease (*path*, pathogenicity; *tox*, toxin; or *vir*, virulence genes) and some do not (*nod* genes). Other types of host specificity genes include those which facilitate the exploitation of the plant as a source of nutrition (host specific parasitism genes). Biotrophic pathogens characteristically possess narrow host ranges whereas non-specialised pathogens (i.e., facultative or opportunist parasites such as *Erwinia* sp.) have wide host-ranges. Gene-for-gene recognition clearly operates in many biotrophic pathogen-plant interactions and this will be treated separately below. Like biotrophic pathogens, most fast-growing rhizobia have relatively narrow host ranges, limited to specific, or closely-related, genera. In contrast, bradyrhizobia and certain fast-growing rhizobia have very extensive host ranges. One hypothesis which is widely accepted is that pathogens with broad host ranges are best able to avoid gene-for-gene recognition and other plant defense responses which limit the colonisation potential. Until recently, there has been little evidence for gene-for-gene recognition operating in the interaction of rhizobia with plants.

In rhizobia, several genes in conjunction with the *nod ABCIJ* and *nod D* genes have been shown to condition host range specificity and act as positive dominant gene traits (Table 1). In *Rhizobium leguminosarum* bv. *trifolii* and *viceae* (which nodulate clover and pea, vetch, and *Lens* sp., respectively), *nod FE(R)L*, *nod T*, *nod MN*, *nod O*, and *nod X* have been identified thus far to condition host range recognition (Canter-Cremers et al., 1989; Davis et al., 1988; De Maagd et al., 1989; Djordjevic et al., 1985b; Economou et al., 1990; Spaink et al., 1989c; Surin and Downie, 1988, 1989). The general organisation of these genes in these two biovars shows a high degree of similarity and some genes such as *nod L* and *nod F* appear to be functionally conserved between the two species (Canter-Cremers et al., 1989; Spaink et al., 1989a). Variations in the genetic content between different strains do exist as, for example, certain biovars of *R. leguminosarum* lack *nod X* (Davis et al., 1988) and others lack *nod T* (Lewis-Henderson and Djordjevic, unpubl. results). The presence and absence of these peripheral *nod* genes is correlated, in part, with their involvement in cultivar specific interactions (see below).

The *nod E* gene has been shown to be a crucial host specificity determinant. The *nod E* gene occurs in an operon coding *nod FE* and *nod L* (Canter-Cremers et al., 1989; Surin and Downie, 1988; Weinman et al., 1988). The *nod E* gene of *R. l.* bv. *trifolii* is important for nodulation of particular clover species but mutation analysis has shown that its presence has a limiting effect for nodulation of pea plants (Djordjevic et al., 1985b; Spaink et al., 1989b). Therefore, the *nod E* gene acts as a "dominant suppressor" for pea nodulation while simultaneously conditioning clover nodulation ability as a positive dominant gene trait (Djordjevic et al., 1985b). Isogenic *R. l.* bv. *trifolii* mutated for *nod E* and differing only by the source of *nod E*, show the host range phenotype of the biovar that the *nod E* was isolated from (Spaink et al., 1989c). The *nod E* proteins of *R. l.* bv. *trifolii* and *viceae* are 78% homologous yet the two differ in function. Using hybrid *nod E* genes, the functional difference was shown to be determined by a central region of 185

Table 1. Nodulation genes identified in rhizobia required for compatibility with plant hosts

nod gene	Estimated M.W. (kDa)	Proposed activity or function (detected homology in parenthesis	Reference/Source
A	22	Nod A and Nod B are cytosolic	Egelhoff and Long, 1985
B	24	proteins essential for nodulation	Schmidt et al., 1986
C	47	transmembrane protein essential for nodulation	John et al., 1988
D	34–39	transcriptional activator and environmental sensor. Membrane- bound protein which responds to inducer and anti-inducer compounds (Lys R regulatory protein family)	Schlaman et al., 1989 Mulligan and Long, 1985
E	42	membrane-bound β-ketoacyl synthase (Fab B condensing enzyme and proteins involved in polyketide synthesis in *Streptomyces*)	Shearman et al., 1986 Bibb et al., 1989
F	10	cytoplasmic protein (acyl carrier protein/transglycosylase)	Shearman et al., 1986
G	27	dehydrogenase (17β-hydroxy-steroid dehydrogenase, ribitol dehydrogenase, glucitol-6-phosphate dehydrogenase and act III reductase)	Debelle and Sharma, 1986 Baker, 1989a, b
H	29	modifies compound(s) produced by Nod ABC activity	Faucher et al., 1989 Banfalvi and Kondorosi, 1989
I	34	(family of ATP binding proteins involved in transport)	Evans and Downie, 1986
J	28	hydrophobic protein, may act in conjunction with Nod I	Evans and Downie, 1986
L	20	acetyl transferase (Lac A and Cys E)	Downie, 1989
M	66	shows positive and negative host range functions (amidophospho-ribosyl transferase)	Surin and Downie, 1988 Weinman et al., 1988
O	30	exported Ca^{2+}-binding protein (Hly A)	De Maagd et al., 1989 Economou et al., 1990
P	35	positive role in host range	Faucher et al., 1989 Schwedock and Long, 1989
Q	71	possible GTP binding protein (EF-Tu)	Schwedock and Long, 1989
R	14	function unknown	this laboratory

Table 1. (continued)

nod gene	Estimated M.W. (kDa)	Proposed activity or function (detected homology in parenthesis	Reference/Source
T	51–58	positive role in cultivar specificity	Djordjevic and Lewis-Henderson, in prep.
V	89	(membrane-bound sensor protein family, e.g., Ntr B)	Gottfert et al., 1989, 1990
W	23	(transcriptional regulatory proteins, e.g., Ntr C)	Gottfert et al., 1989, 1990
X	41	hydrophobic protein, positive role in cultivar specificity	Davis et al., 1988 Djordjevic and Lewis-Henderson, in prep.
Y	16	function unknown	Banfalvi et al., 1988a
Z	35	function unknown	Banfalvi et al., 1988b

amino acids containing only 44 non-identical amino acids (Spaink et al., 1989b). The functional similarity of *nod E* to β-ketoacyl synthases genes involved in polyketide synthesis (Table 1) may offer clues to its real biochemical function, especially since *nod F* and *nod L* are homologous to proteins (acyl-carrier protein and acetyl-transferase, respectively) that are involved in the metabolism of acetate via acetyl-CoA (Downie, 1989; Shearman et al., 1986). A large number of natural products are derived from acetyl-CoA which is one of the building blocks for secondary metabolism.

Genes in the *nod MN* operon are required for the nodulation of certain clover species (Canter-Cremers et al., 1989; Djordjevic et al., 1986; Surin and Downie, 1989). However, in *R. l.* bv. *trifolii* strains containing *nod DABCIJ*, the addition of the *nod MN* operon did not confer host specificity unless the *nod L* gene was also present (Weinman et al., 1988). In *R. l.* bv. *viceae*, the *nod T* gene is located downstream of *nod MN* and is transcribed in the same operon (Canter-Cremers et al., 1989). In contrast, in *R. l.* bv. *trifolii nod T* occurs in the *nod ABCIJ* operon downstream of *nod J* (B. Surin, pers. comm.)

The *nod O* gene occurs on a separate operon in *R. l.* bv. *viceae* and encodes a secreted protein which binds Ca^{2+} (De Maagd et al., 1989; Economou et al., 1990). Mutations in *nod O* affect the timing of onset of nodulation as well as nodule number. Curiously, the *nod O* protein is exported from the cell despite the lack of a conventional N-terminal leader sequence required for processing and extrusion of many other exported proteins (De Maagd et al., 1989; Economou et al., 1990). Nod O is homologous to other proteins (e.g., hemolysin A) which are also exported from bacteria without conventional N-terminal processing. The binding of Ca^{2+}

to Nod O may reflect the crucial role this ion plays in root nodulation (Economou et al., 1990; Vincent, 1974, 1980).

In *R. meliloti*, which nodulates *Melilotus* and *Trigonella* species *nod FEG*, *nod H*, *nod P*, and *nod Q* have been identified as host range recognition factors (Debelle et al., 1988, 1986; Debelle and Sharma, 1986; Faucher et al., 1988, 1989; Fisher et al., 1987; Horvath et al., 1986 1987; Schwendock and Long, 1989). There appears to be no counterpart of *nod G, H, P* or *nod Q* in *R. leguminosarum*. The *nod H* and *nod Q* genes act as dominant positive traits for the nodulation of alfalfa but as dominant suppressors of the nodulation of vetch and clover plants (Debelle et al.,

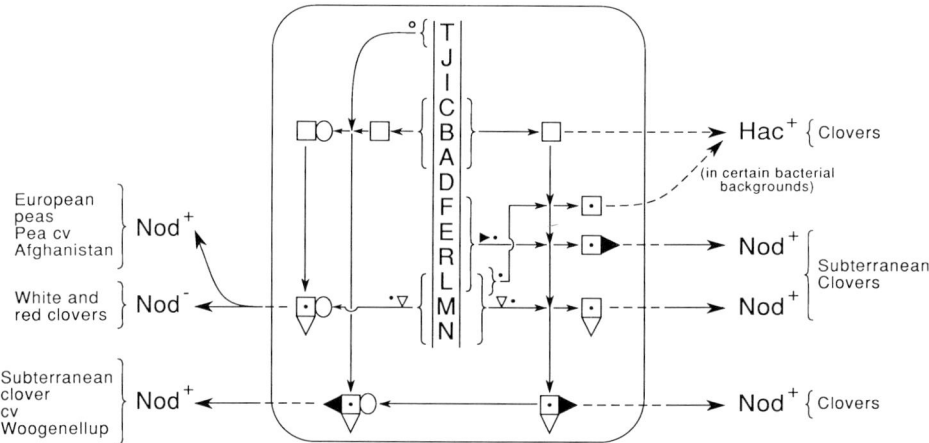

Fig. 2. The effect of the interaction of *R. l.* bv. *trifolii* nodulation genes on the infection of homologous and heterologous hosts. The *Rhizobium* strain induced by plant signals produces extracellular compounds which either cause the root hair curling (Hac$^+$) pheno-type or aid in the infection and nodulation (Nod$^+$) of different host plants. It is proposed that a variety of extracellular products are formed as a result of different parallel pathways, each under the control of the various host-specific nodulation genes, acting on the factors produced after the expression of the *nod A–C* genes. The model provides a mechanistic framework The extracellular factors produced as a result of *nod A–C* gene expression (□) usually cause the Hac$^+$ phenotype on all tested clovers (white, red, and subterranean clovers). However, in certain *R. l.* bv. *trifolii* strains a combination of *nod A–C* plus *nod L* (▣) is required for the Hac$^+$ phenotype. The Sym-plasmid cured *R. l.* bv. *trifolii* strain ANU845, containing combinations of *nod* genes, *nod DABC* plus *nod FERL* or *nod DABC* plus *nod LMN*, can induce the Hac$^+$ Inf$^+$ (infection thread formation) phenotype, however, significant, but relatively poor nodulation results on subterranean clovers. The complete *Nod* gene combination, *Nod DABCFERLMN*, produces a more rapid and efficient nodu-lation of all tested clovers. The host specific nodulation region of *R. l.* bv. *trifolii* also shows the phenomenon of dominance or epistasis. The *nod E$^-$* mutants are Nod$^-$ on white and red clovers, Nod$^+$ on subterranean clovers and have an extended host range, as they are able to nodulate (Nod$^+$) peas but not vetches. Finally, certain strains of *R. l.* bv. *trifolii* exhibit cultivar specificity; the positively acting *nod T* gene counteracting the negatively acting *nod M* gene to enable nodulation of subterranean clover cv. Woogenellup

1988; Faucher et al., 1989). Mutation of *R. meliloti nod FE, nod H*, or *nod Q* results in altered infection and nodulation on the normal host plant (alfalfa) but permits interactions with the root hair cells of heterologous host plants such as clover and vetch. There is evidence for *nod H* and *nod Q* modifying an extracellular, *nod ABC*-dependent factor and thereby conditioning alfalfa nodulation ability (Banfalvi and Kondorosi, 1989; Debelle et al., 1988, 1986; Horvath et al., 1987). The introduction of *R. meliloti nod FE, nod Q*, and *nod H* into *R. leguminosarum* biovars allows transconjugants to nodulate alfalfa but effectively abolishes nodulation ability on their normal host plants. Functional homology of *nod G* exists to several dehygrogenases including the act III protein involved in antibiotic synthesis in *Streptomyces coelicolor* (Baker, 1989a, b; Debelle and Sharma, 1986).

A clearer picture is emerging on the role of nodulation genes in conditioning host range. First, the *nod ABC* genes do not condition host specificity directly but they act in conjunction with other *nod* genes to determine this phenotype (Banfalvi and Kondorosi, 1989; Djordjevic et al., 1986; Faucher et al., 1989, 1988; Surin and Downie, 1989). The functions of the *nod ABC* gene products remain unknown but genetic evidence suggests that they act upstream in a pathway which generates heat-stable low molecular weight compound(s) which can be modified further by genes in each rhizobia that condition host specificity (Fig. 2). Because several gene products act as dominant suppressors (e.g., *nod F, E* in *R. l.* bv. *trifolii*, and *nod H, nod E*, and *nod Q* in *R. meliloti*) it appears that several possible pathways can be generated, leading to the tailoring of different host specific factors. There is further evidence that some genes act sequentially in the same pathway (e.g., the *nod H* and *nod Q* genes in *R. meliloti*) and other genes which act in two parallel, but *nod ABC*-dependent, pathways (e.g., *nod L* with *nod E* and *nod L* in conjunction with the *nod M, N* operon in *R. l.* bv. *trifolii*) (Fig. 2). These new findings explain earlier observations which showed "symbiotic interference" in *Rhizobium* transconjugants carrying two sets of nodulation and host range genes (Debelle et al., 1988; Djordjevic et al., 1982).

E. Evidence of Nodulation Genes in Rhizobium Which Condition Gene-for-Gene Cultivar Specificity

Strain-cultivar specificity is common amongst associations with biotrophic pathogens such as *Pseudomonas* and *Xanthomonas* and their respective plant hosts but there is little evidence for this in *Rhizobium* infections. Race or cultivar specificity is determined by gene-for-gene interactions. The concept of "gene-for-gene" interactions was first applied to plant pathology by Flor (1947, 1955), as discussed further by De Wit (1992). The host codes a single gene trait (R) for resistance and this R gene is matched by a gene(s) in the pathogen for avirulence (*A* for fungi, *avr* for bacteria) (Flor, 1947, 1955). A specific recognition event is mediated, directly or indirectly, by the products of the matching R and *avr* genes. This specific match of R and *avr* genes (e.g., $R_1 \times avr_1$), usually leads to an adverse plant response (e.g., a HR), which limits pathogen colonisation and an avirulent, or incompatible, response results. If there is no match of R and *avr* genes (e.g., $R_2 \times avr_1$), the plant cultivar is unable to establish an incompatible (resistant) response.

Several examples of *Rhizobium*–cultivar specific interactions have emerged. In the best studied example, a single recessive gene in a pea (*Pisum sativum*) cultivar, called Afghanistan, is responsible for poor nodulation by several *R. l.* bv. *viceae* strains (Holl, 1975; Lie, 1984). Naturally-occurring isolates of *R. l.* bv. *viceae* which are capable of nodulating cv. Afghanistan possess, in addition to other *nod* genes, *nod X* (Davis et al., 1988). Mutation of *nod X* in these isolates abolishes nodulation ability on cv. Afghanistan, but does not affect nodulation on European peas (Hombrecher et al., 1984).

The subterranean clover (*Trifolium subterraneum*) cv. Woogenellup is not nodulated by a *R. l.* bv. *trifolii* strain which does not code the *nod T* gene (Lewis-Henderson and Djordjevic., manuscript in prep.). This *R. l.* bv. *trifolii* strain causes root hair curling and infection thread formation but fails to initiate sustained cortical cell division. No evidence for plant cell death can be observed in this association. The *nod T* gene of *R. l.* bv. *trifolii* appears to act as a dominant suppressor of *nod M* (Lewis-Henderson and Djordjevic, manuscript in prep.). Naturally occurring *R. l.* bv. *trifolii* strains which do not code *nod T* can nodulate the restrictive cultivar only (*i*) when the *nod M* gene is mutated or (*ii*) if *nod T* is introduced to this strain from other *R. l.* bv. *trifolii* strains. In this particular strain, *nod M* acts as a cultivar specific avirulence gene and *nod T* as a dominant suppressors of avirulence. A single recessive host gene also appears to be involved in conferring resistance to *Rhizobium* infection in cv. Woogenellup.

A recent report shows that *R. fredii* strains also harbour genes which condition cultivar specific avirulence on soybean. Mutation of these *R. fredii* genes extends the host range of these strains to the normally restrictive soybean cultivars (Heron et al., 1989).

1. Host Specificity Determinants in Plant Pathogens

Due to the agronomic importance of these associations, much of plant pathogen research has been focussed on determinants (*avr* genes) which condition cultivar specific interactions. Little attention has been given to studies on more general host range determinants. Nevertheless, positively and negatively acting host ranges have been recently identified by mutants and transconjugants which exhibit loss or acquisition of pathogenicity on particular host plants (Gabriel, 1989; Keen and Staskawicz, 1988).

The transfer of cosmid clones isolated from *Xanthomonas citri* (the causal agent of citrus canker), to a variety of *X. citrus* pathovars resulted in transconjugants able to induce canker-like lesions (Swarup and Gabriel, 1989). The acquisition of pathogenicity on the new host by the transconjugants was accompanied by a tempering of the pathogenicity effects on their normal host plants. In some cases, suppression of cultivar specific avirulence responses occurred. These results indicate that, like *Rhizobium* *nod* genes, the introduced pathogenicity functions were acting as positive gene traits as well as dominant suppressors of resident pathogenicity and avirulence genes. This suggests that several complex pathways may be acting together to confer pathogenicity.

Genes in pathogens which condition incompatibility on a certain host can be transferred to a compatible strain for that host and this results in reduced pathogenicity in the transconjugant. For example, the transfer of genes from *X. c.* pv. *vitans* (incompatible on *Brassica*) to *X. c.* pv. *campestris* (compatible on *Brassica*) resulted in transconjugants with reduced pathogenicity (Roberts et al., 1987). The treatment of *X. c.* pv. *campestris* transconjugants with antisera prepared against *X. c.* pv. *vitans* resulted in a reduction of the effect of the introduced genes (Roberts et al., 1987). This antisera appeared to bind to surface components of the transconjugant.

Host specific virulence determinants also have been found in *X. c.* pv. *citrumelo* and *X. c.* pv. *translucens* after Tn5 mutagenesis. *X. c.* pv. *citrumelo* causes mild leaf spot disease on *Citrus* and *Phaseolus* species while *X. c.* pv. *translucens* infects wheat, oats, barley, rye, and triticale. Specific classes of mutants were capable of inducing pathogenic symptoms on some hosts but not others (Gabriel, 1989; Kingsley and Gabriel, 1989). Similar examples of host specificity genes have been found in agrobacteria and pseudomonads (Hoekema et al., 1984; Kobayashi and Keen, 1985; Yanofsky and Nester, 1986). These findings parallel those in *R. l.* bv. *trifolii* where specific mutations in *nod L* or *nod E* affect nodulation ability on red clover and white clover, respectively, but have little discernible affect upon nodulation of subterranean clover (Canter-Cremers et al., 1989; Djordjevic et al., 1985b; Spaink et al., 1989c).

2. Gene-for-Gene Interactions in Plant Pathogens

Many examples of cultivar specific associations occur in phytopathogenic bacteria, due to *avr*–*R* gene interactions. Because many avirulence genes in pathogens appear to have no selective advantage they have been thought of as being "gratuitous" in terms of plant interactions, simply serving as molecular targets for resistance genes in the plant (Gabriel, 1989). Mutation of *avr* genes usually results in a compatible interaction with plants that are normally resistant. In contrast to these observations, recent evidence indicates that some *avr* genes code functions which are important for in planta growth and are, therefore, not gratuitous. The *avr Bs2* gene carried by strains of *X. c.* pv. *vesicatoria* is responsible for inducing a HR on pepper plants carrying the Bs2 resistance gene. This *avr Bs2* gene appears widespread in other *X. campestris* pathovars. Mutation of the gene in *X. c.* pv. *vesicatoria* affects the ability of these strains to grow in planta in susceptible pepper plants (Kearney and Staskawicz, 1989) and therefore this gene appears necessary for conferring some level of compatibility. Thus, *avr* genes that are necessary for growth in planta offer significant opportunities to breed, or engineer, plants which show persistence in their ability to resist plant pathogens because the plant resistance gene is targeted to a gene essential for pathogenicity. In this case, the *avrBs2* gene is clearly analogous to *Rhizobium* host range genes in that both condition host range and cultivar specificity (see above).

While several *avr* genes have been cloned in bacterial phytopathogens, recent advances in the understanding of the role of *avr D* in the triggering of a HR in soybeans establishes some broad similarities with low molecular weight factors conditioned by host specific nodulation genes in rhizobia as further discussed by Keen and Dawson (1992). The *avr D* gene—cloned from *P. s.* pv. *tomato* (Kobayashi and Keen, 1985)—can elicit a cultivar-specific HR on soybeans when present in *P. s.* pv. *glycinea* (Kobayashi et al., 1990) and in *Escherichia coli*. The *avr D* product itself is not responsible for the elicitation the plant response. Instead, *avr D* is responsible for the synthesis of a low molecular weight compound which is secreted into the culture fluids of cells containing this gene (Keen et al., 1990; Kobayashi et al., 1990). In interactions where exported bacterial proteins are not directly involved in conditioning plant resistance responses, the involvement of low molecular weight metabolites (e.g., secondary metabolites, toxin-like compounds or oligosaccharides) will be a recurring theme in plant–microbe interactions. Potential plant receptors will only be isolated when the compounds mediating these intercellular communication events are isolated and characterised.

F. Genes in Rhizobia Which Establish Specialised Nutritional Relationships with Plant Hosts

There are several examples of genes in rhizobia which reflect the requirement to establish or exploit a specialized nutritional source in planta. In *R. meliloti*, certain highly competitive isolates are able to synthesise a compound as bacteroids, which provides free-living rhizobia with a specialised source of carbon. These compounds are called rhizopines. The genes required for rhizopine synthesis (*mos*) and catabolism (*moc*) are linked on the nodulation plasmid in some strains of *R. meliloti* (Murphy et al., 1987). The promoter of *mos* is regulated by Nif A and expression is restricted to bacteroids (Murphy et al., 1988). In contrast, the *moc* gene is expressed in free living rhizobia which can occur on the surface of the nodule and this could result in enhanced ability to colonise the rhizosphere. It is not known whether rhizobia inside infection threads can also utilise rhizopines. Utilisation of rhizopines by rhizobia reflects similar strategies used by agrobacteria, where genes located within the T-DNA code for the synthesis of opines and the corresponding catabolism genes occur elsewhere on the Ti plasmid (Hooykaas and Schilperoort, 1986).

In *R. l.* bv. *viceae*, genes for the catabolism of homoserine are located on nodulation plasmids (Johnston et al., 1988). Up to 75% of free amino acids in pea exists as homoserine and this compound can be used as the sole carbon source by many *R. l.* bv. *viceae* (Van Egeraat, 1975). In most cases, these additional genes that satisfy nutritional requirements are not absolutely necessary for infection; however, in other instances the ability to catabolise specific plant compounds is important in establishing virulence on a particular host. The ability to catabolise tartaric acid in certain agrobacteria for example, is important for establishing pathogenicity on grape plants (Hagiya et al., 1985).

G. Genes Conditioning Pathogenicity in Rhizobium

Rhizobium strains rarely elicit pathogenic responses from their plant hosts. However, two examples of fast-growing rhizobia are known which induce a pronounced deformation of the leaves of nodulated pigeonpea, siratro and Desmodium plants (Kumer Rao et al., 1984; Upadhyaya et al., 1985). The relationship between nodulation and leaf deformation has not been established although low molecular weight systemic factors are involved. It appears that additional genes have been acquired by these rhizobia which confer the ability to induce a pathogenic response.

Certain bradyrhizobia are capable of inducing pathogenic responses on plants by secreting a toxin (rhizobitoxine) which causes chlorosis of leaves in soybeans and other plants when added exogenously to seedling culture solutions (Johnson et al., 1959; Owens and Wright, 1965). At low concentrations, the toxin affects biosynthetic pathways not only in plants, but also other genera of bacteria, and results in an inhibition of growth. The presence of rhizobitoxine could provide competitive advantages for root colonisation for these rhizobia when in the rhizosphere.

H. Chromosomal Genes Conditioning Basic Compatibility in Rhizobium

Certain metabolic pathways present in saprophytes may provide integral precursors that are involved in plant interactions. Mutation of these genes may sufficiently affect general cellular metabolism and have pleiotrophic consequences. One clear example of this is provided by studies on an adenine auxotrophic mutant of strain NGR234 induced by transposon Tn5. A distinct pathogenic response is induced after the inoculation of this mutant strain on siratro plants (Djordjevic et al., 1988). Other plants appear to respond in a similar way to inoculation of this mutant. The ade⁻ mutant (called ANU2861) is complemented for growth by the addition to defined growth media of AICAR (5-aminoimidazole-4-carboxamide ribonucleotide – an intermediate in purine biosynthesis). The addition of AICAR to siratro plants inoculated with this mutant also enhances the ability of this strain to induce nodule formation. In addition to the ade⁻

phenotype, ANU2861 overproduces exopolysaccharides and is unable to nodulate a range of legume plants (Chen et al., 1985; Djordjevic et al., 1988). Early infection events are initiated by ANU2861 on the roots of siratro plants including root hair curling, the initiation of plant cortical cell division and the very initiation of infection thread synthesis. However, after 24 h, signs of a HR-like response occurs in the infected root hair cell which, after 48 h, is apparent in surrounding epidermal cells but not sub-epidermal cells (Djordjevic et al., 1988). The timing and localisation of the induction of this HR is comparable to normal incompatible plant–pathogen interactions. There are several possibilities which could explain these observations. First some chemical signal not present in normally growing parental rhizobia, is released that induces this response. Alternatively, *Rhizobium* compatibility may be conditioned by molecules which suppress plant defense responses leading to plant cell death (e.g., a HR) and the synthesis of these compounds may be affected in mutant ANU2861. Finally, a molecule(s) which normally masks the adverse recognition of rhizobia by plant hosts is/are absent in the mutant strain (Rolfe et al., 1989).

IV. Discussion

Rhizobia have clearly developed a subtle mechanism to permit growth in the root tissues of legumes. Several features of this compatible interaction allow clear distinctions to be made compared to plant–pathogen interactions.

Rhizobia perturb root hair growth in unique ways through the elaboration of low molecular weight compounds. A controlled wounding response involving localised degradation and rebuilding of the plant cell wall leads to infection thread synthesis (Rolfe et al., 1989) and possibly the generation of a secondary signal (Long and Cooper, 1988). This wounding reaction may be the trigger for a "repair mechanism" to be initiated in the plant cortex and manifested in the induction of sustained cortical cell divisions (D. C Loschke, pers. comm.). This is consistent with the notion that rhizobia have evolved the capacity to perturb a normal developmental pathway in plants and this is consistent with the existence of plant variants that spontaneously nodulate in the absence of rhizobia (Truchet et al., 1989). These plants are similar in this regard to *les* mutants in maize (Pryor, 1987) which are easily triggered to initiate defense responses in the absence of obvious pathogen colonisation. Finally, a nutritional exchange is established which permits rhizobia to (*i*) exploit available nutrients without overly depriving plant cells and (*ii*) provide a readily available source of fixed nitrogen for the plant. These features require a particular blend of input from chromosomal and plasmid located genes. The identification of the compounds which cause root hair curling and cortical cell division is vital for an understanding of gene function and could have ramifications for understanding plant developmental pathways.

Given the homology that several *nod* genes have to genes which are involved in secondary metabolism and fatty acid synthesis, one distinct possibility is that host range specificity genes, along with *nod A–C*, are involved in the biosynthesis of low molecular weight secondary metabolites or "bioactive" oligosaccharides which affect plant hormonal balances and developmental pathways (Shearman et al., 1986). If this speculation is accurate, then these compounds might be able to bind to key receptors important in plant development. The receptor could be located in the root hair cell only or, in addition, in the cortical tissue. Plant receptors could be preformed transmembrane proteins which can bind gene products or metabolites under *avr* or *nod* gene control.

The utilisation of multidisciplinary approaches to identify the types of molecules which are exchanged will be necessary to establish the biochemical roles of the genes involved in symbiosis and pathogenesis. Gene-for-gene systems provide ideal materials where the small genetic differences between hosts and their interacting microbes will permit the receptors in the plant host to be characterised, if the binding of the signal molecules to the plant receptor is sufficiently strong and the signal compounds can be labelled without affecting activity. The recent identification of probable gene-for-gene(s) systems in *Rhizobium*–plant interactions due to input from host specific nodulation genes is significant. A variety of *Rhizobium* single and double mutant strains should allow the complex epistatic effects observed, to be unravelled. Epistatic effects will be common where a number of host range genes are involved in establishing a particular phenotype. This could explain recent findings where some *avr* genes in *Xanthomonas* appear to be necessary for in planta growth. Here one would predict that several loci contribute to the avirulence/host range phenotype, as is apparent in *Rhizobium*. The *Rhizobium* host range genes can act to suppress or inhibit virulence on some host plants (Debelle et al., 1988; Djordjevic et al., 1985b; Faucher et al., 1988, 1989) but genuine gene-for-gene avirulence should only be seen in host cultivar variants which distinguish between two highly similar isolates of *Rhizobium*.

The identified extracellular molecules, produced as a result of the combined action of several *nod* gene operons (see Fig. 2), will provide plant cell and molecular biologists with a range of signals to probe the underlying principles involved in plant-microbe interactions and, possibly, root morphogenesis and plant cell differentiation. Whatever the mechanism is that underlies root nodule formation, it should involve the alteration of patterns of positional information in cortical, endodermal and pericycle cells. It remains to be shown whether legumes vary only because of the type of receptors which are present or if there is a fundamental difference in the cell biology principles required to form a nodule in each legume.

Finally, while clear differences exist between infections initiated by rhizobia and other phytobacteria, some general principles are shared. Multigenic traits conditions pathogenicity and host specific nodulation ability, in all well studied biotrophic phytobacteria. Similarities exist in the mechanisms by which phytobacteria sense the presence of their respective host plants. The location on plasmids of many of the genes conferring basic compatibility permits the horizontal transfer of these genes through soil microbe populations. A range of factors mediate communication events between the interacting organisms but the identity of most of these are not yet known. Ultimately, the usefulness of studying host-specific nodulation genes as model systems for genes conferring virulence, will depend upon how similar the strategies and mechanisms are in phytopathogens, that condition basic compatibility in a host plant environment.

Acknowledgements

WL-H is a recipient of an Australian Postgraduate Research Award provided by the Australian Government. Research funds provided by the Australian Wool Board (to MAD and BGR) are gratefully acknowledged.

V. References

Agrios GN (1988) Plant pathology, 3rd edn. Academic Press, San Diego

Al-Mousawai AH, Richardson PE, Essenberg M, Johnson WM (1982) Ultrastructural studies of a compatible interaction between *Xanthomonas campestris* pv. *malvacearum* and cotton. Phytopathology 72: 1222–1230

Baker ME (1989a) A common ancestor for human placental 17β-hydroxysteroid dehydrogenase, *Streptomyces coelicolor actIII* protein, and *Drosophila melanogaster* alcohol dehydrogenase. Fed Am Soc Exp Biol J 4: 222–226

Baker ME (1989b) Human placental 17β-hydroxysteroid dehydrogenase is homologous to NodG protein of *Rhizobium meliloti*. Mol Endocrinol 3: 881–884

Banfalvi Z, Kondorosi A (1989) Production of root hair deformation factors by *Rhizobium meliloti* nodulation genes in *Escherichia coli*: HsnD (*NodH*) is involved in the plant host-specific modification of the NodABC factor. Plant Mol Biol 13: 1–12

Banfalvi Z, Sakanyan V, Koncz C, Kiss A, Dusha I, Kondrosi A (1981) Location of nodulation and nitrogen fixation genes on a high molecular weight plasmid of *R. meliloti*. Mol Gen Genet 184: 318–325

Banfalvi Z, Nieuwkoop A, Schell M, Besl L, Stacey G (1988a) Regulation of *nod* gene expression in *Bradyrhizobium japonicum*. Mol Gen Genet 214: 420–424

Banfalvi Z, Schell M, Desmane N, Nieuwkoop A, Kondorosi A, Stacey G (1988b) A *Bradyrhizobium japonicum* gene which directs nodulation of siratro and selected soybean cultivars. In: Bothe H, de Bruijn FJ, Newton WE (eds) Nitrogen fixation: hundred years after. G Fischer, Stuttgart, p 482

Bauer WD, Bhuvaneswari TV, Calvert HE, Law IJ, Malik NSA, Vesper SJ (1985) Recognition and infection by slow-growing rhizobia. In: Evans HJ, Bottomley PJ, Newton WE (eds) Nitrogen fixation research progress. Proceedings of the 6th International Symposium on Nitrogen Fixation. Martinus Nijhoff, Dordrecht, pp 247–253

Bhuvaneswari TV, Turgeon GB, Bauer DW (1980) Early events in the infection of soybean (*Glycine max* L. Merr). Plant Physiol 66: 1027–1031

Bibb MJ, Biro S, Motamedi H, Collins JF, Hutchinson CR (1989) Analysis of the nucleotide sequence of the *Streptomyces glaucescens tcml* genes provides key information about the enzymology of polyketide antibiotic synthesis. EMBO J 8: 2727–2736

Billing E (1987) Bacteria as plant pathogens. Van Nostrand Reinhold, Wokingham (Cole JA, Knowels CJ, Schlessinger D (eds) Aspects of microbiology, vol 14)

Burn JE, Hamilton WD, Wootton JC, Johnston AWB (1989) Single and multiple mutations affecting properties of the regulatory gene *nodD* of *Rhizobium*. Mol Microbiol 3: 1567–1577

Caetano-Anolles G, Crist-Estes DK, Bauer WD (1988) Chemotaxis of *Rhizobium meliloti* to the plant flavone luteolin requires functional nodulation genes. J Bacteriol 170: 3164–3169

Callaham DA, Torrey JG (1981) The structural basis for infection of root hairs of *Trifolium repens* by *Rhizobium*. Can J Bot 59: 1647–1664

Calvert HE, Pence MK, Pierce M, Malik NSA, Bauer WD (1984) Anatomical analysis of the development and distribution of *Rhizobium* infections in soybean roots. Can J Bot 62: 2375–2384

Canter-Cremers HCJ, Spaink HP, Wijfjes AHM, Pees E, Wijffelman CA, Okker RJH, Lugtenberg BJJ (1989) Additional nodulation genes on the Sym plasmid of *Rhizobium leguminosarum* biovar *viciae*. Plant Mol Biol 13: 163–174

Carroll BJ, McNeil DL, Gresshoff PM (1985) Isolation and properties of soybean mutants which nodulate in the presence of high nitrate concentrations. Proc Natl Acad Sci USA 82: 4162–4166

Chen H, Bately M, Redmond JW, Rolfe BG (1985) Alteration of the effective nodulation properties of a fast-growing broad host range *Rhizobium* due to changes in exopolysaccharide synthesis. J Plant Physiol 120: 331–349

Collinge DB, Slusarenko AJ (1987) Plant gene expression in response to pathogens. Plant Mol Biol 9: 389–410

Collmer A, Berman P, Mount MS (1982) Pectate lyase regulation and bacterial soft-rot pathogenesis. In: Mount MS, Lacy EH (eds) Phytopathogenic prokaryotes, vol 1. Academic Press, New York, pp 396–422

Daniels MJ, Dow M (1989) Regulation of gene expression in *Xanthomonas–Brassica* interaction. In: Molecular biology of bacterial plant pathogens, Fallen Leaf Lake, South Lake Tahoe, California, 37

Daniels MJ, Dow JM, Osbourn AE (1988) Molecular genetics of pathogenicity in phytopathogenic bacteria. Annu Rev Phytopathol 26: 285–312

Dart PJ (1974) The infection process. In: Quispal A (ed) The biology of nitrogen fixation. North-Holland, Amsterdam, pp 381–429

Davis EO, Evans IJ, Johnston AWB (1988) Identification of *nodX*, a gene that allows *Rhizobium leguminosarum* biovar *viciae* strain TOM to nodulate Afghanistan peas. Mol Gen Genet 212: 531–535

Dazzo FB, Truchet G (1983) Interactions of lectins and their saccharide receptors in the *Rhizobium* legume symbiosis. J Membrane Biol 73: 1–16

De Maagd RA, Wijfjes AHM, Spaink HP, Ruiz-Sainz JE, Wijffelman CA, Okker RJH, Lugtenberg BJJ (1989) *NodO*, a new *nod* gene of the *Rhizobium leguminosarum* biovar *viciae* Sym plasmid pRL1JI, encodes a secreted protein. J Bacteriol 171: 6764–6770

Debelle F, Sharma SR (1986) Nucleotide sequence of *Rhizobium meliloti* PCR 2011 genes involved in host specificity of nodulation. Nucleic Acids Res 14: 7453–7472

Debelle F, Rosenberg C, Vasse, J, Maillet F, Martinez E, Denarie J, Truchet G (1986) Assignment of symbiotic developmental phenotypes to common and specific nodulation (*nod*) genetic loci of *Rhizobium meliloti*. J Bacteriol 168: 1075–1086

Debelle F, Maillet F, Vasse J, Rosenberg C, De Billy F, Truchet G, Denarie J, Ausubel FM (1988) Interference between *Rhizobium meliloti* and *Rhizobium trifolii* nodulation genes: genetic basis of *R. meliloti* dominance. J Bacteriol 170: 5718–5727

Delves AC, Mathews A, Day DA, Carter AS, Carroll BJ, Gresshoff PM (1986) Regulation of the soybean and *Rhizobium* nodule symbiosis by shoot and root factors. Plant Physiol 82: 588–590

De Wit PJGM (1992) Functional models to explain gene-for-gene relationships in plant–pathogen interactions. In: Boller T, Meins F (eds) Genes involved in plant defense. Springer, Wien New York, pp 25–47 [Dennis ES et al (eds) Plant gene research. Basic knowledge and application]

Diaz CL, Melchers LS, Hooykaas PJJ, Lugtenberg BJJ, Kijne JW (1989) Root lectin as a determinant of host-plant specificty in the *Rhizobium*-legume symbiosis. Nature 338: 579–581

Dickinson CH, Lucas JA (1982) Plant pathology and plant pathogens, 2nd edn. Blackwell, Oxford

Dixon RA (1986) The phytoalexin response: elicitation, signalling and control of host gene expression. Biol Rev 61: 239–291

Djordjevic MA, Zurkowski W, Rolfe BG (1982) Plasmids and stability of symbiotic properties of *Rhizobium trifolii*. J Bacteriol 151: 560–568

Djordjevic MA, Schofield PR, Ridge RW, Morrison NA, Bassam BJ, Plazinski J, Watson JM, Rolfe BG (1985a) *Rhizobium* nodulation genes involved in root hair curling (Hac) are functionally conserved. Plant Mol Biol 4: 147–160

Djordjevic MA, Schofield PR, Rolfe BG (1985b) Tn5 mutagenesis of *Rhizobium trifolii* host-specific nodulation genes result in mutants with altered host range ability. Mol Gen Genet 200: 463–471

Djordjevic MA, Innes RW, Wijffelman CA, Schofield PR, Rolfe BG (1986) Nodulation of specific legumes is controlled by several loci in *Rhizobium trifolii*. Plant Mol Biol 6: 389–403

Djordjevic MA, Gabriel DW, Rolfe BG (1987a) *Rhizobium*: the refined parasite of legumes. Annu Rev Phytopathol 25: 145–168

Djordjevic MA, Redmond JW, Batley M, Rolfe BG (1987b) Clovers secrete specific phenolic compounds which either stimulate or repress *nod* gene expression in *Rhizobium trifolii* EMBO J 6: 1173–1179

Djordjevic SP, Ridge RW, Chen H, Redmond JW, Bately M, Rolfe BG (1988) Induction of pathogenic-like responses in the legume *Macroptilium atropurpureum* by a transposon-induced mutant of the fast-growing, broad-host-range *Rhizobium* strain NGR234. J Bacteriol 170: 1848–1857

Downie JA (1989) The *nodL* gene from *Rhizobium leguminosarum* is homologous to the acetyl transferases encoded by *lacA* and *cysE*. Mol Microbiol 3: 1649–1651

Downie JA, Johnston AWB (1986) Nodulation of legumes by *Rhizobium*: the recognised root? Cell 47: 153–154

Downie JA, Hombrecher G, Ma QS, Knight CD, Wells B, Johnston AWB (1983) Cloned nodulation genes of *Rhizobium leguminosarum* determine host-range specificity. Mol Gen Genet 190: 359–365

Dudley ME, Jacobs TW, Long SR (1987) Microscopic studies of cell divisions induced in alfalfa roots by *Rhizobium meliloti*. Planta 171: 289–301

Dusha I, Bakos A, Kondorosi A, De Bruijn FJ, Schell J (1989) The *Rhizobium meliloti* early nodulation genes (*nodABC*) are nitrogen-regulated: isolation of a mutant strain with efficient nodulation capacity on alfalfa in the presence of ammonium. Mol Gen Genet 219: 89–97

Economou A, Hamilton WDO, Johnston AWB, Downie JA (1990) The *Rhizobium* nodulation gene *nodO* encodes a Ca^{2+}-binding protein that is exported without N-terminal cleavage and is homologous to haemolysin and related proteins. EMBO J 9: 349–354

Egelhoff TT, Long SR (1985) *Rhizobium meliloti* nodulation genes: identification of *nodDABC* gene products, purification of *nodA* protein, and expression of *nodA* in *Rhizobium meliloti*. J Bacteriol 164: 591–599

Evans IJ, Downie JA (1986) The *nodI* product of *Rhizobium leguminosarum* is closely related to ATP-binding bacterial transport proteins; nucleotide sequence analysis of the *nodI* and *nodJ* genes. Gene 43: 95–101

Faucher F, Maillet F, Vasse J, Rosenberg C, van Brussel AAN, Truchet G, Denarie J (1988) *Rhizobium meliloti* host range *nodH* gene determines production of an alfalfa-specific extracellular signal. J Bacteriol 170: 5489–5499

Faucher C, Camut S, Denarie J, Truchet G (1989) The *nodH* and *nodQ* host range genes of *Rhizobium meliloti* behave as avirulence genes in *R. leguminosarum* bv. *viciae* and determine changes in the production of plant-specific extracellular signals. Mol Plant Microbe Interact 2: 291–300

Firmin JL, Wilson KE, Rossen L, Johnston AWB (1986) Flavonoid activation of nodulation genes in *Rhizobium* reversed by other compounds present in plants. Nature 324: 90–92

Fisher RF, Tu JK, Long SR (1985) Conserved nodulation genes in *Rhizobium meliloti* and *Rhizobium trifolii*. Appl Environ Microbiol 49: 1432–1435

Fisher RF, Swanson JA, Mulligan JT, Long SR (1987) Extended region of nodulation genes in *Rhizobium meliloti* 1021. II. Nucleotide sequence, transcription sites and protein products. Genetics 117: 191–201

Fisher RF, Egelhoff TT, Mulligan JT, Long SR (1988) Specific binding of proteins from *Rhizobium meliloti* cell-free extracts containing NodD to DNA sequences upstream of inducible nodulation genes. Genes Dev 2: 282–293

Flor HH (1947) Inheritance of reaction to rust in flax. J Agricult Res 74: 241–262

Flor HH (1955) Host–parasite interaction in flax rust—its genetics and other implications. Phytopathology 45: 680–685

Franssen HJ, Nap J-P, Gloudemans T, Stiekema W, van Dam H, Govers F, Louwerse J, van Kammen A, Bisseling T (1987) Characterization of cDNA for nodulin-75 of soybean: a gene product involved in early stages of root nodule development. Proc Natl Acad Sci USA 84: 4495–4499

Fraser RSS (1985) Some basic concepts and definitions in resistance studies. In: Fraser RSS (ed) Mechanisms of resistance to plant disease. Martinus Nijhoff, Dordrecht, pp 1–12

Gabriel DW (1986) Specificity and gene function in plant–pathogen interactions. Am Soc Microbiol News 52: 19–25

Gabriel DW (1989) The genetics of plant pathogen population structure and host–parasite specificity. In: Kosuge T, Nester EW (eds) Plant–microbe interactions: molecular and genetic perspectives, vol 3. Macmillan, New York, pp 343–379

Gerhold DL, Dazzo FB, Gresshoff PM (1985) Selective removal of seedling root hairs for studies of the *Rhizobium*-legume symbiosis. J Microbiol Methods 4: 95–102

Gloudemans T, Bhuvaneswari TV, Moerman M, Van Brussel T, Van Kammen A, Bisseling T (1989) Involvement of *Rhizobium leguminosarum* nodulation genes in gene expression in plant root hairs. Plant Mol Biol 12: 157–167

Gottfert M, Grob P, Hennecke H (1990) Proposed regulatory pathways encoded by the *nodV* and *nodW* genes, determinants of host specificity in *Bradyrhizobium japonicum*. Proc Natl Acad Sci USA 87: 2680–2684

Gottfert M, Grob P, Rossbach S, Fischer H-M, Thony B, Anthamatten D, Kullik I, Henecke H (1989) Bacterial genes involved in the communication between soybean and its root nodule symbiont, *Bradyrhizobium japonicum*. In: Lugtenberg BJJ (ed) Signal molecules in plants and plant–microbe interactions. Springer, Berlin Heidelberg New York Tokyo, pp 295–301 (NATO ASI Series, series H, vol 36)

Grimm C, Panopoulus NJ (1989) The predicted protein product of a pathogenicity locus from *Pseudomonas syringae* pv. *phaseolicola* is homologous to a highly conserved domain of several procaryotic regulatory proteins. J Bacteriol 171: 5031–5038

Gussin GN, Ronson CW, Ausubel FM (1986) Regulation of nitrogen fixation genes. Annu Rev Genet 20: 567–591

Gyorgypal Z, Iyer N, Kondorosi A (1988) Three regulatory *nodD* alleles of diverged flavonoid-specificity are involved in host-dependent nodulation by *Rhizobium meliloti*. Mol Gen Genet 212: 85–92

Hagiya M, Close TJ, Tait RT, Kado CI (1985) Identification of pTiC58 plasmid-encoded proteins for virulence in *Agrobacterium tumefaciens*. Proc Natl Acad Sci USA 82: 2669–2673

Halverson LJ, Stacey G (1986) Signal exchange in plant-microbe interactions. Microbiol Rev 50: 193–225

Heron DS, Ersek T, Krishnan HB Peuppke SG (1989) Nodulation mutants of *Rhizobium fredii* USDA257. Mol Plant Microbe Interact 2: 4–10

Hirsch AM, Bhuvaneswari TV, Torrey JG, Bisseling T (1989) Early nodulin genes are induced in alfalfa root outgrowths elicited by auxin transport inhibitors. Proc Natl Acad Sci USA 86: 1244–1248

Hoekema A, de Pater BS, Fellinger AJ, Hooykaas PJJ, Schilperoort RA (1984) The limited host range of an *Agrobacterium tumefaciens* strain extended by a cytokinin gene from a wide host range T-region. EMBO J 3: 3043–3047

Holl FB (1975) Host plant control of the inheritance of dinitrogen fixation in the *Pisum-Rhizobium* symbiosis. Euphytica 24: 767–770

Hombrecher G, Gotz R, Dibb NJ, Downie JA, Johnston AWB, Brewin J (1984) Cloning and mutagenesis of nodulation genes from *Rhizobium leguminosarum* TOM, a strain with extended host range. Mol Gen Genet 194: 293–298

Hong G-F, Burn JE, Johnston AWB (1987) Evidence that DNA involved in the expression of nodulation (*nod*) genes in *Rhizobium* binds to the product of the regulatory gene *nodD*. Nucleic Acids Res 15: 9677–9690

Honma MA, Ausubel FM (1987) *Rhizobium meliloti* has three functional copies of the *nodD* symbiotic regulatory gene. Proc Natl Acad Sci USA 84: 8558–8562

Hooykaas PJJ, Schilperoort RA (1986) The molecular basis of the *Agrobacterium*–plant interaction. In: Lugtenberg B (ed) Recognition in microbe–plant symbiotic and pathogenic interactions. Springer, Berlin Heidelberg New York, pp 189–202 (NATO ASI Series, series H, vol 4)

Horvath B, Kondorosi E, John M, Schmidt J, Torok I, Gyorgypal Z, Barabas I, Wieneke U, Schell J, Kondorosi A (1986) Organization, structure and symbiotic function of *Rhizobium meliloti* nodulation genes determining host specificity for alfalfa. Cell 46: 335–343

Horvath, B, Bachem CWB, Schell J, Kondorosi A (1987) Host-specific regulation of nodulation genes in *Rhizobium* is mediated by a plant-signal, interacting with the *nodD* gene product. EMBO J 6: 841–848

Huang SZ, Djordjevic MA, Rolfe BG (1988) Characterisation of aberrant infection events induced on *Trifolium subterraneum* by *Rhizobium trifolii* region II mutants. J Plant Physiol 133: 16–24

Huang Y, Morel P, Powell B, Kado CI (1990) Vir A, a coregulator of Ti-specified virulence genes, is phosphorylated in vitro. J Bacteriol 172: 1142–1144

Innes RW, Kuempel PL, Plazinski J, Canter-Cremers H, Rolfe BG, Djordjevic MA (1985) Plant factors induce expression of nodulation and host range genes in *Rhizobium trifolii*. Mol Gen Genet 201: 426–432

Jin S, Roitsch T, Ankenbauer RG, Gordon MP, Nester EW (1990a) The Vir A protein of *Agrobacterium tumefaciens* is autophosphorylated and is essential for *vir* gene regulation. J Bacteriol 172: 525–530

Jin S, Roitsch T, Christie PJ, Nester EW (1990b) The regulatory Vir G protein specifically binds to a *cis*-acting regulatory sequence involved in transcriptional activation of *Agrobacterium tumefaciens* virulence genes. J Bacteriol 172: 531–537

Johnson HW, Means UM, Clark FE (1959) Responses of seedlings to extracts of soybean nodules bearing selected strains of *Rhizobium japonicum*. Nature 183: 308–309

Johnston AWB, Beynon JL, Buchanan-Wollaston AV, Setchell SM, Hirsch PR, Beringer JE (1978) High frequency transfer of nodulation ability between strains and species of *Rhizobium*. Nature 276: 634–636

Johnston AWB, Burn JE, Economou A, Davis EO, Hawkins FKL, Bibb MJ (1988) Genetic factors affecting host-range in *Rhizobium leguminosarum*. In: Palacios R, Verma DPS (eds) Molecular genetics of plant–microbe interactions. American Phytopathological Society, St. Paul, MN, pp 378–384

Kamoun S, Kado CI (1989) A plant inducible pathogenicity (pth) gene of *X. campestris* pv. *campestris* encodes an exocellular components required for growth in the host. In: Molecular biology of bacterial plant pathogens, Fallen Leaf Lake, South Lake Tahoe, California, 37

Kearney B, Staskawicz BJ (1989) Analysis of *avrBs2* in *Xanthomonas campestris* pathovars: conservation and possible mechanism of stable disease resistance. In: Molecular biology of bacterial plant pathogens, Fallen Leaf Lake, South Lake Tahoe, California, 39

Keen NT (1986) Pathogenic strategies of fungi. In: Lugtenberg BJJ (ed) Recognition in microbe–plant symbiotic and pathogenic interactions. Springer, Berlin Heidelberg New York Tokyo, pp 171–188 (NATO ASI Series, series H, vol 4)

Keen NT, Dawson WO (1992) Pathogen avirulence genes and elicitors of plant defense. In: Boller T, Meins F (eds) Genes involved in plant defense. Springer, Wien New York, pp 85–114 [Dennis ES et al (eds) Plant gene research. Basic knowledge and application]

Keen NT, Staskawicz B (1988) Host range determinants in plant pathogens and symbionts. Annu Rev Microbiol 42: 421–440

Keen NT, Tamaki S, Kobayashi D, Gerhold D, Stayton M, Shen H, Gold S, Lorang J, Thordal-Christensen H, Dahlbeck D, Staskawicz B (1990) Bacteria expressing avirulence gene D produce a specific elicitor of the soybean hypersensitive reaction. Mol Plant Microbe Interact 3: 112–121

Kerr A (1987) The impact of molecular genetics on plant pathology. Annu Rev Phytopathol 25: 87–110

Kingsley MT, Gabriel DW (1989) Tn5-induced mutations in a wide host range xanthomonad affecting general and host-specific virulence. In: Molecular biology of bacterial plant pathogens, Fallen Leaf Lake, South Lake Tahoe, California, 32

Klement Z, Goodman RN (1967) The hypersensitive response to· infection by bacterial pathogens. Annu Rev Phytopathol 5: 17–44

Kobayashi DY, Keen NT (1985) Cloning of a factor from *Pseudomonas syringae* pv. *tomato* responsible for a hypersensitive response on soybean. Phytopathology 75: 1355

Kobayashi DY, Tamaki SJ, Keen NT (1990) Molecular characterisation of avirulence gene D from *Pseudomonas syringae* pv. *tomato*. Mol Plant Microbe Interact 3: 94–102

Kondorosi A (1989) *Rhizobium*–legume interactions: nodulation genes. In: Kosuge T, Nester EW (eds) Plant–microbe interactions: molecular and genetic perspectives, vol 3. McGraw-Hill, New York pp 383–420

Kondorosi E, Gyuris J, Schmidt J, John M, Duda E, Hoffman B, Schell J, Kondorosi A

(1989) Positive and negative control of *nod* expression in *Rhizobium meliloti* is required for optimal nodulation. EMBO J 8: 1331–1340

Kumer Rao JVDK, Dart PJ, Usha Kiran M (1984) *Rhizobium*-induced leaf roll in pigeonpea (*Cajanus cajan* L. Millsp.). Soil Biochem 16: 89–91

Lamb CJ, Lawton MA, Dron M, Dixon RA (1989) Signals and transduction mechanisms for activation of plant defenses against microbial attack. Cell 56: 215–224

Leroux B, Yanofsky MF, Winans SC, Ward JE, Zeigler SF, Nester EW (1987) Characterization of the *vir A* locus of *Agrobacterium tumefaciens*: a transcriptional regulator and host range determinant. EMBO J 4: 849–856

Libbenga KR, Bogers RJ (1974) Root-nodule morphogenesis. In: Quispel A (ed) The biology of nitrogen fixation. North-Holland, Amsterdam, pp 430–472

Lie TA (1984) Host genes in *Pisum sativum* L. conferring resistance to European *Rhizobium leguminosarum* strains. Plant Soil 82: 415–425

Long SR (1984) Genetics of *Rhizobium* nodulation. In: Kosuge T, Nester EW (eds) Plant–microbe interactions. Macmillan, New York, pp 265–306

Long SR (1989) *Rhizobium*-legume nodulation: life together in the underground. Cell 56: 203–214

Long SR, Cooper J (1988) Overview of symbiosis. In: Verma DPS, Palacios R (eds) Molecular genetics of molecular plant–microbe interactions. American Phytopathological Society, St. Paul, MN, pp 163–178

Lynch JM, Whipps JM (1989) Substrate flow in the rhizosphere. In: The rhizosphere and plant growth, Beltsville Symposium XIV, Beltsville, Maryland, 20

Mass JL, Finney MM, Civerolo EL, Sasser M (1985) Association of an unusual strain of *Xanthomonas campestris* with apple. Phytopathology 75: 438–445

Martin JF, Demain AL (1980) Control of antibiotic biosynthesis. Microbial Rev 44: 230–251

Martinez E, Romero D, Palacios R (1990) The *Rhizobium* genome. Crit Rev Plant Sci 9: 59–93

McIver J, Djordjevic MA, Weinman JJ, Bender GL, Rolfe BG (1989) Extension of host range in *Rhizobium leguminosarum* bv. *trifolii* caused by point mutations in *nod D* that result in alterations in regulatory function and recognition of inducer molecules. Mol Plant Microbe Interact 2: 97–106

Morris RO (1986) Genes specifying auxin and cytokinin biosynthesis in phytopathogens. Annu Rev Plant Physiol 37: 509

Morrison NA, Bisseling T, Verma DPS (1988) Development and differentiation of the root nodule. In: Browder LW (ed) Developmental biology, vol 5. Plenum, New York, pp 405–425

Mulligan JT, Long SR (1985) Induction of *Rhizobium meliloti nod C* expression by plant exudate requires *nod D*. Proc Natl Acad Sci USA 82: 6609–6613

Murphy PJ, Heycke N, Banfalfi Z, Tate ME, De Bruijn F, Kondorosi A, Tempe J, Schell J (1987) Genes for the catabolism and synthesis of an opine-like compound in *Rhizobium meliloti* are closely linked on the Sym plasmid. Proc Natl Acad Sci USA 84: 493–497

Murphy PJ, Heycke N, Trenz SP, Ratet P, De Bruijn FJ, Schell J (1988) Synthesis of an opine-like compound, a rhizopine, in alfalfa nodules is symbiotically regulated. Proc Natl Acad Sci USA 85: 9133–9137

Nixon BT, Ronson CW, Ausubel FM (1986) Two component regulatory systems responsive to environmental stimuli share strongly conserved domains with the nitrogen assimilation regulatory genes *ntr B* and *ntr C*. Proc Natl Acad Sci USA 83: 7850–7854

Owens LD, Wright DA (1965) Rhizobial-induced chlorosis in soybeans: isolation, production in nodules and varietal specificity of the toxin. Plant Physiol 40: 927–930

Panopoulos NJ, Peet RC (1985) The molecular genetics of plant pathogenic bacteria and their plasmids. Annu Rev Phytopathol 23: 381–419

Peters NK, Long SR (1988) Alfalfa root exudates and compounds which promote or inhibit induction of *Rhizobium meliloti* nodulation genes. Plant Physiol 88: 396–400

Peters NK, Verma DPS (1990) Phenolic compounds as regulators of gene expression in plant-microbe interactions. Mol Plant Microbe Interact 3: 4–8

Peters NK, Frost JW, Long SR (1986) A plant flavone, luteolin, induces expression of *Rhizobium meliloti* nodulation genes. Science 233: 917–1008

Philip-Hollingsworth S, Hollingsworth RI, Dazzo FB, Djordjevic MA, Rolfe BG (1989) Effects of interspecies transfer of *Rhizobium* host specific nodulation genes on acidic polysaccharide structure and lectin recognition processes. J Biol Chem 264: 5710–5714

Pryor T (1987) The origin and structure of fungal disease resistance genes in plants. Trends Biol Sci 3: 157–161

Rahme LG, Mindrinos MN, Panopoulos NJ (1989) Physiological and genetic control of *hrpABCD* gene expression in *Pseudomonas syringae* pv. *phaseolicola*. In: Molecular biology of bacterial plant pathogens, Fallen Leaf Lake, South Lake Tahoe, California, 22

Redmond JR, Batley M, Djordjevic MA, Innes RW, Keumpel PL, Rolfe BG (1986) Flavones induce expression of *nod* genes in *Rhizobium*. Nature 323: 632–635

Ridge RW, Rolfe BG (1985) *Rhizobium* sp. degradation of legume root hair cell wall at the site of infection thread origin. Appl Environ Microbiol 50: 717–720

Ridge RW, Rolfe BG (1986) Sequence of events during the infection of the tropical legume *Macroptilium atropurpureum* Urb. by the broad-host-range, fast-growing *Rhizobium* ANU240. J Plant Physiol 122: 121–137

Roberts IN, Dow JM, Lum KY, Scofield G, Barber CE, Daniels MJ (1987) Antiserum against *Xanthomonas* phytopathogen inhibits host-pathogen interaction in seedlings of *Brassica campestris*. FEMS Microbiol Lett 44: 383–387

Rolfe BG (1988) Flavones and isoflavones as inducing substances of legume nodulation. Biofactors 1: 3–10

Rolfe BG, Gresshoff PM (1988) Genetic analysis of legume nodule initiation. Annu Rev Plant Physiol Plant Mol Biol 39: 297–319

Rolfe BG, Gresshoff PM, Shine J, Vincent JM (1980) Interaction between a non-nodulating and an ineffective mutant of *Rhizobium trifolii* resulting in effective (nitrogen-fixing) nodulation. Appl Environ Microbiol 39: 449–452

Rolfe BG, Batley M, Redmond JW, Richardson AE, Simpson RJ, Bassam BG, Sargent CL, Weinman JJ, Djordjevic MA, Dazzo FB (1988) Phenolic compounds secreted by legumes. In: Bothe H, de Bruijn FJ, Newton WE (eds) Nitrogen fixation: hundred years after. Proceedings of the 7th International Congress on Nitrogen Fixation. G Fischer, Stuttgart, pp 405–409

Rolfe BG, Weinman JJ, Djordjevic MA (1989) Host-specificity in *Rhizobium*. In: Kosuge T, Nester EW (eds) Plant–microbe interactions: molecular and genetic perspectives, vol 3. McGraw-Hill, New York, pp 421–456

Rossen L, Johnston AWB, Downie JA (1984) DNA sequence of the *Rhizobium leguminosarum* nodulation genes *nodAB* and *C* required for root hair curling. Nucleic Acids Res 12: 9497–9508

Ryan CA (1988) Oligosaccharides as recognition signals for the expression of defensive genes in plants. Biochemistry 27: 8879–8883

Sargent L, Huang SZ, Rolfe BG, Djordjevic MA (1987) Split root assays using *Trifolium subterraneum* show that *Rhizobium* infection induces a systemic response that can inhibit nodulation of another invasive *Rhizobium* strain. Appl Environ Microbiol 53: 1611–1619

Scheres B, Van De Wiel C, Zalensky A, Horvath B, Spaink H, Van Eck H, Zwartkruis F, Wolters A-M, Gloudemans T, Van Kammen A, Bisseling T (1990) The ENOD12 gene product is involved in the infection process during the pea–*Rhizobium* interaction. Cell 60: 284–291

Schlaman HRM, Spaink HP, Okker RJH, Lugtenberg BJJ (1989) Subcellular localisation of the *nod D* gene product in *Rhizobium leguminosarum*. J Bacteriol 171: 4686–4693

Schmidt J, John M, Weineke U, Krussmann H-D, Schell J (1986) Expression of the nodulation gene *nod A* in *Rhizobium meliloti* and the localisation of the gene product in the cytosol. Proc Natl Acad Sci USA 83: 9581–9585

Schofield PR, Djordjevic MA, Rolfe BG, Shine J, Watson JM (1983) A molecular linkage map of nitrogenase and nodulation genes in *Rhizobium trifolii*. Mol Gen Genet 192: 459–465

Schofield PR, Ridge RW, Rolfe BG, Shine J, Watson JM (1984) Host-specific nodulation is encoded on a 14 kb DNA fragment in *Rhizobium trifolii*. Plant Mol Biol 3: 3–11

Schwedock J, Long SR (1989) Nucleotide sequence and protein products of two new nodulation genes of *Rhizobium meliloti*, *nod P* and *nod Q*. Mol Plant Microbe Interact 2: 184–194

Sequeira F (1983) Mechanisms of induced resistance in plants. Annu Rev Microbiol 37: 51–79

Shearman CA, Rossen L, Johnston AWB, Downie JA (1986) The *Rhizobium leguminosarum* nodulation gene *nod F* encodes a polypeptide similar to acyl-carrier protein and is regulated by *nod D* and a factor in pea root exudate. EMBO J 5: 647–652

Spaink HP, Okker RJH, Wijffelman CA, Pees E, Lugtenberg BJJ (1987a) Promoters in the nodulation region of the *Rhizobium leguminosarum* Sym plasmid pRL1JI. Plant Mol Biol 9: 27–39

Spaink HP, Wijffelman CA, Pees E, Okker PJH, Lugtenberg BJJ (1987b) *Rhizobium* nodulation gene *nod D* as a determinant of host specificity. Nature 328: 337–340

Spaink HP, Okker RJH, Wijffelman CA, Tak T, Goosen-De Roo L, Pees E, Van Brussel AAN, Lugtenberg BJJ (1989a) Symbiotic properties of rhizobia containing a flavonoid-independent hybrid *nod D* product. J Bacteriol 171: 4045–4053

Spaink HP, Weinman JJ, Djordjevic MA, Wijffelman CA, Okker RJH, Lugtenberg BJJ (1989b) Genetic analysis and cellular localization of the *Rhizobium* host specificity-determining Nod E protein. EMBO J 8: 2811–2818

Spaink HP, Wijffelman CA, Okker RJH, Lugtenberg BJJ (1989c) Localisation of functional regions of the *Rhizobium nod D* product using hybrid *nod D* genes. Plant Mol Biol 12: 59–73

Stachel SE, Messens E, Van Montague M, Zambryski P (1985) Identification of signal molecules produced by wounded plant cells that activate T-DNA transfer in *Agrobacterium tumefaciens*. Nature 318: 624–629

Surin BP, Downie JA (1988) Characterisation of the *Rhizobium leguminosarum* genes *nod LMN* involved in efficient host specific nodulation. Mol Microbiol 2: 173–183

Surin BP, Downie JA (1989) *Rhizobium leguminosarum* genes required for expression and transfer of host specific nodulation. Plant Mol Biol 12: 19–29

Suslow TV (1982) Role of root-colonizing bacteria in plant growth. In: Mount MS, Lacy GH (eds) Phytopathogenic prokaryotes, vol 2. Academic Press, New York, pp 187–224

Swarup S, Gabriel DW (1989) Host-specific virulence genes from *Xanthomonas citri* enable several pathovars to induce canker-like lesions on citrus. In: Molecular biology of bacterial plant pathogens, Fallen Leaf Lake, South Lake Tahoe, California, 32

Triplett EW, Roberts GP, Ludden PW, Handelsman J (1989) What's new in nitrogen fixation. Am Soc Microbiol News 55: 15–21

Truchet G, Barker DG, Camut S, De Billy F, Vasse J, Huguet T (1989) Alfalfa nodulation in the absence of *Rhizobium*. Mol Gen Genet 219: 65–68

Upadhyaya NM, Tucker WT, Kumer Rao JVDK, Dart PJ (1985) Analysis of a leaf-curl phenomenon in pigeonpea (*Cajanas cajan* (L.) Millsp.) induced by *Rhizobium* nodulation.

In: Evans HJ, Bottomley PJ, Newton WE (eds) Nitrogen fixation research progress. Martinus Nijhoff, Dordrecht, p 145

Van Den Bosch KA, Newcomb EH (1986) Immunogold localisation of nodule-specific uricase in developing soybean root nodules. Planta 167: 425–436

Van Egeraat AWSM (1975) The possible role of homoserine in the development of *R. leguminoarum* in the rhizosphere of pea seedlings. Plant Soil 42: 387–396

Van Kammen A (1984) Suggested nomenclature for plant genes involved in nodulation and symboisis. Plant Mol Biol Rep 2: 103–131

Vance CP (1983) *Rhizobium* infection and nodulation: a beneficial plant disease? Annu Rev Microbiol 37: 399–424

Vance CP, Egli MA, Girffith SM, Miller SS (1988) Plant regulated aspects of nodulation and nitrogen fixation. Plant Cell Environ 11: 413–427

Vasse JM, Truchet G (1984) The *Rhizobium*-legume symbiosis: observation of root infection by bright-field microscopy after staining with methylene blue. Planta 161: 487–489

Vidaver AK (1983) Bacteriocins: the lure and the reality. Plant Dis 67: 471–472

Vincent J (1974) Root nodule symbioses with *Rhizobium*. In: Quispel A (ed) The biology of nitrogen fixation. North-Holland, Amsterdam, pp 265–341

Vincent JM (1980) Factors controlling the legume-*Rhizobium* symbiosis. In: Newton WE, Orme-Johnson WH (eds) Nitrogen fixation, vol 2. University Park Press, Baltimore, pp 103–129

Weinman JJ, Djordjevic MA, Sargent CL, Dazzo FB, Rolfe BG (1988) A molecular analysis of host range genes of *Rhizobium trifolii*. In: Palacios R, Verma DPS (eds) Molecular genetics of plant–microbe interactions. American Phytopathological Society, St. Paul, MN, pp 33–34

Yanofsky MF, Nester EW (1986) Molecular characterisation of a host-range determining locus from *Agrobacterium tumefaciens*. J Bacteriol 168: 244–250

Zaat SAJ, Schripsema J, Wijffelman CA, van Brussel AAN, Lugtenberg BJJ (1989) Analysis of the major inducers of the *Rhizobium nod A* promoter from *Vicia sativa* root exudate and their activity with different *nod D* genes. Plant Mol Biol 13: 175–188

Zhou JC, Tchan YT, Vincent JM (1985) Reproductive capacity of bacteroids in nodules of *Trifolium repens* L. and *Glycine max* (L.). Planta 163: 473–482

Chapter 4

Pathogen Avirulence Genes and Elicitors of Plant Defense

Noel T. Keen and William O. Dawson

Department of Plant Pathology, University of California Riverside, Riverside, CA 92521, U.S.A.

With 3 Figures

Contents

I. Introduction

The hypersensitive response (HR) is an active, inducible defense reaction employed by plants against pathogens. Its main superficial feature is the rapid necrosis of plant cells surrounding the infection site of an invading pathogen. Of more functional importance, 30 or more plant genes are specifically derepressed following recognition of a pathogen, some of which account for the inhibition of subsequent pathogen invasion. These 'defense response genes' control a wide range of expressive mechanisms associated with the HR, including the production of chitinases, antiviral proteins, protease inhibitors, β-1,3 glucanases and hydroxyproline-rich glycoproteins as well as encoding enzymes involved in the production of lignin, callose and phytoalexins at the infection site.

The HR is initiated by highly specific surveillance systems utilized by plants that detect some but not all pathogens during initial infection and thereby prevent disease. Thus, pathogen strains that avoid recognition by the HR may cause disease while detected biotypes generally do not. This review will consider recent information on the mechanics of pathogen recognition that is establishing a framework for understanding how pathogen surveillance and active disease defense function in plants.

The HR operates on various hierarchal levels in plant–pathogen interactions. One of these is termed *general resistance*, in which an entire plant species is uniformly resistant to an entire pathogen taxon. Another level, called *specific resistance* or race–cultivar resistance, involves the resistance of certain cultivars of a single plant species to only certain biotypes or races of a single pathogen taxon. Specific resistance is generally controlled by single plant *disease resistance genes* and single complementary genes in the pathogen, called *avirulence genes* (Day, 1974; Ellingboe, 1981; Flor, 1942; Keen and Staskawicz, 1988). The functional alleles in both plant and pathogen are generally inherited as dominant characters, both of which must be present for plant recognition to occur. This gene-for-gene relationship has been known for some 50 years but only recently have we begun to understand the underlying mechanisms.

The fact that active plant defense is invoked in response to a certain pathogen avirulence gene a priori implies the function of a specific plant recognitional system, presumably a specific receptor. These receptors must detect pathogen elicitors, which we will call general or specific (Fig. 1). In specific resistance, avirulence genes in the pathogen direct production of a particular ligand (called a specific elicitor) that is recognized only by plants carrying a complementary or matching disease resistance gene. This has been formally proposed as the *elicitor–receptor model* to explain gene-for-gene specificity (Callow, 1977; Day, 1974; Dixon, 1986; Gabriel et al., 1988; Keen and Bruegger, 1977). It predicts that the primary avirulence gene protein product or a metabolite resulting from its catalytic activity is recognized by a specific plant receptor encoded totally or in part by a disease resistance gene. Such a system has features in common with some genes of the IG superfamily in vertebrates, particularly those encoding Fc receptors and interleukin receptors (Kinet, 1989; Sims et al., 1989). Some evidence also suggests that closely linked plant disease resistance genes may recombine to develop new recognitional specificities (Bennetzen et al., 1988; Ellis et al., 1988; Islam et al., 1989; Pryor, 1987; Shepherd and Mayo, 1972), again similar to the well known properties of antibody genes (Alt et al., 1987; Williams and Barclay, 1988). However, plant disease resistance genes

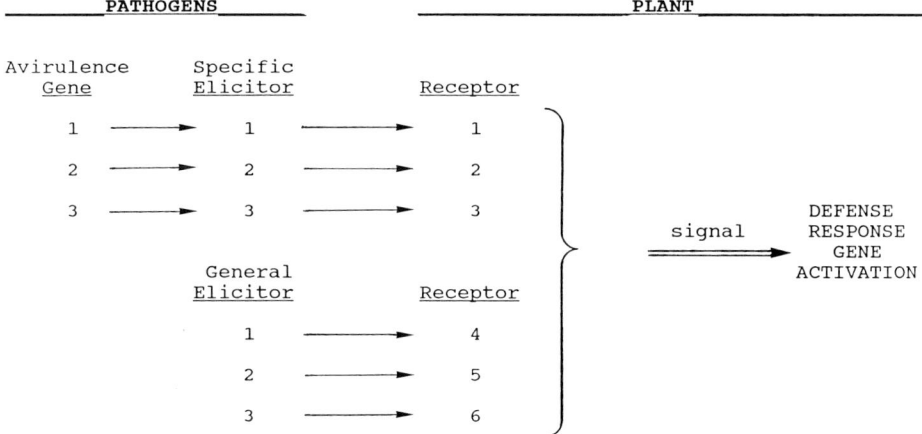

Fig. 1. Conceptual framework whereby diverse pathogen-produced general elicitors and specific elicitors resulting from avirulence gene activity are recognized by unique plant receptors. These receptors all invoke a common signal transduction pathway that de-represses expression of an array of disease response genes, as discussed in the text

have not yet been cloned and characterized, thus precluding critical tests of these possibilities.

Regardless of the elicitor, the executional events of the HR in a certain plant are similar (e.g., Daniels et al., 1987; Lamb et al., 1989). Thus, different elicitor signals appear to activate a common signal transduction mechanism which in turn modulates the derepression of many different plant defense response genes (Fig. 1).

II. Host Range Determinants

A. Positive-Acting Factors

The pattern is emerging that the host range of pathogenic bacteria and fungi may be determined by pathogen genes which operate positively to extend the range of plants affected or negatively (avirulence genes) to restrict the host range. Djordjevic et al. (1992) address the host range genes of *Rhizobium*, some but not all of which confer positive-acting factors. In bacterial pathogens, mutational and cloning studies have shown the existence of specific genetic loci that are required for virulence on certain plant species (Gabriel, 1989; Mellano and Cooksey, 1988). *Xanthomonas campestris* pv. *translucens* occurs in nature as several strains that can infect various grass plants. Mellano and Cooksey (1988) studied a strain that normally infected rye, barley, wheat and triticale and induced several mutants that were non-pathogenic on some but not all of these plants. Since all the tested mutants retained virulence on at least one of the host plants, none of the mutations involved genes conferring basic pathogenicity. Instead, the mutated genes must be required for virulence on the various grass plants. The functions of these *X. campestris* pv. *translucens* host-range genes, however, are not yet known.

Recent work with two fungal pathogens has provided evidence that single genes may expand the host range of a pathogen to include additional plant species. In the first case, Schäfer et al. (1989) reported that introduction of a gene encoding the enzyme pisatin demethylase into *Cochliobolus maydis* enabled this fungal pathogen of corn plants to cause disease on pea plants, not a normal host. The pisatin demethylase gene was originally cloned from the pea pathogen, *Nectria haematococca* (Weltring et al., 1988) and had been shown to be important in its virulence by detoxifying the pea phytoalexin, pisatin (see VanEtten et al., 1989). Surprisingly, transformation of the cloned pisatin demethylase gene into *Cochliobolus maydis* permitted this organism to exhibit significant virulence on pea plants. Thus, at least under experimental conditions, production of pisatin demethylase extended the host range of *C. maydis* to include pea.

A second example of host range expansion by a defined pathogen gene was reported by Dickman et al. (1989). These workers transformed a gene encoding the enzyme, cutinase, into the pathogen *Mycosphaerella* sp. and noted that, unlike the wildtype fungus, transformants were able to directly penetrate the cuticle of papaya fruits and cause disease. Normally *Mycosphaerella* is only able to attack papaya fruits through wounds in the fruit surface. On the other hand, several pathogenic fungi such as *Fusarium solani* elaborate cutinase enzymes that permit breaching the barrier of the plant cuticle (see Kolattukudy, 1985). The work of Dickman et al. (1989) therefore illustrates an expansion of the host range of *Mycosphaerella* to include unwounded papaya fruits by the sole expedient of introducing a gene encoding cutinase.

B. Avirulence Genes

Since this topic was discussed in more depth in another review (Keen, 1990), only coverage germane to this paper will be included here. Avirulence (*avr*) genes were first described by genetic crossing of pathogen biotypes that differed in their reactions on a resistant plant cultivar (see Day, 1974; Flor, 1942). Hundreds of such genes have been defined in this way. Essentially all of them are inherited as dominant genetic characters. Avirulence genes have recently been defined in fungal, bacterial, and viral pathogens and several of them have been sequenced and characterized. The first bacterial avirulence gene was cloned from race 6 of *Pseudomonas syringae* pv. *glycinea* and called *avr A* (Staskawicz et al., 1984). Avirulence genes were subsequently cloned by several laboratories from various *P. syringae* and *Xanthomonas campestris* pathovars (e.g., Bonas et al., 1989; Gabriel et al., 1986; Hitchin et al., 1989; Huang et al., 1988; Kelemu and Leach, 1990; Kobayashi et al., 1989; Minsavage et al., 1990; Ronald et al., 1988; Shintaku et al., 1989; Swanson et al., 1988; Vivian et al., 1989; Whalen et al., 1988). All of the bacterial *avr* genes thus far characterized specify the avirulence phenotype when expressed in appropriate bacterial hosts, including pathogens other than that from which the gene was cloned and, in some cases, even saprophytic organisms like *E. coli* (S. Beer, pers. comm.; Huang et al., 1988; Keen et al., 1990). Recent work has also indicated that certain viral genes may be regarded as avirulence genes (Knorr and Dawson, 1988; Saito et al., 1987; Schoelz et al., 1986). All of the avirulence genes thus far sequenced from both viral and bacterial pathogens encode single protein products. With a few noteworthy exceptions, these protein products have not exhibited significant homology to those in data banks, but certain *avr* genes share significant homology with each other. However, the sequencing and expression data have generally not solved the key questions of (*i*) how do *avr* genes function to cause plant recognition of the pathogen? and (*ii*) what is the function of *avr* genes in the pathogen? However, recent work with avirulence genes from tobacco mosaic virus (TMV) and bacterial pathogens as well as findings with several microbial elicitors have illuminated these questions.

III. General Elicitors

Several recent reviews (e.g., Anderson, 1989; Collinge and Slusarenko, 1987; Dixon, 1986; Kuc and Rush, 1985) have covered the topic sufficiently well that this paper will only emphasize recent developments. Two types of elicitors have been detected from plant pathogens that elicit the HR in the absence of a pathogen. One group includes general elicitors. All members of a pathogen taxon produce the elicitor and it uniformly affects all members of a plant species. As such, these elicitors may be involved in the general resistance of plants to broad taxonomic groups of potential pathogens. This is in contrast to specific elicitors, which mirror the specificity of gene-for-gene interactions—that is, the elicitor is produced only by pathogen biotypes containing a certain avirulence gene and is effective only in plants carrying the complementary disease resistance gene. Since general elicitors from pathogens include structural components of the cell wall or other essential metabolites that cannot be dispensed with, it is unlikely that mutations for their synthesis will be isolated from native populations of pathogens.

General elicitors include abiotic and biotic factors. Among the former are heavy metal salts, certain synthetic detergents and UV light. The effects of these agents are not well understood, but may involve either mimicry of or interference with normal plant signal transduction mechanisms or inhibition of enzymes responsible for phytoalexin turnover (Yoshikawa, 1978; and unpubl. data). Among the biotic elicitors made by plant pathogens, oligosaccharides, proteins and glycoproteins have been the most studied (see Ryan, 1988; Stone, 1989; West, 1981; Ebel and Scheel, 1992). These elicitors are frequently present on or near the cell surface of the pathogen, often as components of the cell wall. While general elicitors may show marked differences when tested on various plant species (Parker et al., 1988), they behave relatively uniformly when tested on several cultivars of a single plant species. The array of factors which can be obtained from a single pathogen that possess elicitor activity (e.g., Hamdan and Dixon, 1987) indicates that a diverse array of molecules may function. Unfortunately, the role of general elicitors as physiologically important agents determining specificity in plant–pathogen interactions has not yet been critically tested. It is, however, probable that general elicitors may account for the general resistance of certain plant species to particular pathogens. Thus, along with specific elicitors, general elicitors are part of the diverse group of pathogen metabolites that are specifically recognized by plants to elicit the HR (Fig. 1).

A. Glucan Elicitors

One class of general elicitors is the oligoglucans present in *Phytophthora* spp. and other fungi. Elicitor-active glucans were recovered from cell walls of *P. megasperma* f. sp. *glycinea* following partial hydrolysis with acid or autoclaving (Ayers et al., 1976). The smallest elicitor-active molecule was isolated and characterized as a heptaglucan (Fig. 2, n = 1) consisting of a five-membered β-1,6 linked glucose chain with β-1,3 linked glucose molecules on the second and fourth residues (Sharp et al., 1984). The heptaglucan was highly active as an elicitor of the hypersensitive response and phytoalexin production in a range of soybean cultivars. Significantly, however, glucan elicitors are inefficient in several other plants such as potato (Bostock et al., 1981) and parsley (Parker et al., 1988).

Yoshikawa and collaborators found that extracellular β-1,3-endoglucanases present in soybean tissues may attack the cell walls of invading fungal pathogens, liberating glucan and glucomannan elicitors (Yoshikawa et al., 1981; Keen and Yoshikawa, 1983). Unlike certain other plants, such as tobacco, where glucanases behave as defense response genes and are significantly expressed after infection (Kauffman et al. 1987), soybean tissues contain a high constitutive level of endoglucanase activity (Keen and Yoshikawa, 1983). The soybean endoglucanases released the heptaglucan elicitor as well as higher homologues from cell walls of *P. megasperma* f. sp. *glycinea* (Yoshikawa, 1988). The homologue glucans had one to four additional glucose residues in the central spacer region (Fig. 2, n = 2–5) and exhibited markedly greater elicitor activity in soybean tissues than the heptaglucan. The more active glucans were apparently missed previously because acid hydrolysis, used in the work of Sharp et al. (1984), cleaves the longer β-1,6-linked glucose backbones. Enzymatic extraction with β-1,3-

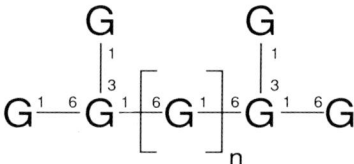

Fig. 2. Homologue polyglucan elicitors isolated from cell walls of *Phytophthora megasperma* f. sp. *glycinea*. G, D-glucose residues and the linkages shown are all β; n = 1 to 5 glucose residues, as discussed in the text

endoglucanase, however, preserves the β-1,6 linked backbones as well as the β-1,3-linked glucose branch residues. It will be of interest to more fully test the possible role of the soybean endoglucanases and the glucanase-related elicitors in the general resistance of soybean plants. The gene encoding the soybean β-1,3-endoglucanase has recently been cloned (Takeuchi and Yoshikawa, pers. comm.) and this should permit incisive tests of its role in resistance.

B. Chitin and Chitosan

Hadwiger and collaborators showed that enzymes present in pea tissues released chitosan oligomers from the cell walls of *Fusarium solani* which elicited defense responses in pea (for review see Kendra and Hadwiger, 1987). In addition, relatively low concentrations of chitosan have been shown to directly inhibit the growth of certain pathogenic fungi (Kendra et al., 1989), raising the possibility that liberation of chitosan oligomers may directly cause disease resistance in the plant. Oligomers containing 6 to 11 residues of glucosamine were most active as phytoalexin elicitors in pea, while glucosamine itself and short chain oligomers were inactive. This work initiated considerable interest in the possible role of chitin and chitosan oligomers as general elicitors of plant defense reactions. The compounds are also of interest since plant chitinases are known to be synthesized in increased quantities during active defense responses (Legrand et al., 1987), as will be discussed later.

Chitin and chitosan are active elicitors of defense responses in pea as well as other plants such as wheat (Ride and Barber, 1987) and rice (C. West, pers. comm.), but are relatively inactive as phytoalexin elicitors in other plants such as soybean (Keen and Yoshikawa, 1983). However, Köhle et al. (1985) reported that chitosan elicited callose synthesis in soybean cells. Barber et al. (1989) recently showed that chitin oligomers with four to six N-acetyl glucosamine residues exhibited maximal activity to elicit lignification in wounded wheat leaves. This is a somewhat shorter size requirement than that noted above for chitosan in the pea system. While definitive proof for a physiologic role is yet lacking, chitin and chitosan oligomers released from pathogen cell walls may be important as primary elicitors of plant defense reactions in certain plants. It is also possible that they may function as secondary elicitors formed following the elaboration of plant chitinases during an initial defense reaction which then re-enforce the resistance response.

C. Arachidonic Acid and Related Compounds as Elicitors

Bostock et al. (1981) showed that arachidonic acid and the related eicosapentaenoic acid were the major elicitors detectable from *Phytophthora infestans* that were active in potato tuber tissue. These unsaturated fatty acids are clearly very different from other currently recognized classes of elicitors. The *Phytophthora* fatty acids, however, functioned as efficient elicitors of sesquiterpenoid phytoalexin accumulation in potato tuber tissue but not in several other plant species (Bloch et al., 1984). The ability of the fatty acids to function as elicitors is very structure dependent, since only 20 and 22 carbon fatty acids with unsaturations at certain positions in the carbon chain have significant elicitor activity. The fatty acid elicitors have also been shown to stimulate increases in enzymes involved in production of sesquiterpenoid phytoalexins and lignification (Stermer and Bostock, 1987). However, it was recently observed that arachidonic acid stimulated a general increase in protein synthesis in potato tuber tissues (Stermer and Bostock, 1989). This is considerably different from the activity of most other elicitors, which derepress discrete sets of plant defense response genes (see Lamb et al., 1989).

D. Protein and Glycoprotein Elicitors

Several proteins and glycoproteins have been identified as general elicitors (e.g., Farmer and Helgeson, 1987), although their physiologic importance in disease defense is not known in most cases. Sutherland et al. (1989) recently extracted elicitors from germ tubes of *Puccinia graminis* f. sp. *tritici* or *P. recondita* and from wheat leaves infected with these fungi. The major elicitor in germ tube extracts had previously been identified as a 67 kDa glycoprotein (Kogel et al., 1988; Moerschbacher et al., 1986). All of the preparations from germ tubes or from infected leaves elicited the hypersensitive response in leaves of some but not all wheat cultivars. The similarity of response suggested that the glycoprotein elicitor was present in all of the preparations. Of particular interest, sensitivity to the elicitor preparations mapped to wheat chromosome 5A, but this gene(s) could not be identified as any known disease resistance gene against the rust fungi (Sutherland et al., 1989). Thus, the glycoprotein elicitor exhibits specificity on a series of host cultivars. However, it does not appear to be a specific elicitor because it has not yet been linked to a defined resistance gene/ avirulence gene set.

Ricci et al. (1989, and pers. comm.) isolated novel proteins of ca. 10 kDa from culture fluids of *Phytophthora cryptogeae*, *P. capsici*, and certain isolates of *P. parasitica*. These proteins were efficient elicitors of the hyper-sensitive reaction in tobacco plants. Isolates of *P. parasitica* that were pathogenic on tobacco plants did not elaborate the protein elicitor, but it was produced by non-pathogenic *P. parasitica* isolates. Application of the proteins to tobacco plants protected them against subsequent inoculation with a normally pathogenic isolate of *P. parasitica* Ricci et al. (1989). While the proteins from the three *Phytophthora* species were highly homologous, sequencing showed that those from *P. capsici* and *P. cyrptogeae* had distinct differences. The protein from *P. capsici* was also only about 1/50 as active as an elicitor in tobacco as that from *P. cryptogeae* or non-patho-genic *P. parasitica* isolates. The results of Ricci and collaborators accord-ingly raise the possibility that these proteins may function as elicitors of the HR in tobacco and play an important role in determining the host range of *Phytophthora* sp.

E. Oligogalacturonide Elicitors

Another class of general elicitor is the oligogalacturonides. These are linear, alpha-linked oligomers of D-galacturonic acid and are of particular interest because they arise from plant cell walls. Oligogalacturonides are accord-ingly sometimes referred to as 'endogenous elicitors'. They are released from the pectic fraction of the cell wall by depolymerizing enzymes of pathogen (Jin and West, 1984) or possibly plant (Hargreaves and Bailey, 1978) origin. While not as active as certain other elicitors, the oligogalactur-onides have been shown to act synergistically with glucan elicitors in soybean (Davis et al., 1986). Oligogalacturonides elicit several components of the HR, including phytoalexins and lignification. Oligomers of about 10–12 galacturonic acid residues exhibited optimal activity in studied plants, with dimers/trimers as well as longer polymers exhibiting little or no activity (Bruce and West, 1989; Nothnagel et al., 1983).

Invading pathogens frequently employ enzymes which attack the pectic fraction of plant cell walls as virulence factors to gain ingress into plant tissue. However, such enzymes may be counter-productive if they liberate significant levels of elicitor-active oligogalacturonides in the infection site. Then the plant may respond with an effective hypersensitive response that precludes further development of the pathogen. Indeed, recent evidence suggests that the efficiency of attack on plant pectic polymers may deter-mine the success of an invading pathogen (Cervone et al., 1989)—if the pathogen efficiently degrades the pectate fraction to oligomers smaller than

10–12 residues, defense will not occur; however, if the pectate polymers are less efficiently degraded and significant levels of elicitor-active oligogalacturonides accumulate, the HR will be invoked.

IV. Suppressors

There has been considerable interest in the possibility that pathogens may produce suppressor molecules which either specifically or non-specifically block the action of general elicitors in certain species or cultivars of plant. Some evidence suggests the involvement of suppressors with specific resistance (Doke et al., 1979), but no corroboration of such a role has as yet been obtained from genetic experiments with bacterial or viral pathogens. Thus, attempts to identify cloned genes modulating the production of specific suppressors have thus far failed (Culver and Dawson, unpubl. data; Kelemu and Leach, 1990; J. Mansfield, pers. comm.; Staskawicz et al., 1984).

Unlike the case with specific resistance, several pathogens have been shown to produce suppressors that function to overcome general resistance. For example, Heath (1980) reported that injection of extracts from plant tissue infected with a compatible rust species into uninfected tissue rendered it susceptible to infection by a normally non-pathogenic rust fungus. Oku et al. (1987) have studied a small peptide from *Mycosphaerella* spp. that specifically suppresses active defense in pea plants. Sutherland et al. (1989) have also speculated that rust fungi may produce suppressor substances which block the activity of the *Puccinia* glycoprotein elicitor discussed above. Peever and Higgins (1989b) reported that intercellular fluids from tomato leaves infected with *Cladosporium fulvum* suppressed the activity of general glycoprotein elicitors produced by the fungus. The tomato suppressor(s) was not identified, but it was suggested that enzymes present in the plant might attack the glycoprotein elicitor. Of importance, the suppressors did not reduce the activity of a race-specific peptide elicitor obtained from *C. fulvum*, to be discussed in the next section. It is therefore appealing to speculate that successful pathogens have evolved mechanisms to mask, degrade or otherwise deal with their general elicitors in order not to elicit plant defense.

V. Specific Elicitors

Elicitors that mirror the specificities dictated by pathogen avirulence genes have been detected in several systems (Bruegger and Keen, 1979; Culver and Dawson, 1989b; De Wit et al., 1985; Keen et al., 1990; Keen and Legrand, 1980; Mayama et al., 1986; Tepper and Anderson, 1986), but only a few of them have been isolated and characterized. One is the classic host selective toxin, victorin, produced by *Helminthosporium victoriae* (Wolpert et al., 1985). Although long believed to function as a toxin, victorin in fact behaved as a specific elicitor of phytoalexins only in oat plants carrying the *Pc-2* allele for sensitivity to victorin and resistance to *Puccinia coronata* (Mayama et al., 1986). It is therefore speculated that races of *P. coronata* carrying the avirulence gene complementary to *Pc-2* must also produce victorin or a structural analog. The findings with victorin and the *avr D* elicitor, to be described later, raise the possibility that other metabolites heretofore considered as host selective toxins may in reality be elicitors. For example, Traylor et al. (1987) reported that a host selective peptide toxin from *Periconia circinata* led to the production of apparent defense response proteins by toxin sensitive but not insensitive sorghum cultivars. Although further tests have not yet been run, it is possible that the *Periconia* peptide in fact behaves as a specific elicitor.

A small linear peptide produced only by *Cladosporium fulvum* races carrying the putative *ACf 9* avirulence gene was identified that functioned as a race specific elicitor in tomato plants carrying the *Cf 9* resistance gene (De Wit et al., 1985; Scholtens-Toma and De Wit, 1988). Since the peptide elicitor is described fully by De Wit (1992), we will not discuss it in detail here.

Tepper and Anderson (1986) have partially characterized a polysaccharide specific elicitor from the alpha race of *Colletotrichum lindemuthianum*. A fraction from the extracellular culture fluids of the alpha race of *C. lindemuthianum* gave high elicitor activity and stimulated defense response genes in a bean cultivar that was resistant to the alpha race but little or no response was observed in compatible cultivars (Tepper et al., 1989). Comparable fractions from a beta race isolate gave little or no elicitor activity on any of these bean cultivars. Although the alpha race elicitor was size heterogenous, a ca. 60 kDa active fraction consisting almost entirely of carbohydrate (45% mannose, 17% galactose and 38% glucose) was isolated.

Recent work with cloned avirulence genes and their associated specific elicitors from viral and bacterial pathogens has provided new insight into how specific resistance functions. We will discuss a few of these cases in detail.

A. The Gene VI Protein of Cauliflower Mosaic Virus

Shepherd and collaborators have shown that gene VI of cauliflower mosaic virus (CaMV) may function as an avirulence gene when the virus is inoculated into certain resistant solanaceous hosts. CaMV is normally pathogenic on cruciferous hosts, but certain strains infect a few solanaceous plants such as *Datura stramonium* and *Nicotiana bigelovii*. The gene VI protein forms extensive arrays in infected plant cells, called inclusions, but recent results (Bonneville et al., 1989; Gowda et al., 1989) have indicated that the viral function of the gene VI protein may be as a *trans*-activator for the expression of other CaMV genes. Schoelz et al. (1986, 1987) interchanged portions of the genomes of CaMV strains which either infected *Datura* or elicited the local lesion HR. In this way, they identified a 496 bp segment within the first half of the open reading frame of gene VI from the local lesion strain that accounted for elicitation of the HR. The gene VI protein is therefore a likely candidate for the specific elicitor of the *Datura* HR. This possibility has been supported by the work of Takahashi et al. (1989), who found that transformation of gene VI from a crucifer strain of CaMV into tobacco resulted in transgenic plants that exhibited autogenous necrotic spots and expression of defense response genes. This experiment therefore also suggests that the gene VI protein product may function as a specific elicitor of active defense responses.

B. The Coat Protein of Tobacco Mosaic Virus

Certain species of *Nicotiana* respond to specific strains or variants of tobacco mosaic virus (TMV) with a HR that confines the virus to the area of initial infection. All naturally occurring strains of TMV induce the HR in plants with the *N* resistance gene from *N. glutinosa*. Other species of *Nicotiana*, including *N. sylvestris* and some varieties of *N. tabacum*, contain a similar gene, N', that conditions HR upon infection by only some isolates of TMV. Plants with the nn, n'n' genotype fail to respond with the HR upon infection by essentially all isolates of TMV. Therefore, the tobacco–TMV interactions appear to be gene-for-gene systems.

The best examined system is the interaction of TMV with plants of the N'N' genotype where the viral coat protein has been shown to be a specific elicitor of the HR. Strain OM of TMV does not induce HR in these plants, whereas TMV-L, the tomato strain, induces the HR. Hybrids of these viruses were constructed in vitro and examined in N'N' plants (Saito et al., 1987). Hybrids in which only the coat protein gene of TMV-OM was substituted into the genome of TMV-L lost the ability to induce the HR, suggesting that the coat protein gene was responsible for induction. TMV-U1 also fails to induce the HR in N'N' plants, but mutants of this virus which elicit the HR can be isolated at a relatively high frequency. One such mutant resulted from the change of a single nucleotide in the coat protein gene (Knorr and Dawson, 1988). This nucleotide alteration resulted in an amino acid substitution of phenylalanine for serine at position 148 in the coat protein. By coincidence, this amino acid substitution had previously been identified in a TMV mutant that was selected to induce the HR in *N. sylvestris* (Funatsu and Fraenkel-Conrat, 1964). Several other mutants with similar phenotypes and different amino acid alterations in the coat protein also had been described (Funatsu and Fraenkel-Conrat, 1964; Wittmann and Wittmann-Liebold, 1966), but the technology of the time did not allow exclusion of the possibility that mutations had occurred in other parts of the genome.

To examine whether other single nucleotide mutations throughout the TMV coat protein gene were responsible for induction of the HR, Culver and

Dawson (1989a) introduced point mutations at different regions of the coat protein gene, resulting in amino acid alterations at positions 11, 20, 25, and 46 of the coat protein. Each of these mutations resulted in induction of HR. Additionally, these mutants varied in the intensity of HR produced, similar to that observed with other types of pathogens and other disease resistance genes (e.g., McIntosh, 1976). Mutants TMV 20 and TMV 25 induced distinct necrotic local lesions within three days post-inoculation and were classified as strong elicitors. Mutant TMV 11 induced lesions that developed four to five days after inoculation; the virus often continued to slowly spread and led to the collapse of large portions of inoculated leaves, as would be expected of a weak elicitor. Mutant 46 presented yet another pattern of HR timing, which was intermediate between the other mutants studied.

The experiments above demonstrated that the coat protein gene was involved in induction of the HR in N'N' genotype plants. Although all of the mutations that induced the HR resulted in amino acid changes in the coat protein, this was not sufficient to demonstrate that the protein rather than the RNA was the elicitor of the HR. To examine whether the altered protein or the altered coat protein gene was the elicitor of the HR, TMV mutants that induced HR were altered by removal of the coat protein gene start codon, leaving production of RNA intact but preventing production of the coat protein (Culver and Dawson, 1989b). The absence of coat protein synthesis resulted in loss of the ability to induce HR, demonstrating that the coat protein, not the m-RNA, is an elicitor of the HR in N'N' plants. This finding has been further substantiated by the production of transgenic N'N' tobacco plants expressing wild type and mutant coat proteins. Plants expressing the coat protein from strains of TMV which do not elicit the HR appeared normal, but those expressing the coat proteins from HR-inducing strains developed necrotic leaf spots on all except the youngest leaves that coalesced into large areas of dead tissue (Culver and Dawson, unpubl. results). The results dramatically demonstrate that viral replication is not required for induction of the HR and strongly indicate that certain coat proteins are elicitors of the HR in N' tobacco plants. The findings also indicate that the coat protein gene of TMV can be regarded as an avirulence gene which interacts specifically with the N' resistance gene in a gene-for-gene fashion. As such, it is the first avirulence gene identified from a plant pathogen with a known function in the pathogen.

Additional questions remain, however, that are being addressed with the TMV-tobacco system. What form of the TMV coat protein elicits the HR? Assembly into virions is not required for induction of HR since a TMV mutant with an alteration in the origin of assembly that prevents the production of virions is still able to induce the HR (Culver et al., unpubl.). However, does the plant recognize an epitope-like area of the coat protein such that only certain regions of the coat protein are active? Recent data suggest that almost the entire coat protein, probably folded into its normal tertiary or quaternary structure, is required for activity. Saito et al. (1989) observed that the 5' half of the TMV-L gene was sufficient for induction of the HR when fused to the 3' half of the TMV-OM gene. However, when the 5' half of the TMV-L gene was expressed as a truncated protein containing only the elicitor-active portion, the HR was not induced. Also it appears that the individual coat protein may not be the active molecule, but that aggregates of coat proteins elicit the HR. In a collaboration between the laboratories of W. Dawson and G. Stubbs at Vanderbilt University, it was recently shown that coat protein mutations which elicit the HR all occurred in a cluster or 'foot print' on the surface of coat protein aggregates; on the other hand, mutations thus far studied which alter non-surface amino acids do not elicit the HR (unpubl. data).

TMV is unusual in that only a relatively unique amino acid sequence of the coat protein fails to elicit HR, but several mutations cause the HR in N' tobacco plants. This behavior is somewhat at odds with the elicitor-receptor model, which would predict that only a single coat protein structure should be recognized. However, all the strains of TMV or related tobamoviruses that replicate in N'N' plants induce HR except the type or tobacco strains. Also, mutants of the type strains that induce HR can be isolated at relatively high frequencies, 10^{-3}–10^{-2}, whereas revertants of these mutants occur at a much lower frequency, less than 10^{-5} (Aldaoud et al., 1989). This suggests that many changes can elicit the HR in N' tobacco, but only mutations that mimic the original structural configuration fail to elicit. It therefore appears that small structural alterations within the 'footprint' on coat protein aggregates are recognized by the host to initiate the resistance reaction.

The interaction of the TMV coat protein with N' tobacco plants has similarities to the activity of tumor necrosis factor (TNF) in mammalian cells. TNF is a cytokine produced by T lymphocytes and macrophages that induces tissue destruction involved in host defense against pathogens in a manner analogous to the plant HR. TNF is a 17 kDa protein whose structure recently was found to be similar to that of the coat protein of satellite tobacco necrosis virus (Jones et al., 1989). TNF is active as an aggregate of three molecules which binds to a glycoprotein receptor on the cell membrane. It is of interest that the TMV coat protein and TNF are of similar size and both are active as macromolecular aggregates which elicit active defense responses of the HR type. Does the coat protein bind to a specific cell receptor encoded by the N' gene? Do coat proteins that fail to induce the HR also fail to bind to the receptor? Does an inactive coat protein interfere with the ability of an elicitor-active coat protein to elicit the HR? Another question is what are the threshold amounts of elicitor-active coat protein required for induction of the HR? For instance, it is noteworthy that in transgenic plants expressing an elicitor-active coat protein, some areas of the leaves contain the coat protein but are not necrotic (J. Culver and W. Dawson, unpubl. results).

In contrast to the interaction of TMV with N'N' tobacco plants, its interaction with other plants is different. In NN genotype plants the viral elicitor has not been identified. It is known not to involve the coat protein gene, however, because mutants with this gene deleted still induced the HR (Dawson et al., 1988; Takamatsu et al., 1987). Additionally, there are numerous other hosts in which tobamoviruses elicit the HR. It is therefore possible that various parts of the viral genome are involved in eliciting the HR, depending on the host plant and the particular resistance gene involved.

C. Barley Stripe Mosaic Virus and Chenopodium amaranticolor

Considerable complexity occurs in the interaction of barley stripe mosaic virus (BSMV) with *Chenopodium amaranticolor* plants (A. O. Jackson, pers. comm.). The type strain of BSMV induces chlorotic lesions while the ND18 strain elicits necrotic HR lesions. Of the three plus-strand genomic viral RNAs (Jackson et al., 1989), the smallest (called the γ-RNA, containing two large open reading frames (ORFs) called a and b) also contains the genetic information that determines whether the HR is elicited in *C. amaranticolor*. At least three regions in this RNA have been shown to result in the production of necrotic lesions (I.T.D. Petty and A.O. Jackson, unpubl. data). A point mutation in the γb gene that results in a leucine to proline alteration in the b protein of the type strain changes the plant reaction from compatible to incompatible. Deletion of the entire γb gene along with deletion of a repeated sequence in the 5' region of the γa gene result in pinpoint-sized necrotic lesions, demonstrating that the protein is not necessarily required for induction of the HR. However, a change that is dominant over both of these alterations and results in necrotic lesions that are identical in size and timing to those induced by the native ND18 strain is the deletion of a small ORF in the leader preceding γa gene. This modification increases the production of the γa protein to a level similar to that of the ND18 strain.

All of the results discussed in this section indicate that plant viruses may elicit the HR by several different methods. However, because of the small size of viral genomes and recent progress in manipulating them, we should soon have a much clearer understanding of how different viruses elicit the plant HR.

D. A Low Molecular Weight Elicitor from Bacteria Expressing Avirulence Gene D of Pseudomonas syringae pv. tomato

Kobayashi et al. (1989) cloned three different avirulence genes from *Pseudomonas syringae* pv. *tomato* (*Pst*) that functioned in the related bacterium, *P. s. glycinea* (*Psg*), causing it to elicit a hypersensitive reaction on some but not all soybean cultivars. Surprisingly, one of these genes was identical to *avr A*, previously cloned from *Psg* race 6 (Napoli and Staskawicz, 1987; Staskawicz et al., 1984). A second *Pst* cosmid clone that elicited the HR in soybean has recently been shown to include a portion of the *Pst hrp* gene cluster required for pathogenicity (J. Lorang et al., unpubl. obs.). The third *Pst* avirulence gene, called *avr D*, was subcloned, sequenced and studied in attempts to determine how it initiates the soybean HR (Kobayashi et al., 1990a). A functional copy of this gene was not present in any tested race of *Psg*, but all *Psg* races contained DNA sequences with high homology to *avr D* from *Pst* despite the fact that they do not exhibit the avirulence phenotype (Kobayashi et al., 1990b). This apparent recessive allele of *avr D* present in *Psg* is of evolutionary interest and was discussed in detail elsewhere (Keen, 1990).

Avirulence gene D from *Pst* was located on a ca. 75 kb indigenous plasmid in *Pst* (Kobayashi et al., 1990a). The *Pst avr D* gene encoded a 34 kDa protein with 311 amino acids that did not contain a signal peptide sequence and appeared to remain in the bacterial cytoplasm (Kobayashi et al., 1990a). Surprisingly, *E. coli* cells expressing the cloned *Pst avr D* gene elicited the HR on precisely the same soybean cultivars as *Psg* race 4 carrying the cloned *avr D* gene (Keen et al., 1990). The *avr D* protein extracted from overproducing *E. coli* cells did not cause the HR when infiltrated into soybean leaves, but culture fluids from *E. coli* or *Psg* cells expressing *avr D* contained a low molecular weight race specific elicitor of the soybean HR (Keen et al., 1990). The elicitor was also produced by *Xanthomonas campestris* pv. *glycines*, *Rhizobium fredii*, and *Erwinia chrysanthemi* cells expressing *avr D*, but none of them normally produce elicitor activity. The *avr D* elicitor has been purified and partially characterized (M. Stayton et al., unpubl. data), but its structure has not yet been determined. However, genetic experiments have proven that it mediates the gene-for-gene interaction between bacteria harboring *avr D* and soybean plants carrying the complementary resistance gene, called *Rpg 4* (Kobayashi et al., 1990a; Keen et al., 1990; Keen and Buzzell, 1991). The *avr D* elicitor is relatively potent, producing HR on sensitive soybean cultivars carrying *Rpg 4*, but not on insensitive cultivars lacking *Rpg 4* at 1000 × or higher concentrations (M. Stayton et al., unpubl. results). The elicitor also produces a spreading, systemic necrosis in *Rpg 4* soybean leaves when infiltrated in relatively high concentrations. Thus, the *avr D* elicitor represents a very different kind of signal molecule than other specific elicitors such as the TMV coat protein. While not fully established, it appears that the *avr D* protein catalyzes the conversion of a metabolite normally present in several Gram-negative bacteria into the *avr D* elicitor (see Fig. 3).

It would have been impossible to detect the presence of the putative soybean disease resistance gene complementing *avr D* before availability of the cloned *Pst avr D* gene because biotypes of *P. s. glycinea* do not contain a functional *avr D* allele (Kobayashi et al., 1990b). We therefore analyzed soybean cross progeny in order to test whether a classical resistance gene in fact interacts with the *avr D* gene (Keen and Buzzell, 1991). The experiment also permitted testing whether resistance to *Psg* carrying *avr D* and sensitivity to the *avr D* elicitor co-segregated. Crosses of the soybean cultivars Flambeau and Merit showed that a single dominant genetic character,

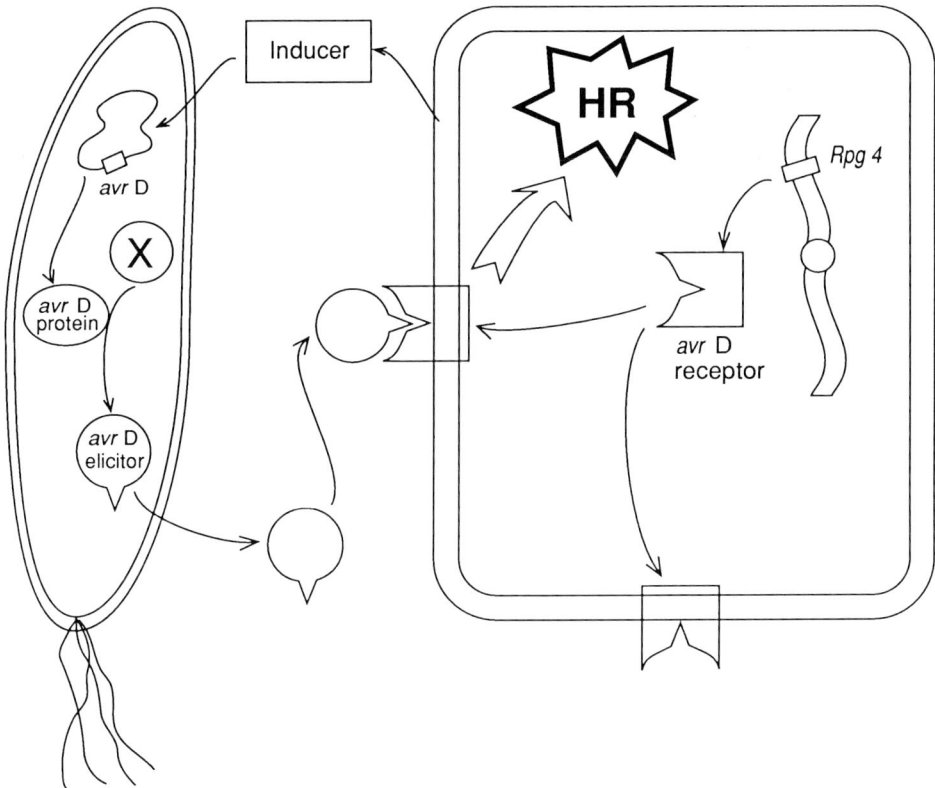

Fig. 3. Summary of the interaction of avirulence gene D and its complementary soybean disease resistance gene, *Rpg 4*. The chromosomal resistance gene in the plant cell (right) is hypothesized to encode a receptor protein that may be associated with the host plasmalemma, but no critical data is yet available on this point. Avirulence gene D in the bacterial pathogen (left) encodes an enzyme that converts a normal cellular metabolite (X) into an elicitor-active compound (the *avr* D elicitor). The *avr* D elicitor is then secreted from the bacterium into the plant intercellular space where it interacts with the plant resistance gene-encoded receptor. Elicitor-receptor binding initiates the plant HR as shown in Fig. 1. A plant inducing substance or perhaps a unique feature of the plant environment also stimulates expression of the *avr* D gene in the bacterium

called *Rpg 4*, segregated in a classical 3:1 ratio, indicating that it indeed constituted a disease resistance gene. All of 552 tested F2 and F3 cross progeny which were resistant to *Psg* race 4 cells carrying *avr* D also produced a necrotic HR in response to the isolated *avr* D elicitor. On the other hand, all of the 190 F2 and F3 segregants of the Flambeau × Merit cross which were susceptible to *Psg* race 4 carrying *avr* D also were insensitive to the *avr* D elicitor (Keen and Buzzell, 1991). These data therefore prove that the *avr* D elicitor mediates the phenotype of the *Pst avr D* gene. They also suggest but do not conclusively prove that the *avr D* elicitor interacts only with plants carrying the *Rpg 4* disease resistance gene to elicit the soybean HR. Thus, the elicitor produced by bacteria expressing *avr D* represents the first case in which the signal molecule has been identified which confers gene-for-gene specificity involving a defined microbial avirulence gene and its complementary disease resistance gene. It is also the first case in which a microbial specific elicitor has been definitively linked to the activity of an avirulence gene.

The *avr D* gene in *Pst* was the first of five tandem open reading frames (ORFs) (Kobayashi et al., 1990a). Based on co-regulation observed when various of the ORFs were used as probes in RNA slot blot experiments with bacteria grown in culture and in planta, it appears that the five ORFs constitute a functional operon (D. Kobayashi and N. Keen, unpubl. data). These results were confirmed when the DNA 5′ to *avr D* was used in a *lux* promoter probing system (H. Shen and N. Keen, unpubl. data). Promoter activity was relatively low when *P. s. glycinea* cells carrying the *lux-avr D* promoter were grown on laboratory culture media, but expression increased by ca. $100 \times$ when the cells were inoculated into soybean, tomato or tobacco leaves. It is not yet known whether a discrete plant factor(s) induces expression of the *avr D* operon or whether the specific plant environment is sufficient. The full significance of these observations is also not clear, pending identification of the *avr D* elicitor and determination of the biochemical functions of the *avr D* protein. The results do indicate, however, that the *avr D* operon may encode enzymes comprising a secondary metabolic pathway in bacterial pathogens which is important for their multiplication in plants.

Taken collectively, the data with *avr D* and the *avr D* elicitor strongly support the elicitor-receptor model (Fig. 3). The genetic evidence demonstrates that the *avr D* elicitor is the signal molecule which is recognized by soybean plants carrying the *Rpg 4* resistance allele. The data also establish that production of the *avr D* elicitor by bacterial hosts is dependent on the *avr D* gene. The *avr D* elicitor is therefore the only specific elicitor currently known to result from activity of a defined avirulence gene and which functions when exogenously supplied to plants of the appropriate resistance genotype.

VI. Do Plants Contain Receptors That Recognize Elicitors and Initiate the HR?

This topic will be dealt with only summarily here since it is explored by Ebel and Scheel (1992) in more depth. Elicitors that are active at low concentrations, such as the β-glucans, victorin or the *avr D* elicitor, would a priori be expected to have highly specific and high affinity receptors in sensitive plants. Experience with vertebrate systems has shown that such active, relatively polar ligands are sensed by specific receptors, usually located on the plasma membrane. The traditional approach for detection and isolation of these receptors is to radioactively or otherwise label the ligand to high specific activity and perform binding experiments with whole target cells or cell fractions such as isolated plasma membranes. Although few such experiments have been conducted with well characterized, labelled elicitors, limited evidence suggests that specific receptors may indeed occur in plants.

Yoshikawa et al. (1983) obtained the first experimental evidence suggesting a plasma membrane-bound receptor for an elicitor. Mycolaminaran, a relatively homogeneous β-1,3;β-1,6-linked glucan from *Phytophthora* species, was employed. ^{14}C-labelled mycolaminaran specifically bound to soybean membranes with a moderate dissociation constant. This was in keeping with the fact that mycolaminaran is only a moderately active phytoalexin elicitor in soybean. Kinetic data also indicated that only a single affinity class of mycolaminaran binding site was present. The putative receptor for mycolaminaran also appeared to be physiologically involved in phytoalexin elicitation, since mycolaminaran derivatives with greater or lesser elicitor activity exhibited greater or lesser affinity, respectively, for the receptor (Yoshikawa et al., 1983). The evidence therefore indicated that ligand binding to the putative receptor was correlated with the physiological response. J. Ebel's group in Freiburg confirmed these results using partially purified but heterogenous *Phytophthora* cell wall glucan preparations and provided evidence that the putative glucan receptor indeed occurred on the plasma membrane (Schmidt and Ebel, 1987; Cosio et al., 1988; see Ebel and Scheel, 1992). However, since mycolaminaran competed with binding of the cell wall glucan preparations, both glucans appear to be recognized by the same receptor.

As noted earlier, victorin causes necrosis and symptoms of the hypersensitive reaction only in oat plants carrying the *Pc-2* disease resistance gene (Litzenberger, 1949; Mayama et al., 1986). Victorin occurs as at least three chemically related cyclic peptides with unusual structures (Wolpert et al., 1985). While victorin has long been considered to be a host specific toxin against *Pc-2* oat plants (Scheffer and Livingston, 1984), more recent results (Mayama et al., 1986) indicate that it in fact is a specific elicitor of the HR.

Wolpert and Macko (1989) presented evidence that a specific, membrane bound victorin receptor occurs in *Pc-2* oat plants. ^{125}I-labeled victorin was applied to leaf tissues of *Pc-2* or *pc-2* oat plants and, following electrophoresis, a 100 kDa protein was found to be labeled only in the case of *Pc-2* leaves. This in vivo binding was competitively displaced by a nontoxic, chemically reduced victorin analogue, thereby indicating that binding was associated with biological activity. However, in vitro binding of the labelled victorin to leaf extracts required the presence of a reducing agent. Further, in vitro binding of the labelled victorin occurred to a 100 kDa protein present in both the *Pc-2* and *pc-2* genotypes. Of particular interest, the binding of victorin to the 100 kDa protein appeared to be covalent, involving a terminal glyoxylic acid residue present on the native molecule.

This functional group was previously shown to be essential for the activity of victorin because reduction of the aldehyde of the glyoxylate residue to a primary alcohol eliminated activity (Wolpert et al., 1988). However, the reduced derivative nevertheless competed with binding of the native molecule to the 100 kDa receptor, as noted above, indicating that it does associate with the putative binding site on the receptor. The work of Wolpert and Macko (1989) therefore indicates that the 100 kDa protein from oat tissues may be involved as a specific receptor for victorin. However, it is not yet clear whether the 100 kDa protein is the primary gene product of the *Pc-2* resistance gene. This would require cloning of the gene encoding the 100 kDa protein, transforming it into an oat cultivar lacking the *Pc-2* allele and testing the resultant regenerated transgenic plants for sensitivity to victorin and resistance to *Puccinia coronata* races carrying the avirulence gene complementary to *Pc-2*.

VII. Signal Transduction and Defense Gene Activation

This topic will be covered in more depth by Ebel and Scheel (1992). One interesting aspect is that inoculation of plant tissues with incompatible pathogens or treatment with elicitors can cause rapid changes in plasma membrane integrity (Keppler and Novacky, 1987; Pavlovkin et al., 1986; Pelissier et al., 1986) and the rapid generation of active oxygen species at the site of the plasma membrane. These include superoxide anion and hydrogen peroxide (Apostol et al., 1989; Doke, 1985; Doke and Ohashi, 1988; Keppler et al., 1989; Lindner et al., 1988; Ocampo et al., 1986; Rogers et al., 1988). P. Low and colleagues have shown that hydrogen peroxide may be a key component of the HR in soybean cells (Apostol et al., 1989, and pers. comm.). A burst of H_2O_2 release occurs relatively rapidly after elicitor treatment and has a duration of ca. 1 h. This oxidative burst is inhibited by citrate and several lines of evidence indicate that it is involved with subsequent phytoalexin production. Peever and Higgins (1989a) also reported that treatment of tomato leaves with the race specific peptide elicitor from *Cladosporium fulvum* caused the activation of lipoxygenase activity and lipid peroxidation both in the dark and light; however, necrosis typical of the HR was only observed when plants were grown in the light. This indicates that the necrosis classically associated with the HR may be a photoactivation artefact associated with the oxidative events but not necessarily a key element in the expression of resistance.

In human cells, levels of activated oxygen species are modulated by enzymes such as superoxide dismutase (SOD) and it has been shown to affect the cellular response to tumor necrosis factor, discussed earlier (Wong et al., 1989). Levels of SOD have also been shown to be stimulated during certain hypersensitive and/or stress reactions in plants (viz. Bowler et al., 1989; Buonaurio et al., 1987; Keppler and Novacky, 1989). However, critical experiments have not yet been done in which cloned sense and anti-sense plant SODs have been used in transgenic plants to modulate activity of the enzyme and determine whether expression of the HR can be correspondingly altered.

Despite indications that activated oxygen species may play a role in elicitor signalling, the precise transductional factors are not known that ultimately result in derepression of plant disease response genes (see Lamb et al., 1989). Candidates for this second messenger role include altered calcium flux (Atkinson et al., 1990; Gabriel et al., 1988; Kurosaki et al., 1987a; Stäb and Ebel, 1987), phosphoinositide formation (Kurosaki, 1987b), altered protein kinase activity (Grab et al., 1989) or perhaps a direct role of hydrogen peroxide as a second messenger (Apostol et al., 1989). The next few years should see considerable progress in this area, particularly since well characterized specific and general elicitors are now available.

Whatever the nature of intermediate signals, elicitor treatment of sensitive plant cells causes a rapid and massive transient derepression of several plant genes. These so-called 'defense response genes' encode bio-synthetic enzymes in phytoalexin and lignin pathways as well as hydroxy-proline-rich glycoproteins, chitinases and glucanases. In situ hybridization experiments have recently yielded a precise picture regarding expression of these genes (Schmelzer et al., 1989). The expression and regulation of defense response genes has recently been reviewed (Anderson, 1989; Collinge and Slusarenko, 1987; Ebel and Grisebach, 1988; Hutcheson et al., 1989; Lamb et al., 1989) and will therefore not be considered in detail here. It is worthy of note, however, that Dron et al. (1988), Cramer et al. (1989) and Lois et al. (1989) identified *cis*-acting elements upstream of the coding regions in two classes of defense response genes (phenylalanine ammonia lyase and chalcone synthase) from *Phaseolus* and parsley that functioned as specific silencer and activator elements important in elicitor derepression of these genes. C. Lamb and associates (unpubl. data) have also recently observed the occurrence in bean tissues of a protein that may function as a *trans*-acting element which may regulate defense response genes by inter-acting with their *cis*-acting activator elements. This development is similar to the recent work of Holdsworth and Laties (1989) who identified a carrot root nuclear factor that bound to the 5' DNA of an extension gene and inhibited expression. During wounding, levels of the binding factor de-creased, thus accounting for increased transcription of the extensin gene. These findings may accordingly parallel those of Lamb and collaborators with the silencer elements of disease response genes. The results are particularly significant because they may permit new approaches to under-standing the signal transduction machinery of plants.

VIII. Conclusions

The cloning of avirulence genes has greatly aided our understanding of plant–pathogen specificity. It has proven that the gene-for-gene relation-ship first noted by Flor (1942) is correct—avirulence genes encode protein products that indeed are the agents which either directly or indirectly interact with plant cells carrying complementary disease resistance genes. However, it is clear that the molecular mechanisms for these interactions may be very diverse, as illustrated by structural differences in the TMV coat protein and victorin. Furthermore, genetic evidence now mechanistically links two very different cloned avirulence gene products (the TMV coat gene and the *avr D* elicitor from *P. syringae* pv. *tomato*) with initiation of the plant hypersensitive response. These results indicate that the coat protein of TMV functions as a specific elicitor per se, but that the *avr D* protein does not; instead, it appears to catalyze the conversion of a normal bacterial metabolite into the *avr D* elicitor. The results therefore strongly support the elicitor-receptor model for recognition of incompatible pathogen races by plants. As such, these specific elicitors are the first pathogen signal molecules proven to initiate active defense in plants, similar to antigens in vertebrates. In both viral and bacterial pathogens, we are also beginning to understand how avirulence genes may be altered by mutation to confound plant recognition of the pathogen. In addition, the ability to express avirulence genes in heterologous pathogens has raised the specter that these genes may control higher level plant–pathogen specificities in addition to specific resistance. The next few years should yield a wealth of additional information on avirulence gene structure as well as illuminate the important questions of their function in the pathogen and the mechanism of elicitor recognition by plant cells.

The marked successes in cloning and characterizing avirulence genes underscore more forcefully the pressing need to clone and characterize plant disease resistance genes. With the major strides that have recently occurred in understanding vertebrate immune systems, it is frustrating that plant molecular biology has not pursued the cloning and characterization of plant disease resistance genes more forcefully. Certainly the cloning of these genes is required before we fully understand the mechanics of pathogen surveillance and active defense in plants. The availability of cloned disease resistance genes, when finally accomplished, will also permit their manipulation for dramatically improved control of plant diseases in practical agriculture.

Acknowledgements

The authors' research is supported by grants from the National Science Foundation and the USDA Competitive Grants Program.

IX. References

Aldaoud R, Dawson WO, Jones GE (1989) Rapid, random evolution of the genetic structure of replicating tobacco mosaic virus populations. Intervirology 30: 227–233
Alt FW, Blackwell TK, Yancopoulos GD (1987) Development of the primary antibody repertoire. Science 238: 1079–1087

Anderson AJ (1989) The biology of glycoporteins as elicitors. In: Kosuge T, Nester EW (eds) Plant–microbe interactions. Molecular and genetic perspectives, vol 3. Macmillan, New York, pp 87–130

Apostol I, Heinstein PF, Low PS (1989) Rapid stimulation of an oxidative burst during elicitation of cultured plant cells. Role in defense and signal transduction. Plant Physiol 90: 109–116

Atkinson MM, Keppler LD, Orlandi EW, Baker CJ, Mischke CF (1990) Involvement of plasma membrane calcium influx in bacterial induction of the K^+/H^+ and hypersensitive responses in tobacco. Plant Physiol 92: 215–221

Ayers AR, Ebel J, Valent B, Albersheim P (1976) Host-pathogen interactions. X. Fractionation and biological activity of an elictior isolated from the mycelial walls of *Phytophthora megasperma* var. *sojae*. Plant Physiol 57: 760–765

Barber MS, Bertram RE, Ride JP (1989) Chitin oligosaccharides elicit lignification in wounded wheat leaves. Physiol Mol Plant Pathol 34: 3–12

Bennetzen JL, Qin M-M, Ingels S, Ellingboe AH (1988) Allele-specific and *mutator*-associated instability at the *Rp1* disease-resistance locus of maize. Nature 332: 369–370

Bloch CB, De Wit PJGM, Kuc J (1984) Elicitation of phytoalexins by arachidonic and eicosapentaenoic acids: a host survey. Physiol Plant Pathol 25: 199–208

Bonas U, Stall RE, Staskawicz B (1989) Molecular and structural characterization of the avirulence gene *avr Bs3* from *Xanthomonas campestris* pv. *vesicatoria*. Mol Gen Genet 218: 127–136

Bonneville JM, Sanfacon H, Fütterer J, Hohn T (1989) Posttranscriptional *trans*-activation in cauliflower mosaic virus. Cell 59: 1135–1143

Bostock RM, Kuć JA, Laine RA (1981) Eicosapentaenoic and arachidonic acids from *Phytophthora infestans* elicit fungitoxic sesquiterpenes in the potato. Science 212: 67–69

Bowler C, Alliotte T, DeLoose M, van Montagu M, Inze D (1989) The induction of manganese superoxide dismutase in response to stress in *Nicotiana plumbaginifolia*. EMBO J 8: 31–38

Bruce RJ, West CA (1989) Elicitation of lignin biosynthesis and isoperoxidase activity by pectic fragments in suspension cultures of castor bean. Plant Physiol 91: 889–897

Bruegger BB, Keen NT (1979) Specific elicitors of glyceollin accumulation in the *Pseudomonas glycinea*–soybean host–parasite system. Physiol Plant Pathol 15: 43–51

Buonaurio R, Torre GD, Montalbini P (1987) Soluble superoxide dismutase (SOD) in susceptible and resistant host–parasite complexes of *Phaseolus vulgaris* and *Uromyces phaseoli*. Physiol Mol Plant Pathol 31: 173–184

Callow JA (1977) Recognition, resistance and the role of plant lectins in host–parasite interactions. Adv Bot Res 4: 1–49

Cervone F, Hahn MG, DeLorenzo G, Darvill A, Albersheim P (1989) Host–pathogen interactions. XXXIII. A plant protein converts a fungal pathogenicity factor into an elicitor of plant defense response. Plant Physiol 90: 542–548

Collinge DB, Slusarenko AJ (1987) Plant gene expression in response to pathogens. Plant Mol Biol 9: 389–410

Cosio EG, Popperl H, Schmidt WE, Ebel J (1988) High-affinity binding of fungal β-glucan fragments to soybean (*Glycine max* L.) microsomal fractions and protoplasts. Eur J Biochem 175: 309–315

Cramer CL, Edwards K, Dron M, Liang X, Dildine SL, Bolwell GP, Dixon RA, Lamb CJ, Schuch W (1989) Phenylalanine ammonia-lyase gene organization and structure. Plant Mol Biol 12: 367–383

Culver JN, Dawson WO (1989a) Point mutations in the coat protein gene of tobacco mosaic virus induce hypersensitivity in *Nicotiana sylvestris*. Mol Plant Microbe Interact 2: 209–213

Culver JN, Dawson WO (1989b) Tobacco mosaic virus coat protein: an elicitor of the hypersensitive reaction but not required for the development of mosaic symptoms in *Nicotiana sylvestris*. Virology 173: 755–758

Daniels CH, Fritensky B, Wagoner W, Hadwiger LA (1987) Pea genes associated with non-host disease resistance to *Fusarium* are also active in race-specific disease resistance to *Pseudomonas*. Plant Mol Biol 8: 309–316

Davis KR, Darvill AG, Albersheim P (1986) Host–pathogen interactions. XXXI. Several biotic and abiotic elicitors act synergistically in the induction of phytoalexin accumulation in soybean. Plant Mol Biol 6: 23–32

Dawson WO, Bubrick, P, Grantham, GL (1988) Modifications of the tobacco mosaic virus coat protein gene affecting replication, movement, and symptomatology. Phytopathology 78: 783–789

Day P (1974) Genetics of host–parasite interactions. Freeman, San Francisco

De Wit PJGM (1992) Functional models to explain gene-for-gene relationships in plant–pathogen interactions. In: Boller T, Meins F (eds) Genes involved in plant defense. Springer, Wien New York, pp 25–47 [Dennis ES et al (eds) Plant gene research. Basic knowledge and application]

De Wit PJGM, Hoffman AE, Velthuis GCM, Kuć JA (1985) Isolation and characterization of an elicitor of necrosis isolated from intercellular fluids of compatible interactions of *Cladosporium fulvum* (Syn. *Fulvia fulva*) and tomato. Plant Physiol 77: 642–647

Dickman MB, Podila GK, Kolattukudy PE (1989) Insertion of cutinase gene into a wound pathogen enables it to infect intact host. Nature 342 446–448

Dixon RA (1986) The phytoalexin response: elicitation, signalling and control of host gene expression. Biol Rev 61: 239–291

Djordjevic MA, Rolfe BG, Lewis-Henderson W (1992) An analysis of host range specificity of *Rhizobium* as a model system for virulence genes in phytobacteria. In: Boller T, Meins F (eds) Genes involved in plant defense. Springer, Wien New York, pp 51–83 [Dennis ES et al (eds) Plant gene research. Basic knowledge and application]

Doke N (1985) NADPH-dependent O_2^- generation in membrane fractions isolated from wounded potato tubers inoculated with *Phytophthora infestans*. Physiol Plant Pathol 27: 311–322

Doke N, Ohashi Y (1988) Involvement of an O_2^- generating system in the induction of necrotic lesions on tobacco leaves infected with tobacco mosaic virus. Physiol Mol Plant Pathol 32: 163–175

Doke N, Garas NA, Kuć J (1979) Partial characterization and aspects of the mode of action of a hypersensitivity-inhibiting factor (HIF) isolated from *Phytophthora infestans*. Physiol Plant Pathol 15: 127–140

Dron M, Clouse SD, Dixon RA, Lawton MA, Lamb CJ (1988) Glutathione and fungal elicitor regulation of a plant defense gene promoter in electroporated protoplasts. Proc Natl Acad Sci USA 85: 6738–6742

Ebel J, Grisebach G (1988) Defense strategies of soybean against the fungus *Phytophthora megasperma* f. sp. *glycinea*: a molecular analysis. Trends Biochem Sci 13: 23–27

Ebel J, Scheel D (1992) Elicitor recognition and signal transduction. In: Boller T, Meins F (eds) Genes involved in plant defense. Springer, Wien New York, pp 183–205 [Dennis ES et al (eds) Plant gene research. Basic knowledge and application]

Ellingboe AH (1981) Changing concepts in host-pathogen genetics. Annu Rev Phytopathol 19: 125–143

Ellis JG, Lawrence GJ, Peacock WJ, Pryor AJ (1988) Approaches to cloning plant genes conferring resistance to fungal pathogens. Annu Rev Phytopathol 26: 245–263

Farmer EE, Helgeson JP (1987) An extracellular protein from *Phytophthora parasitica* var. *nicotianae* is associated with stress metabolite accumulation in tobacco callus. Plant Physiol 85: 733–740

Flor HH (1942) Inheritance of pathogenicity in *Melampsora lini*. Phytopathology 32: 653–669

Funatsu G, Fraenkel-Conrat H (1964) Location of amino acid exchanges in chemically evoked mutants of tobacco mosaic virus. Biochemistry 3: 1356–1362

Gabriel DW (1989) Genetics of plant parasite populations and host–parasite specificity. In: Kosuge T, Nester EW (eds) Plant–microbe interactions. Molecular and genetic perspectives, vol 3. McGraw-Hill, New York, pp 343–379

Gabriel DW, Burges A, Lazo GR (1986) Gene-for-gene interactions of five cloned avirulence genes from *Xanthomonas campestris* pv. *malvacearum* with specific resistance genes in cotton. Proc Natl Acad Sci USA 83: 6415–6419

Gabriel DW, Loschke DC, Rolfe BG (1988) Gene-for-gene recognition: the ion channel defense model. In: Palacios R, Verma DPS (eds) Molecular genetics of plant–microbe interactions 1988. American Phytopathological Society, St. Paul MN, pp 3–14

Gowda S, Wu FC, Scholthof HB, Shepherd RJ (1989) Gene VI of figwort mosaic virus (caulimovirus group) functions in posttranscriptional expression of genes on the full-length RNA transcript. Proc Natl Acad Sci USA 86: 9203–9207

Grab D, Feger M, Ebel J (1989) An endogenous factor from soybean (*Glycine max* L.) cell cultures activates phosphorylation of a protein which is dephosphorylated in vivo in elicitor-challenged cells. Planta 179: 340–348

Hamdan MAMS, Dixon RA (1987) Fractionation and properties of elicitors of the phenylpropanoid pathway from culture filtrates of *Colletotrichum lindemuthianum*. Physiol Mol Plant Pathol 31: 91–103

Hargreaves JA, Bailey JA (1978) Phytoalexin production by hypocotyls of *Phaseolus vulgaris* in response to constitutive metabolites released by damaged bean cells. Physiol Plant Pathol 13: 89–100

Heath MC (1980) Effects of infection by compatible species or injection of tissue extracts on the susceptibility of nonhost plants to rust fungi. Phytopathology 70: 356–360

Hitchin FE, Jenner CE, Harper S, Mansfield JW, Barber, CE, Daniels MJ (1989) Determinant of cultivar specific avirulence cloned from *Pseudomonas syringae* pv. *phaseolicola* race 3. Physiol Mol Plant Pathol 34: 309–322

Holdsworth MJ, Laties GG (1989) Identification of a wound-induced inhibitor of a nuclear factor that binds the carrot extensin gene. Planta 180: 74–81

Huang H-C, Schuurink R, Denny TP, Atkinson MM, Baker CJ, Yucel I, Hutcheson SW, Collmer A (1988) Molecular cloning of *Pseudomonas syringae* pv. *syringae* gene cluster that enables *Pseudomonas fluorescens* to elicit the hypersensitive response in tobacco plants. J Bacteriol 170: 4748–4756

Hutcheson SW, Collmer A, Baker CJ (1989) Elicitation of the hypersensitive response by *Pseudomonas syringae*. Physiol Plant 76: 155–163

Islam MR, Shepherd KW, Mayo GME (1989) Recombination among genes at the L group in flax conferring resistance to rust. Theor Appl Genet 77: 540–546

Jackson AO, Hunter BG, Gustafson GD (1989) Hordeivirus relationships and genome organization. Annu Rev Phytopathol 27: 95–121

Jin D-F, West CA (1984) Characteristics of galacturonic acid oligomers as elicitors of casbene synthetase activity in castor bean seedlings. Plant Physiol 74: 989–992

Jones EY, Stuart DI, Walker NPC (1989) Structure of tumour necrosis factor. Nature 338: 225–228

Kauffman S, Legrand M, Geoffrey P, Fritig B (1987) Biological function of pathogenesis-related proteins: four PR proteins of tobacco have 1,3-β-glucanase activity. EMBO J 6: 3209–3212.

Keen NT (1990) Gene-for-gene complementary in plant–pathogen interactions. Annu Rev Genetics 24: 441–463

Keen NT, Bruegger B (1977) Phytoalexins and chemicals that elicit their production in plants. ACS Symp Ser 62: 1–26

Keen NT, Buzzell RI (1991) New disease resistance genes in soybean against *Pseudomonas syringae* pv. *glycinea*: evidence that one of them interacts with a bacterial elicitor. Theor Appl Genet 81: 133–138

Keen NT, Legrand M (1980) Surface glycoproteins: evidence that they may function as the race specific phytoalexin elicitors of *Phytophthora megasperma* f. sp. *glycinea*. Physiol Plant Pathol 17: 175–192

Keen NT, Staskawicz B (1988) Host range determinants in plant pathogens and symbionts. Annu Rev Microbiol 42: 421–440

Keen NT, Yoshikawa M (1983) β-1,3-endoglucanase from soybean releases elicitor-active carbohydrates from fungus cell walls. Plant Physiol 71: 46–465

Keen NT, Tamaki S, Kobayashi DY, Gerhold D, Stayton M, Shen H, Gold S, Lorang J, Thordal-Christensen H, Dahlbeck D, Staskawicz B (1990) Bacteria expressing avirulence gene D produce a specific elicitor of the soybean hypersensitive reaction. Mol Plant Microbe Interact 3: 112–121

Kelemu S, Leach JE (1990) Cloning and characterization of an avirulence gene from *Xanthomonas campestris* pv. *oryzae*. Mol Plant Microbe Interact 3: 59–65

Kendra DF, Hadwiger LA (1987) Cell death and membrane leakage not associated with the induction of disease resistance in peas by chitosan or *Fusarium solani* f. sp. *phaseoli* Phytopathology 77: 100–106

Kendra DF, Christian D, Hadwiger LA (1989) Chitosan oligomers from *Fusarium solani*/pea interactions, chitinase/B-glucanase digestion of sporelings and from fungal wall chitin actively inhibit fungal growth and enhance disease resistance. Physiol Mol Plant Pathol 35: 215–230

Keppler LD, Novacky A (1987) The initiation of membrane lipid peroxidation during bacteria-induced hypersensitive reaction. Physiol Mol Plant Pathol 30: 233–245

Keppler LD, Novacky A (1989) Changes in cucumber cotyledon membrane lipid fatty acids during paraquat treatment and a bacteria-induced hypersensitive reaction. Phytopathology 79: 705–708

Keppler LD, Baker CJ, Atkinson MM (1989) Active oxygen production during a bacteria-induced hypersensitive reaction in tobacco suspension cells. Phytopathology 79: 974–978

Kinet J-P (1989) Antibody-cell interactions: Fc receptors. Cell 57: 351–354

Knorr DA, Dawson WO (1988) A point mutation in the tobacco mosaic virus capsid protein gene induces hypersensitivity in *Nicotiana sylvestris*. Proc Natl Acad Sci USA 85: 170–174

Kobayashi DY, Tamaki SJ, Keen NT (1989) Cloned avirulence genes from the tomato pathogen, *Pseudomonas syringae* pv. *tomato* confer cultivar specificity on soybean. Proc Natl Acad Sci USA 86: 157–161

Kobayashi DY, Tamaki SJ, Keen NT (1990a) Molecular characterization of avirulence gene D from *Pseudomonas syringae* pv. *tomato*. Mol Plant Microbe Interact 3: 94–102

Kobayashi DY, Tamaki SJ, Trollinger DJ, Gold S, Keen NT (1990b) A gene from *Pseudomonas syringae* pv. *glycinea* with homology to avirulence gene D from *P. s.* pv. *tomato* but devoid of the avirulence phenotype. Mol Plant Microbe Interact 3: 103–111

Kogel G, Beissmann B, Reisener HJ, Kogel KH (1988) A single glycoprotein from *Puccinia graminis* f. sp. *tritici* cell walls elicits the hypersensitive lignification response in wheat. Physiol Mol Plant Pathol 33: 173–185

Köhle H, Jeblick W, Poten F, Blaschek W, Kauss H (1985) Chitosan-elicited callose synthesis in soybean cells as a Ca^{2+}-dependent process. Plant Physiol 77: 544–551

Kolattukudy PE (1985) Enzymatic penetration of the plant cuticle by fungal pathogens. Annu Rev Phytopathol 23: 223–250

Kuć J, Rush JS (1985) Phytoalexins. Arch Biochem Biophys 236: 455–472

Kurosaki F, Tsurusawa Y, Nishi A (1987a) The elicitation of phytoalexins by Ca^{++} and cyclic AMP in carrot cells. Phytochemistry 26: 1919–1923

Kurosaki F, Tsurusawa Y, Nishi A (1987b) Breakdown of phosphatidylinositol during the elicitation of phytoalexins produced in cultured carrot cells. Plant Physiol 85: 601–604

Lamb CJ, Lawton MA, Dron M, Dixon RA (1989) Signals and transduction mechanisms for activation of plant defenses against microbial attack. Cell 56: 215–224

Legrand M, Kauffmann S, Geoffroy P, Fritig B (1987) Biological function of pathogenesis-related proteins: four tobacco pathogenesis-related proteins are chitinases. Proc Natl Acad Sci USA 84: 6750–6754

Lindner WA, Hoffmann C, Grisebach H (1988) Rapid elicitor-induced chemiluminescence in soybean cell suspension cultures. Phytochemistry 27: 2501–2503

Litzenberger SC (1949) Nature of susceptibility to *Helminthosporium victoriae* and resistance to *Puccinia coronata* in Victoria oats. Phytopathology 39: 300–318

Lois R, Dietrich A, Hahlbrock K, Schulz W (1989) A phenylalanine ammonia-lyase gene from parsley: structure, regulation and identification of elicitor and light responsive *cis*-acting elements. EMBO J 8: 1641–1648

McIntosh RA (1976) Genetics of wheat and wheat rusts since Farrer. J Austr Inst Agricult Sci: 203–216

Mayama S, Tani T, Ueno T, Midland SL, Sims JJ, Keen NT (1986) The purification of victorin and its phytoalexin elicitor activity in oat leaves. Physiol Mol Plant Pathol 29: 1–18

Mellano VJ, Cooksey DA (1988) Development of host range mutants of *Xanthomonas campestris* pv. *translucens*. Appl Environ Microbiol 54: 884–889

Minsavage GV, Dahlbeck D, Whalen MC, Kearney B, Bonas U, Staskawicz BJ, Stall RE (1990) Gene-for-gene relationships specifying disease resistance in *Xanthomonas campestris* pv. *vesicatoria*-pepper interactions. Mol Plant Microbe Interact 3: 41–47

Moerschbacher B, Kogel KH, Noll U, Reisener HJ (1986) An elicitor of the hypersensitive lignification response in wheat leaves isolated from the rust fungus *Puccinia graminis* f. sp. *tritici*. I. Partial purification and characterization. Z Naturforsch 41c: 830–838

Napoli C, Staskawicz B (1987) Molecular characterization and nucleic acid sequence of an avirulence gene from race 6 of *Pseudomonas syringae* pv. *glycinea*. J Bacteriol 169: 572–578

Nothnagel EA, McNeil M, Albersheim P, Dell A (1983) Host–pathogen interactions XXII. A galacturonic acid oligosaccharide from plant cell walls elicits phytoalexins. Plant Physiol 71: 916–926

Ocampo CA, Moerschbacher B, Grambow HJ (1986) Increased lipoxygenase activity is involved in the hypersensitive response of wheat leaf cells infected with virulent rust fungi or treated with fungal elicitor. Z Naturforsch 41c: 559–563

Oku H, Shiraishi T, Ouchi S (1987) Role of specific suppressors in pathogenesis of *Mycosphaerella* species. In: Nishimura S, Vance CP, Doke N (eds) Molecular determinants of plant diseases. Springer, Tokyo Berlin Heidelberg New York, pp 145–156

Parker JE, Hahlbrock K, Scheel D (1988) Different cell-wall components from *Phytophthora megasperma* f. sp. *glycinea* elicit phytoalexin production in soybean and parsley. Planta 176: 75–82

Pavlovkin J, Novacky A, Ulrich-Eberius CI (1986) Membrane potential changes during bacteria-induced hypersensitive reaction. Physiol Mol Plant Pathol 28: 125–135

Peever TL, Higgins VJ (1989a) Electrolyte leakage, lipoxygenase, and lipid peroxidation induced in tomato leaf tissue by specific and nonspecific elicitors from *Cladosporium fulvum*. Plant Physiol 90: 867–875

Peever TL, Higgins VJ (1989b) Suppression of the activity of non-specific elicitor from *Cladosporium fulvum* by intercellular fluids from tomato leaves. Physiol Mol Plant Pathol 34: 471–482

Pelissier B, Thibaud JB, Grignon C, Esquerre-Tugaye MT (1986) Cell surfaces in plant–microorganism interactions. VII. Elicitor preparations from two fungal pathogens depolarize plant membranes. Plant Sci 46: 103–109.

Pryor T (1987) The origin and structure of fungal disease resistance genes in plants. Trends Genet 3: 157–161

Ricci P, Bonnet P, Huet J-C, Sallantin M, Beauvais-Cante F, Bruneteau M, Billard V, Michel G, Pernollet J-C (1989) Structure and activity of proteins from pathogenic fungi *Phytophthora* eliciting necrosis and acquired resistance in tobacco. Eur J Biochem 183: 555–563

Ride JP, Barber MS (1987) The effects of various treatments on induced lignification and the resistance of wheat to fungi. Physiol Mol Plant Pathol 31: 349–360

Rogers KR, Albert F, Anderson AJ (1988) Lipid peroxidation is a consequence of elicitor activity. Plant Physiol. 86: 547–553

Ronald PC, Staskawicz BJ (1988) The avirulence gene *avr Bs1* from *Xanthomonas campestris* pv. *vesicatoria* encodes a 50 kDa protein. Mol Plant Microbe Interact 1: 191–198

Ryan CA (1988) Oligosaccharides as recognition signals for the expression of defensive genes in plants. Biochemistry 27: 8879–8883

Saito T, Meshi T, Takamatsu N, Okada Y (1987) Coat protein gene sequence of tobacco mosaic virus encodes a host response determinant. Proc Natl Acad Sci USA 84: 6074–6077

Saito T, Yamanaka K, Watanabe Y, Takamatsu N, Meshi T, Okada Y (1989) Mutational analysis of the coat protein gene of tobacco mosaic virus in relation to hypersensitive response in tobacco plants with the N′ gene. Virology 173: 11–20

Schäfer W, Straney D, Ciuffetti L, VanEtten HD, Yoder OC (1989) One enzyme makes a fungal pathogen, but not a saprophyte, virulent on a new host plant. Science 246: 247–249

Scheffer RP, Livingston RS (1984) Host-selective toxins and their role in plant diseases. Science 223: 17–21

Schmelzer E, Kruger-Lebus S, Hahlbrock K (1989) Temporal and spatial patterns of gene expression around sites of attempted fungal infection in parsley leaves. Plant Cell 1: 993–1001

Schmidt WE, Ebel J (1987) Specific binding of a fungal glucan phytoalexin elicitor to membrane fragments from soybean *Glycine max* Proc Natl Acad Sci USA 84: 4117–4121

Schoelz J, Shepherd RJ, Daubert S (1986) Region VI of cauliflower mosaic virus encodes a host range determinant. Mol Cell Biol 6: 2632–2637

Schoelz J, Shepherd RJ, Daubert SD (1987) Host response to cauliflower mosaic virus (CaMV) in solanaceous plants is determined by a 496 bp DNA sequence within gene VI. In: Arntzen CJ, Ryan CA (eds) Molecular strategies for crop protection. AR Liss, New York, pp 253–265

Scholtens-Toma IMJ, De Wit PJGM (1988) Purification and primary structure of a necrosis-inducing peptide from the apoplastic fluids of tomato infected with *Cladosporium fulvum* (syn *Fulvia fulva*). Physiol Mol Plant Pathol 33: 59–67

Sharp JK, McNeil M, Albersheim P (1984) The primary structures of one elicitor-active and seven elicitor-inactive hexa (β-D-glucopyranosyl)-D-glucitols isolated from the mycelial walls of *Phytophthora megasperma* f. sp. *glycinea*. J Biol Chem 259: 11321–11336

Shepherd KW, Mayo GME (1972) Genes conferring specific plant disease resistance. Science 175: 375–380

Shintaku MH, Kluepfel DA, Yacoub A, Patil SS (1989) Cloning and partial characterization of an avirulence determinant from race 1 of *Pseudomonas syringae* pv. *phaseolicola*. Physiol Mol Plant Pathol 35: 313–322

Sims JE, Acres RB, Grubin CE, McMahan CJ, Wignall JM, March CJ, Dower SK (1989) Cloning the interleukin 1 receptor from human T cells. Proc Natl Acad Sci USA 86: 8946–8950

Stäb MR, Ebel J (1987) Effects of Ca^{2+} on phytoalexin induction by fungal elicitor in soybean cells. Arch Biochem Biophys 257: 416–423

Staskawicz BJ, Dahlbeck D, Keen NT (1984) Cloned avirulence gene of *Pseudomonas syringae* pv. *glycinea* determines race-specific incompatibility on *Glycine max* (L.) Merr Proc Natl Acad Sci USA 81: 6024–6028

Stermer BA, Bostock RM (1987) Involvement of 3-hydroxy-3-methylglutaryl coenzyme A reductase in the regulation of sesquiterpenoid phytoalexin synthesis in potato. Plant Physiol 84: 404–408

Stermer BA, Bostock RM (1989) Rapid changes in protein synthesis after application of arachidonic acid to potato tuber tissue. Physiol Mol Plant Pathol 35: 347–356

Stone BA (1989) Cell walls in plant–microbe associations. Aust J Plant Physiol 16: 5–17

Sutherland MW, Deverall BJ, Moerschbacher BM, Reisener H-J (1989) Wheat cultivar and chromosomal selectivity of two types of eliciting preparations from rust pathogens. Physiol Mol Plant Pathol 35: 535–541

Swanson J, Kearney B, Dahlbeck D, Staskawicz B (1988) Cloned avirulence gene of *Xanthomonas campestris* pv. *vesicatoria* complements spontaneous race-change mutants. Mol Plant Microbe Interact 1: 5–9

Takahashi H, Shimamoto K, Ehara Y (1989) Cauliflower mosaic virus gene VI causes growth suppression, development of necrotic spots and expression of defence-related genes in transgenic tobacco plants. Mol Gen Genet 216: 188–194

Takamatsu N, Ishikawa M, Meshi T, Okada Y (1987) Expression of bacterial chloramphenicol acetyltransferase gene in tobacco plants mediated by TMV-RNA. EMBO J 6: 307–311

Tepper CS, Anderson AJ (1986) Two cultivars of bean display a differential response to extracellular components from *Colletotrichum lindemuthianum*. Physiol Mol Plant Pathol 29: 411–420

Tepper CS, Albert FG, Anderson AJ (1989) Differential mRNA accumulation in three cultivars of bean in response to elicitors from *Colletotrichum lindemuthianum*. Physiol Mol Plant Pathol 34: 85–98

Traylor EA, Shore SH, Ransom RF, Dunkle LD (1987) Pathotoxin effects in sorghum are also produced by mercuric chloride treatment. Plant Physiol 84: 975–978

VanEtten HD, Matthews DE, Matthews PS (1989) Phytoalexin detoxification: importance for pathogenicity and practical implications. Annu Rev Phytopathol 27: 143–164

Vivian A, Atherton GT, Bevan JR, Crute IR, Mur, LAJ Taylor JD (1989) Isolation and characterization of cloned DNA conferring specific avirulence in *Pseudomonas syringae* pv. *pisi* to pea (*Pisum sativum*) cultivars, which possess the resistance allele, R2. Physiol Mol Plant Pathol 34: 335–344

Weltring K-M, Turgeon BG, Yoder OC, VanEtten HD (1988) Isolation of a phytoalexin-detoxification gene from the plant pathogenic fungus *Nectria haematococca* by detecting its expression in *Aspergillus nidulans*. Gene 68: 335–344

West CA (1981) Fungal elicitors of the phytoalexin response in higher plants. Naturwissenschaften 68: 447–457

Whalen MC, Stall RE, Staskawicz BJ (1988) Characterization of a gene from a tomato pathogen determining hypersensitive resistance in non-host species and genetic analysis of this resistance in bean. Proc Natl Acad Sci USA 85: 6743–6747

Williams AF, Barclay AN (1988) The immunoglobulin superfamily—domains for cell surface recognition. Annu Rev Immunol 6: 381–405

Wittmann HG, Wittmann-Liebold B (1966) Protein chemical studies of two RNA viruses and their mutants. Cold Spring Harbor Symp Quant Biol 31: 163–172

Wolpert TJ, Macko V (1989) Specific binding of victorin to a 100 kDa protein from oats. Proc Natl Acad Sci USA 86: 4092–4096

Wolpert, TJ, Macko V, Acklin W, Juan B, Seibl J, Meili J, Arigoni D (1985) Structure of victorin C, the major host-selective toxin from *Cochliobolus victoriae*. Experientia 41: 1524–1529

Wolpert TJ, Macko V, Acklin W, Arigoni D (1988) Molecular features affecting the biological activity of the host-selective toxins from *Cochliobolus victoriae*. Plant Physiol 88: 37–41

Wong GH, Elwell JH, Oberley LW, Goeddel DV (1989) Manganous superoxide dismutase is essential for cellular resistance to cytotoxicity of tumor necrosis factor. Cell 58: 923–931

Yoshikawa M (1978) Diverse modes of action of biotic and abiotic phytoalexin elicitors. Nature 275: 546–547

Yoshikawa M (1988) Molecular mechanisms for induction of host defense in fungal diseases. In: Molecular strategies for pathogenicity and host defense in viral, bacterial and fungal diseases. Abstracts of satellite meeting of the 5th International Congress of Plant Pathology, Kyoto, Japan, pp 3–7

Yoshikawa M, Matama M, Masago H (1981) Release of a soluble phytoalexin elicitor from mycelial walls of *Phytophthora megasperma* var. *sojae* by soybean tissues. Plant Physiol 67: 1032–1035

Yoshikawa M, Keen NT, Wang M-C (1983) A receptor on soybean membranes for a fungal elicitor of phytoalexin accumulation. Plant Physiol 73: 497–506

Chapter 5

Pathogenicity Determinants in the Smut Fungi of Cereals

Flora Banuett and Ira Herskowitz

Department of Biochemistry and Biophysics, School of Medicine, University of California
at San Francisco, San Francisco, CA 94143-0448, U.S.A.

With 3 Figures

Contents

I. Introduction

The smut fungi are a large group of plant pathogenic fungi and belong to the order Ustilaginales of the Basidiomycetes (Kenaga, 1972; Agrios, 1988). Most attack cereal grains and grasses, though they also affect broad leaf plants. Smut fungi can cause serious grain losses because they develop in the kernels, replacing the kernel contents with black masses of sooty spores, from which the term smut is derived (Agrios, 1988). Some of the most common smut fungi of cereals and the diseases they incite are *Ustilago maydis* (corn smut disease), *U. nuda* (loose smut of barley), *U. avenae* (loose smut of oats), *U. tritici* (loose smut of wheat), and *U. hordei* (covered smut of barley).

The focus of this review will be *Ustilago maydis*, although relevant aspects of other smut fungi will be included. *U. violacea*, though not a cereal smut fungus (it causes anther smut in the Caryophyllaceae), has been one of the best characterized smut fungi and references to it will also be made throughout this chapter. For more details on smut fungi, the reader is referred to reviews by Thomas (1988), Day and Garber (1988), Nielsen (1988), and Banuett and Herskowitz (1988).

In *U. maydis*, two master regulatory loci, *a* and *b*, govern many steps in the life cycle and are thus determinants of pathogenicity (Rowell and DeVay, 1954; Rowell, 1955; Holliday, 1961; Puhalla, 1968, 1970; Banuett and Herskowitz, 1988). The other smut fungi have a single regulatory locus, *a*, governing their life cycle and consequently pathogenicity (reviewed in Nielsen, 1988). Additional genes that affect pathogenicity, *fuz 1*, fuz 2, and *rtf 1*, have recently been identified in *U. maydis* (Banuett, 1991) and may be the targets of action for the master regulatory loci. Genes involved in compatible interactions with the host (*vir* genes) have been identified in some of the smut fungi and will be discussed at the end of the chapter. Molecular techniques will make it possible to determine if the *fuz* and *rtf* genes correspond to any of the *vir* genes.

The smut fungi all share basic features in their life cycles. Therefore, we will first describe general aspects of this life cycle and then address those features unique to *U. maydis*.

II. General Aspects of the Life Cycle of the Smut Fungi With Emphasis on *U. maydis*

Several of the smut fungi are dimorphic, that is, they exhibit two distinct forms (a yeast-like and a filamentous form) in their life cycles that differ in several aspects: the yeast-like form divides by budding, is haploid and non-pathogenic, and the filamentous form is dikaryotic and pathogenic. The filamentous dikaryon arises after cell fusion of two haploid strains. This filamentous form can grow only in the plant and is responsible for disease symptoms. All smut fungi that have been characterized, except *Ustilago maydis*, have a single mating type or sexual incompatibility locus that governs the generation of the two forms and other aspects of the life cycle (see Nielsen, 1988). This locus, the *a* locus, has two alleles. In contrast, *Ustilago maydis* has two mating type loci, *a* and *b* (see Holliday, 1974; Banuett and Herskowitz, 1988). For all smut fungi, formation of the filamentous pathogenic dikaryon requires that the two haploid strains differ at the *a* locus. For *U. maydis*, the mating partners must also differ at *b*.

A. U. maydis

The filamentous dikaryon of *U. maydis* induces the formation of galls (or tumors) in stems, leaves, tassels, and kernels (Christensen, 1963) (Fig. 1), a symptom characteristic only of corn smut disease. The term tumor is used here to denote the outgrowths resulting from increased cell division and enlargement. Whether the continued presence of the fungus is required for

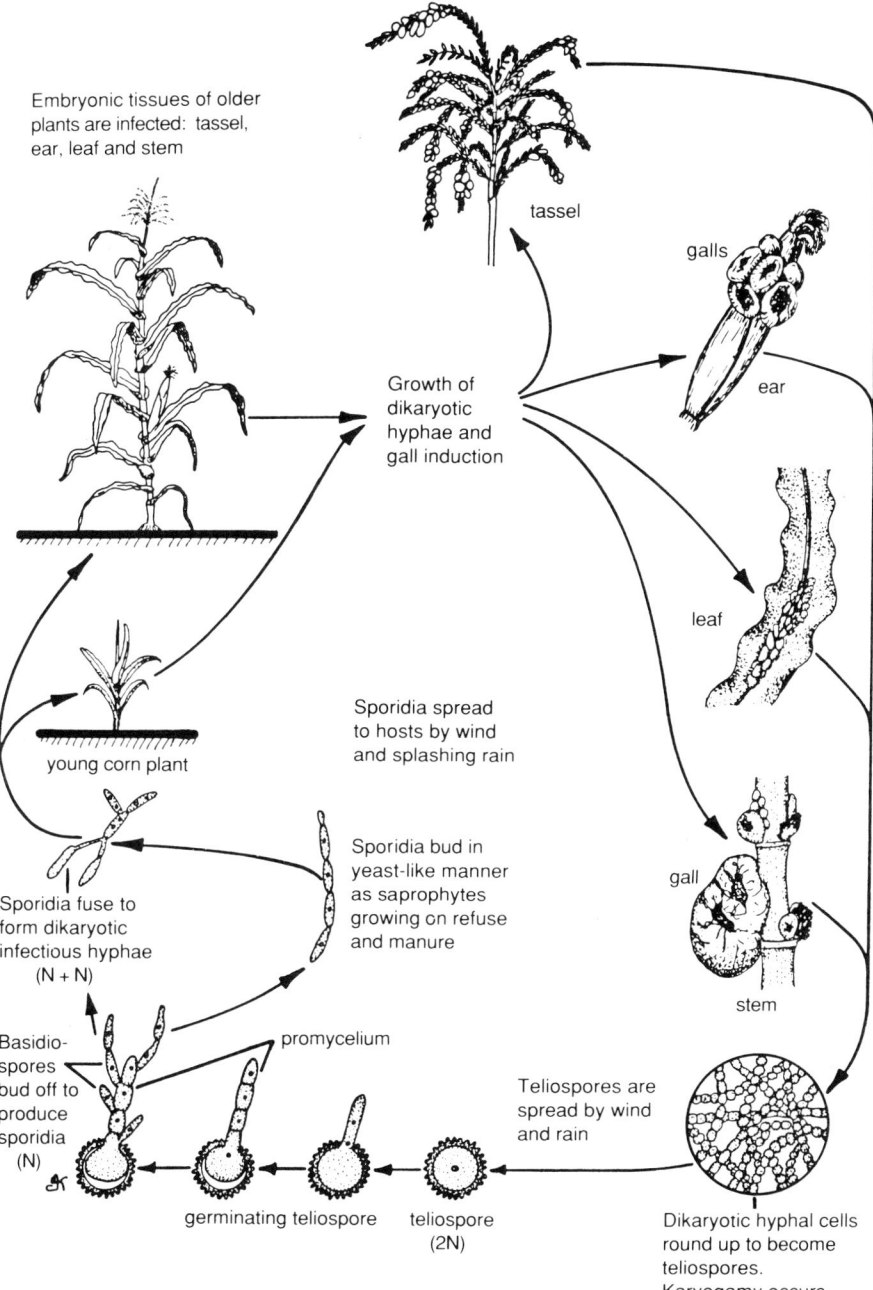

Embryonic tissues of older plants are infected: tassel, ear, leaf and stem

tassel

galls

ear

Growth of dikaryotic hyphae and gall induction

leaf

young corn plant

Sporidia spread to hosts by wind and splashing rain

Sporidia bud in yeast-like manner as saprophytes growing on refuse and manure

gall

stem

Sporidia fuse to form dikaryotic infectious hyphae (N + N)

Basidiospores bud off to produce sporidia (N)

promycelium

Teliospores are spread by wind and rain

germinating teliospore

teliospore (2N)

Dikaryotic hyphal cells round up to become teliospores. Karyogamy occurs.

Fig. 1. Life cycle of *Ustilago maydis*

alteration of growth control in the host is not known. The fungus might alter the host hormonal balance by producing plant hormones or transfer some of its DNA to the host. The fungal filaments undergo differentiation within the tumors: the hyphal cells enlarge, then separate and gelatinize, and a thick cell wall is deposited. Karyogamy takes place resulting in production of diploid spores known as teliospores. Teliospores can be removed from the tumors and, when placed on nutrient media, germinate by forming a short filament, the promycelium, where the diploid nucleus migrates and undergoes meiosis (O'Donnell and MacLaughlin, 1984a, b). Even though meiosis itself can occur outside the plant, *competence* to undergo meiosis is acquired only by passage through the plant. The plant signals that induce this competence and the fungal genes that respond to such signals have not yet been identified. The components of this signalling process may prove to be important in pathogenesis.

B. Other Smut Fungi

For most smut fungi, the filamentous dikaryon grows or is carried along in the meristematic region of the plant and produces no symptoms, except for stunting of plants and early emergence of flowering spikes in infected plants (Kenaga, 1972; Agrios 1988). It should be noted that there are few studies of the infectious process and growth of the fungus within its host. Immuno-cytology with fungal-specific antibodies should allow a better understanding of the fungal growth process in the host. Overt symptoms of infection begin to appear at onset of flowering; fungal differentiation within the developing seeds results in replacement of the seed tissue with sooty masses of diploid teliospores. In the case of *U. violacea*, the dikaryotic mycelium penetrates the anthers, undergoes differentiation and produces purple-brown teliospores conferring this distinctive color to the anthers, which is the most obvious symptom of the disease (Day and Garber, 1988). Another consequence of infection by some of these fungi that has not been studied is their ability to induce their host to undergo sexual transformation: development of anthers instead of ovaries. For all these fungi, competence for meiosis is acquired only after passage through the plant.

Clearly, completion of the life cycle for most of these fungi requires that the host reach sexual maturity (i.e., undergo floral development) which takes several months for wheat, barley, and oats. In contrast, *U. maydis* does not require sexual maturity of its host for completion of its life cycle because of its ability to induce tumors and produce teliospores in very young plants. Its life cycle can be completed in as short a time as two weeks in laboratory conditions (Holliday, 1961, 1974; Banuett and Herskowitz, 1988). Obviously, such a short life cycle is an asset in studies of this organism.

In general, the haploid form can be grown on a variety of laboratory media whereas the dikaryon is difficult or impossible to grow under similar conditions (Day and Anagnostakis, 1971). Thus, the transition from one form to the other entails decreased ability to grow on nutrient media in addition to acquisition of pathogenic ability.

Several properties of the smut fungi make them an ideal group of fungal pathogens for the study of regulation of the pathogenic form and for host–pathogen interactions: (1) Most of these fungi have a haploid unicellular form that can be manipulated by standard microbiological techniques, permitting the isolation of different types of mutants (auxotrophic, drug resistance, UV sensitive) as has been done in *U. maydis* (Holliday, 1961; F. Banuett, unpubl.), in *U. violacea* (Day and Garber, 1985; see Day and Garber, 1988), and in *U. hordei* (Kozar, 1969; J. Sherwood, pers. comm.). It should be possible to isolate mutants that affect pathogenicity, filamentous growth, germination of teliospores, etc. This has been accomplished only for *U. maydis* (Banuett, 1991). (2) The ease of manipulating meiotic products under laboratory conditions facilitates genetic analysis, e.g., construction of strains of known genotype, allelism tests, etc. (3) It is possible to construct stable diploids for *U. maydis* (Holliday, 1961; Day et al., 1971; Banuett and Herskowitz, 1989) and *U. violacea* (Day and Jones, 1968, 1969). (4) Introduction of exogenous DNA into *U. maydis* by DNA transformation (Wang et al., 1988; Tsukuda et al., 1988) makes it possible to clone genes by complementation and also to introduce mutations constructed in vitro into the genome for phenotypic analysis. Several *U. maydis* genes have already been cloned (Kronstad et al., 1989; Fotheringham and Holloman, 1989), including the *b* locus (see below). Recently, *U. hordei* was transformed with high efficiency using the *U. maydis* autonomously replicating (ARS) vector developed by Tsukuda et al. (1988) (D. Pope, pers. comm.; J. Sherwood, pers. comm.). Low efficiency transformation of *U. hordei* and *U. nigra* has been described by Holden et al. (1988).

III. The Mating Type Loci of *U. maydis*

Ustilago maydis has been the subject of genetic and biochemical studies on recombination, and more recently it has become the subject of intensive studies of the mating type loci (Kronstad and Leong, 1989; Banuett and Herskowitz, 1989; Schulz et al., 1990). Two master regulatory loci, *a* and *b*, govern life cycle transitions in *U. maydis*. These loci reside on different chromosomes (Holliday, 1974). The *a* locus has two alleles (Rowell and DeVay, 1954), and thus in this respect it is like the single mating type locus of the other smut fungi. Whether this similarity will be true at the molecular level remains to be determined. The two alleles of the *a* locus have been cloned (Froeliger and Leong, 1991) and sequenced (M. Bölker and R. Kahmann, pers. comm). They can now be used as molecular probes to analyze other smut fungi. The *b* locus is estimated to have at least 25 naturally occurring alleles (Puhalla, 1970). A *b* locus has not been described for any of the other smut fungi. There are, however, other Basidiomycetes that have two unlinked genetic determinants of sexual development. For example, in *Schizophyllum commune* (Novotny et al., 1991) and *Coprinus cinereus* (Casselton, 1978) the two determinants are complex, each consisting of two closely linked loci α and β, which are multiallelic. In *Tremella mesenterica*, a member of the jelly fungi, the situation is like that in *U. maydis*: one locus is multiallelic and the other is not (Wong and Wells, 1985).

As noted earlier, fusion of two *U. maydis* haploid strains carrying different *a* and *b* alleles leads to formation of the filamentous pathogenic dikaryon. The *a* locus has been proposed to govern cell fusion (Rowell and DeVay, 1954; Rowell, 1955), although this has not been established conclusively. The *a* and *b* loci are both necessary for filamentous growth (Rowell and DeVay, 1954; Rowell, 1955; Holliday, 1961; Puhalla, 1968; Day et al., 1971; Banuett and Herskowitz, 1989), and the *b* locus is the key pathogenicity determinant (Holliday, 1961; Puhalla, 1968; Banuett and Herskowitz, 1989). These loci most likely regulate expression of other genes more intimately involved in determining the specialized properties of this form.

A. Diploid Strains: Formation and Use in Studying the Pathogenicity Determinants

Diploid strains have been invaluable in demonstrating different processes regulated by the *a* and *b* loci. They have been used extensively in genetic analysis of *U. maydis* (Holliday, 1961) and *U. violacea* (Garber and Day, 1985; see also Day and Garber, 1988), for example, for determining dominance or recessiveness of mutations. Attempting to determine dominance relationships by analysis of dikaryons is cumbersome and potentially misleading because dikaryons are subject to instability, to variation in number of nuclei per cell, and to nucleus-limited gene products.

Diploid strains capable of vegetative growth are not part of the normal life cycle of the smut fungi but can be constructed rather easily in the laboratory by mating two haploid strains carrying complementing auxotrophic markers and then selecting for prototrophs (Holliday, 1961, 1974; Day et al., 1971; Day and Jones, 1968; F. Banuett, unpubl.). Diploids of *U. maydis* and *U. violacea* are very stable, except that those of *U. violacea* appear to undergo high levels of mitotic recombination at the mating type locus (see references in Day and Garber, 1988). Haploidization, a phenomenon in which whole sets of chromosomes or individual chromosomes are lost, does not occur spontaneously (Holliday, 1961; Day and Jones, 1969), but it can be induced in *U. violacea* by treatment with *p*-fluorophenylalanine (Day and Jones, 1969). This treatment is not effective in *U. maydis* diploids. Construction of stable vegetative diploids has not been described for other smut fungi but should be possible, for example, in *U. hordei*, where auxotrophic mutants have been isolated (Kozar, 1969; J. Sherwood, pers. comm.).

U. maydis diploids heterozygous for both *a* and *b* exhibit properties of the dikaryon: these diploids form mycelial or fuzzy colonies (Rowell and DeVay, 1954; Holliday, 1961; Puhalla, 1968; Day et al., 1971; Banuett and Herskowitz, 1989) and are thus said to exhibit the Fuz$^+$ phenotype. They are also solopathogenic, that is, a single culture induces tumor formation (Tum$^+$ phenotype) (Rowell and DeVay, 1954; Holliday, 1961; Puhalla, 1968; Day et al., 1971; Banuett and Herskowitz, 1989). A convenient property of the diploids is that, unlike the dikaryon, they can be maintained in culture in nutrient media for indefinite periods of time. Their filamentous mode of growth is evoked when these cells are exposed to media containing charcoal (see Sect. III.C). Diploids have also been obtained that are heterozygous at *a* but homozygous at *b*, either by mating strains that differ only at the *a* locus (Holliday, 1961; Day et al., 1971) or as mitotic recombinant derivatives of diploid strains heterozygous at both mating type loci (Banuett and Herskowitz, 1989). Strains homozygous at *a* and heterozygous at *b* have been obtained as mitotic recombinant derivatives of diploids heterozygous at both loci (Banuett and Herskowitz, 1989).

Comparison of isogenic diploids that differ in composition at *a* or *b* or at both loci has allowed examination of the requirements of these loci in maintenance of filamentous growth, tumor-inducing ability, teliospore formation, and meiosis (Banuett and Herskowitz, 1989). Diploid strains that are homozygous at *b* or at *a* are not capable of mycelial growth; they exhibit the Fuz$^-$ phenotype. Thus, the presence of different alleles at *a* and *b* is necessary for filamentous growth. Even though these two types of diploid strains are Fuz$^-$, they do differ with respect to pathogenicity: *a1/a2 b1/b1* and *a1/a2 b2/b2* strains are not pathogenic, whereas *a1/a1 b1/b2* and *a2/a2 b1/b2* strains are pathogenic. The latter diploids are also capable of forming teliospores that undergo normal meiosis indicating that the *a* locus is not necessary for these processes (Banuett and Herskowitz, 1989). In contrast, the presence of different *b* alleles is necessary and sufficient for pathogenicity (once cell fusion has taken place to produce the cell with different *b* alleles). In summary, the presence of different *b* alleles is necessary for maintenance of the pathogenic state itself whereas the presence of different *a* alleles is necessary to establish this state.

B. Molecular Analysis of the b Locus

An understanding of the molecular mechanisms by which *a* and *b* govern filamentous growth and pathogenicity can be obtained by cloning these loci and determining the type of product they encode. The *b* locus has attracted considerable attention (1) because it is the major pathogenicity determinant of *U. maydis*, and thus it is likely to regulate genes for tumor induction; and (2) because the *b* locus is multiallelic, with 25 alleles. Any combination of two different alleles results in pathogenic development, which raises the fascinating problem of the molecular basis for self-nonself recognition in this organism. Understanding this discrimination at the molecular level

```
                 20           40           60           80          100
     b2  MSNYPNFSLTSFVECLNEIEHEFLRDKLENRPVLVRKLQELRRKTPNNVASLSYDPGTIHQIHQTTHRIKVAAKAFIRIDQSFVSLHSDAVEDTSKALKK
     b3  --RD-KL--SK-L-----------V-H-----------QQ--KH--K-FHE-EM-Q----AA---DIV-Q---F--K----C-EI-H--T-VMQE
     b4  --SD-KI--L---L---S---EK-P---S----QQ---I-N-DH--K--QK------N--V---C---T----R----A-R---
     b1  --SD-----I--L--------G--Y------R--QQ-I-DI-N-PR--E--Q-----RAV-Q---F--K---C-EV-HG---VMQE
     b5  --SD-----L-------E---II---R--QQ---H--D-AH-SK--E----A---V-T------K---C-EV-HG---VMEE
     b6  --RD-KL--SK-L-----------V-H------QQ--KH-TN-PN--E--Q---IA--LE--V-LH--RK--R--V----QE
     b7  --SD-----L---SQ--------V--P------QQ---GH----LH--E--Q----A---E--V-V--H---K-T-GC-EV-HG---VMQE

                                                        ** * * *
                120          140          160          180          200
     b2  ADASSPVVGCRDLSEDLPAYHMRKHFLHTLDNPYPTQEEKEGLVRLTNESTARVGLSKANRPPLEVHQLTLWFINARRSGWSHILKKFAREDRSRMKHL
     b3  -NVV--GE--N-----------N------E------T-----IRP-N-I-------Q--------------R-
     b4  --N---------E------------E------T-----Q-IV-
     b1  FNVV--D--N-----------L-----------T--Q-SV---------HV-
     b5  FNVV--A-V-N-----------L-----------QN----V--S-NPA-
     b6  VNVA--A-EY-N----------S-----------T--------V--SV-
     b7  VNVG--A---N----------L--S---------T-------N-T------E----R-

                220          240          260          280          300
     b2  VRAKLSSSNQSTPPSSTDSLSNNLDDVLSDNLGREPLTPVDKQQFEDDWASMISWIKYGVKEKVGDWVYDLCAASKKTPKPGMRPVTTVAKRHPARKTK
     b3  ------------L--EKP-DD--V----------LA--------------------------Q--
     b4  --------------P-EYP------F------A---------------------N----Q-
     b1  -------------Y---------F------A-----------------N-----T--Q-
     b5  ------------L--EYP----F-----------A-----------------Q-
     b6  ------------L--EKP-DD--V-----------LA----------------T--Q-
     b7  ------------T-S--PMPEYP----NI-----A-----------------Q-

                320          340          360          380          400
     b2  PAAKPKSRTANPRASTTPSIDSTLDSSKLESTPELSMCSTADTSFSTFGSSLSMSHYNPFQDGNDILQSPTVKARGNRKVKALPKRAGKQQPDEVENGKI
     b3  -----------------------------------D-
     b4  ---------------------------D--Y----F------D-
     b1  ---------------------------------D----
     b5  ---------------------------------D----F------D-
     b6  ---------------------------D--Y----F------ID-
     b7  --N-

                410
     b2  PFLCLSVAFV
     b3  ----------
     b4  ----------
     b1  ----------
     b5  ----------
     b6  --F--I---
     b7  ----------
```

may provide important information on the functioning of an ubiquitous class of regulatory proteins, the homeodomain proteins.

Because diploids homozygous for *b* are non-fuzzy, the *b* locus could be cloned by a simple functional assay in which one screens for fuzzy colonies after introduction into cells of the appropriate library (see Kronstad and Leong, 1989; Schulz et al., 1990). A similar strategy was used to clone the a locus (M. Bölker and R. Kahmann, pers. comm.). An *a1/a2 b1/b1* diploid was transformed with a library from an *a2 b2* strain constructed in a plasmid vector that confers hygromycin resistance. Hygromycin resistant transformants were selected and then screened for the fuzzy phenotype on charcoal plates. A transformant carrying a plasmid with an insert responsible for the fuzzy phenotype and ability to induce tumors was identified. The region within the insert responsible for the phenotype was sequenced and shown to contain an open reading frame (ORF) of 410 amino acids (Schulz et al., 1990). Three other *b* alleles (*b1, b3, b4*) were identified by nucleic acid hybridization using the *b2* clone as a probe. All four alleles contain an ORF of 410 amino acid residues. Comparison of the amino acid sequences revealed that the ORFs consist of a variable region in the amino terminal 110 amino acid residues and a constant region for the remainder of the ORF (Fig. 2). Notably, the constant region contains a sequence (WF-N-R) found in all homeodomain proteins of higher eukaryotes, suggesting that the b-polypeptides are DNA-binding proteins (see Schulz et al., 1990). The b-polypeptides also share extensive similarity with known yeast DNA-binding proteins that contain a divergent homeodomain (Schulz et al., 1990) (Fig. 3). Nucleotide sequence analysis of other *b* alleles, *b5, b6, b7* (Fig. 2) (Kronstad and Leong, 1990), further demonstrates the conservation of the organization of the *b* locus and the presence of the homeodomain motif. Because the biological observations indicate that different *b* alleles are necessary for pathogenic development, Schulz et al. (1990) proposed that self-nonself recognition occurs at the level of polypeptides, which are capable of interacting to form a multimeric regulatory protein. This protein is proposed to govern target genes for filamentous growth (*fuz* genes) and pathogenicity (*tum* genes). Thus, the *b* locus is a master regulator that encodes a DNA-binding protein responsible for the developmental switch that results in a pathogenic dikaryon.

Many questions about the *b* locus remain. What is the functional species produced by interaction of b-monomers? It might be a repressor or an activator. What residues determine allele specificity? Analysis of the behavior of chimeric *b* alleles and of the *b* null mutant should provide answers

Fig. 2. Alignment of the deduced amino acid sequence of 7 different *b* alleles. Dashes indicate amino acid identity. Asterisks denote the four invariant amino acids, WF-N-R, found in all multicellular eukaryotic homeodomain proteins (Scott et al., 1989). Data for *b1, b2, b3,* and *b4* are from Schulz et al. (1990) and for *b5, b6,* and *b7* from Kronstad and Leong (1990)

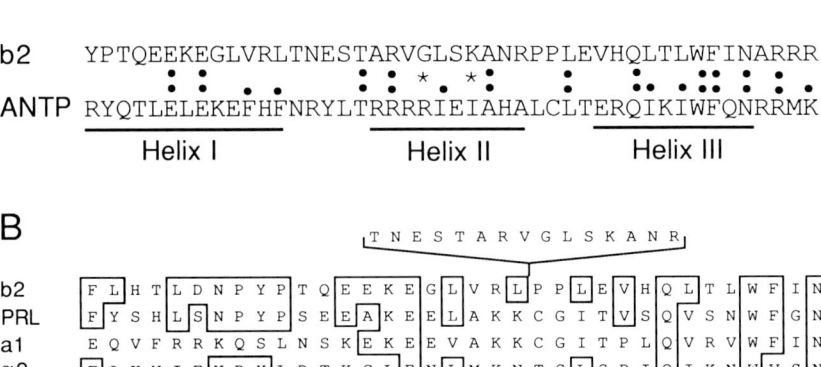

Fig. 3. The homeodomain-related region of *b*. The homeodomain-related region of *b* is compared with the homeodomain of the *Drosophila* Antennapedia protein (see Scott et al., 1989) (A) and with the homeodomain-related proteins from humans (*prl*; Kamps et al., 1990; Nourse et al., 1990) and yeast (a1, α2, matPi, PHO2; see Schulz et al., 1990) (B). The α2 protein of yeast and other members of the fungal family of divergent homeodomain proteins (see Novotny et al., 1991) contain only 3 of the 4 invariant amino acids. Helix II and helix III refer to proposed helical regions of the homeodomain (Kissinger et al., 1990). Identical amino acids are indicated with a colon; conservative changes with a single dot. In B, 15 amino acids have been looped out from *b2* in order to maximize identities

to some of these questions. What domain(s) of the polypeptides govern association and activity of the multimer formed? Two broad categories of models were proposed by Schulz et al. (1990). These models differ in whether it is the variable or the constant region that determines association of the monomers. These models consider the functional consequences of whether the homomultimer or the heteromultimer is the active species.

C. Other Genes Necessary for Filamentous Growth and Pathogenicity in U. maydis

The above discussion has focused on the putative *b* regulatory protein. What are its target genes? One approach to identify such genes is to isolate mutants defective in filamentous growth and pathogenicity. The next step is to clone the wildtype versions of these genes, and then to determine whether their transcription is regulated by *b*. Such mutants can be identified using charcoal nutrient media to assay production of filaments. On this medium, a mixture of haploid strains that differ at both *a* and *b* produces filaments (the Fuz$^+$ phenotype; Banuett and Herskowitz, 1988). If the strains carry identical *a* or identical *b* alleles, no filaments are formed (the Fuz$^-$ phenotype). The charcoal plate assay allows one to screen for mutants that are unable to produce filaments with the appropriate tester strain (Banuett and Herskowitz, 1988). For example, mutagenesis of an *a1 b1* strain, followed by mating with an *a2 b2* strain led to the identification of mutants that exhibit altered fuzz reaction: some mutants give no reaction, others exhibit reduced fuzziness or very altered filamentous structures (Banuett, 1991). Because the process of filament formation involves cell fusion and subsequent growth of the fusion product as a filament, and because the mating type loci govern these processes, the mutants may have mutations in *a* or *b*, or in other genes necessary for cell fusion or for filamentous growth. Subsequent genetic analysis of some of these mutants has led to the identification of three new genes distinct from *a* and *b* designated *rtf 1, fuz 1* and *fuz 2* (Banuett, 1991). The behavior of the mutants suggests that they alter steps after cell fusion. Some of them have marked effects on pathogenicity: *fuz 1*$^-$ strains lead to production of very small tumors devoid of teliospores; *fuz 2*$^-$ strains induce tumors that produce teliospores, but these teliospores are unable to germinate. The *fuz 1* and *fuz 2* genes are thus necessary for filament formation, full tumor induction, and teliospore production and germination (Banuett, 1991).

Mutants defective in *rtf* exhibit a remarkable behavior: co-inoculation of plants with two haploids having the same *b* allele and the *rft*$^-$ mutation induces tumors that are identical to those obtained in inoculations with wildtype strains carrying different *b* alleles, for example, *b1 + b2*. Thus, the *rtf*$^-$ mutation bypasses the need for different *b* alleles. The wildtype *rtf* product is proposed to inhibit tumor formation (Banuett, 1991). This inhibitory action of *rtf* is antagonized by the combinatorial activity proposed to result from interaction of two different b-monomers.

Using the approach described above one can begin to dissect the pathway of filamentous growth and pathogenicity and identify the targets for *b* and *a*. The genes identified will provide a better understanding of what determines filamentous growth and pathogenicity. In addition, if these genes are regulated by *b*, they will also provide useful reagents in biochemical studies on the mode of action of the b-protein.

IV. Host–Pathogen Interactions in the Smut Fungi

Genes that condition whether a pathogen will produce an incompatible reaction (no disease) in a given cultivar of its host have been described for a number of fungal and bacterial pathogens (Day, 1974). These genes are designated avirulence (*avr*) and are usually dominant, although in the case of many fungal pathogens, it is not clear how dominance was conclusively determined. For every such gene, there is a corresponding dominant resistance gene in the host, such that if the pathogen carries an *avr* gene and the host the corresponding resistance gene, there is an incompatible reaction. Any other combination between host and pathogen results in a compatible reaction or disease. This is known as the gene-for-gene hypothesis (Flor, 1956).

The interaction between *U. hordei* and its host, barley, is one of the best studied in the smut fungi. Several dominant avirulence genes (*avr*) and the corresponding resistance genes have been described for the *U. hordei*–barley interaction (Sidhu and Person, 1971, 1972; see Thomas, 1988, for a review). Avirulence genes in *U. tritici* and *U. nuda* that condition interaction with their respective hosts, wheat and barley, have also been described (see Nielsen, 1988). No such genes have been described in *U. maydis* or *U. violacea*.

Construction of diploid strains in *U. hordei* should make it possible to test conclusively the dominance or recessiveness of the avirulence genes in this species. Furthermore, the ability to transform *U. hordei* with high efficiency using the *U. maydis* autonomously replicating vector opens the way for cloning these genes and to begin their molecular analysis.

V. Conclusion

In this review we have described salient features of the smut fungi that make them amenable for analyzing the development of the pathogenic dikaryon and its ability to incite disease. Among these fungi, *U. maydis* offers unique opportunities for such studies because of the extensive genetic and molecular techniques available for cloning and subsequent molecular characterization of the cloned genes, and because certain aspects of its life cycle make it easier to work with. *Fuz* genes identified in *U. maydis* might play similar roles in other smut fungi as well as in other fungal pathogens. Once *fuz* genes have been cloned, they can be used to identify their counterparts in other fungal pathogens by nucleic acid hybridization. Such studies should provide insights as to the molecular mechanisms underlying fungal pathogenicity.

Acknowledgements

We would like to thank John Sherwood, Dave Pope, Regine Kahmann, and Charles Novotny for communicating unpublished results. FB has been supported by the Weingart Program in Developmental Genetics at UCSF and by a Research Grant (to IH) from the National Institutes of Health (AI 18738).

VI. References

Agrios GN (1988) Plant pathology, 3rd edn. Academic Press, London
Banuett F (1991) Identification of genes necessary for filamentous growth and tumor induction of the plant pathogen *Ustilago maydis*. Proc Natl Acad Sci USA 88: 3922–3926

Banuett F, Herskowitz I (1988) *Ustilago maydis*, smut of maize. In: Sidhu GS (ed) Advances in plant pathology, vol 6, genetics of plant pathogenic fungi. Academic Press, London, pp 427–455

Banuett F, Herskowitz I (1989) Different *a* alleles of *Ustilago maydis* are necessary for maintenance of filamentous growth but not for meiosis. Proc Natl Acad Sci USA 86: 5878–5882

Casselton LA (1978) Dikaryon formation in higher Basidiomycetes. In: Smith JE, Berry DR (eds) The filamentous fungi, vol 3. Arnold, London, pp 275–297

Christensen JJ (1963) Corn smut caused by *Ustilago maydis*. Am Phytopathol Soc Monogr 2

Day AW, Garber ED (1988) *Ustilago violacea*, anther smut of the Caryophyllaceae. In: Sidhu GS (ed) Advances in plant pathology, vol 6, genetics of plant pathogenic fungi. Academic Press, London, pp 457–482

Day AW, Jones JK (1968) The production and characteristics of diploids in *Ustilago violacea*. Genet Res (Camb) 11: 63–81

Day AW, Jones JK (1969) Sexual and parasexual analysis of *Ustilago violacea*. Genet Res (Camb) 14: 195–221

Day PR (1974) Genetics of host–parasite interactions. Freeman, San Francisco

Day PR Anagnostakis SL (1971) Corn smut dikaryon in culture. Nature (New Biol) 231: 19–20

Day PR, Anagnostakis SL, Puhalla JE (1971) Pathogenicity resulting from mutation at the *b* locus of *Ustilago maydis*. Proc Natl Acad Sci USA 68: 533–535

Flor HH (1956) The complementary genetic systems in flax and flax rust. Adv Genet 8: 29–54

Fotheringham S, Holloman WK (1989) Cloning and disruption of *Ustilago maydis* genes. Mol Cell Biol 9: 4052–4055

Garber ED, Day AW (1985) Genetic mapping of a phytopathogenic basidiomycete, *Ustilago violacea*. Bot Gaz 146: 449–459.

Holden DW, Wang J, Leong SA (1988) DNA-mediated transformation of *Ustilago hordei* and *Ustilago nigra*. Physiol Mol Plant Pathol 33: 235–239

Holliday R (1961) The genetics of *Ustilago maydis*. Genet Res 2: 204–230

Holliday R (1974) *Ustilago maydis*. In: King RC (ed) Handbook of genetics, vol 1. Plenum, New York, pp 575–595

Kamps MP, Murre C, Sun X-h, Baltimore D (1990) A new homeobox gene contributes the DNA binding domain of the t(1:19) translocation protein in pre-B ALL. Cell 60: 547–555

Kenaga CB (1972) Principles of phytopathology, 2nd edn. Balt, Lafayette, IN

Kissinger CR, Beishan L, Martin-Blanco E, Kornberg TB, Pabo CO (1990) Crystal structure of an engrailed homeodomain-DNA complex at 2.8 Å resolution: a framework for understanding homeodomain-DNA interactions. Cell 63: 579–590

Kozar F (1969) Mitotic recombiantion in biochemical mutants of *Ustilago hordei*. Can J Genet Cytol 11: 961–966

Kronstad JW, Leong SA (1989) Isolation of two alleles of the *b* locus of *Ustilago maydis*. Proc Natl Acad Sci USA 86: 978–982

Kronstad JW, Leong SA (1990) The *b* mating type locus of *Ustilago maydis* contains variable and constant regions. Genes Dev 4: 1384–1395

Kronstad JW, Wang J, Covert SF, Holden DW, McKnight GL, Leong SA (1989) Isolation of metabolic genes and demonstration of gene disruption in the phytopathogenic fungus *Ustilago maydis*. Gene 79: 97–106

Nielsen J (1988) *Ustilago* spp., smuts. In: Sidhu GS (ed) Advances in plant pathology, vol 6, genetics of plant pathogenic fungi. Academic Press, London, pp 483–490

Nourse J, Mellentin JD, Galili N, Wilkinson J, Stanbridge E, Smith SD, Cleary ML (1990) Chromosomal translocation t(1:19) results in synthesis of a homeobox fusion mRNA that codes for a potential chimeric transcription factor. Cell 60: 535–545.

Novotny CP, Stankis MM, Specht CA, Yang H, Giasson L, Ullrich R (1991) The $A\alpha$ mating type locus of *Schizophyllum commune*. In: Bennett J, Lasure L (eds) More gene manipulations in fungi. Academic Press, London, pp 234–257

O'Donnell KL, McLaughlin DJ (1984a) Ultrastructure of meiosis in *Ustilago maydis*. Mycologia 76: 468–485

O'Donnell KL, McLaughlin DJ (1984b) Postmeiotic mitosis, basidiospore development, and septation in *Ustilago maydis*. Mycologia 76: 486–502

Puhalla JE (1968) Compatibility reactions on solid medium and interstrain inhibition in *Ustilago maydis*. Genetics 60: 461–474

Puhalla JE (1970) Genetic studies of the *b* incompatibility locus of *Ustilago maydis*. Genet Res 16: 229–232

Raper CA (1983) Controls for development and differentiation of the dikaryon in Basidiomycetes. In: Bennett JW, Ciegler A (eds) Secondary metabolism and differentiation in fungi. Marcel Dekker, New York, pp 195–238

Rowell JB (1955) Functional role of compatibility factors and an in vitro test for sexual compatibility with haploid lines of *Ustilago zeae*. Phytopathology 45: 370–374

Rowell JB, DeVay JE (1954) Genetics of *Ustilago zeae* in relation to basic problems of its pathogenicity. Phytopathology 44: 356–362

Scott MP, Tamkun JW, Hartzell GW, III (1989) The structure and function of the homeodomain. Biochim Biophys Acta 989: 25–48

Sidhu G, Person C (1971) Genetic control of virulence in *Ustilago hordei*. II. Segregation for higher levels of virulence. Can J Genet Cytol 13: 173–178

Sidhu G, Person C (1972) Genetic control of virulence in *Ustilago hordei*. III. Identification of genes for host resistance and demonstration of gene-for-gene relations. Can J Genet Cytol 14: 209–213

Thomas PL (1988) *Ustilago hordei*, covered smut of barley and *Ustilago nigra*, false loose smut of barley. In: Sidhu GS (ed) Advances in plant pathology, vol 6, genetics of plant pathogenic fungi. Academic Press, London, pp 415–425

Tsukuda T, Carleton S, Fotheringham S, Holloman WK (1988) Isolation and characterization of an autonomously replicating sequence from *Ustilago maydis*. Mol Cell Biol 8: 3703–3709

Wang J, Holden DW, Leong SA (1988) Gene transfer system for the phytopathogenic fungus *Ustilago maydis*. Proc Natl Acad Sci USA 85: 865–869

Wong GJ, Wells K (1985) Modified bifactorial incompatibility in *Tremella mesenterica*. Trans Br Mycol Soc 84: 95–109

Chapter 6

Identification of Fungal Genes Involved in Plant Pathogenesis and Host Range

Willi Schäfer, Dietmar Stahl, and Enrico Mönke

Institut für Genbiologische Forschung Berlin, D-W-1000 Berlin 33,
Federal Republic of Germany

With 2 Figures

Contents

I. Introduction

The application of molecular genetic techniques to study the interactions of a fungal pathogen with its host plant is just beginning. Nevertheless a number of encouraging results have already emerged and they highlight the

power of this approach. In this chapter we will describe some of these
techniques and demonstrate with a few selected examples how the application of molecular biology can provide answers to long-standing questions.

II. Transformation of Phytopathogenic Fungi

The development of a transformation system is a prerequisite for a molecular genetic analysis of fungal plant pathogens. Since the first reports of DNA
mediated transformation of yeast (Hinnen et al., 1978) and the filamentous
fungi *Neurospora crassa* (Case et al., 1979) and *Aspergillus nidulans* (Tilburn
et al., 1983), gene transfer systems for an increasing number of filamentous
fungal species including phytopathogens have been developed (for a review
see Fincham, 1989). Geneticists are encountered with two main problems in
generating a transformation procedure: Cells have to be rendered into a
transformable state able to take up DNA and subsequently regenerated.
Furthermore, a selection marker for detection of transformed colonies has
to be found.

A. Transformation Procedure

In principle two kinds of protocols for DNA uptake have been applied.
Transformable protoplasts are produced by enzymatic digestion of cell
walls from fungal mycelium or germinating conidia in the presence of an
osmotic stabilizer. DNA uptake can be achieved by incubation in the
presence of 40% polyethylene glycol (PEG) and 10–50 mM $CaCl_2$ (Yelton
et al., 1984). As a rapid but generally less efficient alternative, intact cells or
germinating spores can take up DNA upon treatment with high concentrations of alkali cations, followed by application of DNA in 0.1 M lithium
acetate (Ito et al., 1983). After plating on regeneration medium, selection
pressure is applied usually after 16 h. In general resistant colonies are
visible after 3–9 days (Turgeon et al., 1985; Kistler and Benny, 1988; Wang
et al., 1988). Since transformed cells of many fungi are heterokaryons,
transformants are often purified by a single conidiation step to get genetically uniform clones. The yield of transformants varies in the range of 1 to
1000 transformants/µg transforming DNA.

Recently two promising new transformation procedures of filamentous
fungi have been reported. Electroporation of fungal protoplasts (Goldmann
et al., 1990), which is already widely used for bacteria, is a fast transformation method, omitting the use of PEG which is toxic to some fungi. The
biolistic gene transfer system is using DNA coated microprojectiles (Armaleo
et al., 1990). These microprojectiles carrying the transforming DNA are
transferred with a "particle gun" into intact mycelium. The DNA is stably
integrated into the genome and transformed cells can be regenerated under
selection pressure. The biolistic gene transfer system is very fast and is an

alternative way for delivering DNA to fungi which are unamenable to protoplast preparation or sensitive to PEG.

B. Selection Markers

Selection of transformants relies on the expression of a marker gene conferred by the transforming DNA. The expression of the marker gene is controlled by fungal promoter and terminator sequences of homologous or heterologous origin. Regulatory DNA sequences of *A. nidulans* (Punt et al., 1987) and *N. crassa* (Orbach et al., 1986) have been used successfully for transformation of many filamentous fungi, including plant pathogens.

Selection based on complementation of auxotrophic mutants with a wild type gene was widely used in the beginning of fungal transformation. This technique has been applied only in rare cases to pathogenic fungi (Rambosek and Leach, 1987; Parsons et al., 1987). The advantage of a higher number of transformed fungal colonies by selection against auxotrophy is countered by a number of disadvantages. Defined nutritional mutations are often not available for fungi of interest. The construction of auxotrophic strains by chemical or UV mutagenesis requires a series of backcrosses to elimate additional undetected mutations. The auxotropic mutation has to be crossed into each desired recipient strain. This is an obstacle particularly for fungi without a sexual stage or for which difficulties in meiotic manipulation exist.

The use of antibiotic or toxin resistance genes of bacterial or fungal origin as dominant selection markers has been applied for the transformation of a large number of pathogenic fungi (Table 1). The antibiotic resistance genes *hyg B* and *APH II* of *E. coli* confer resistance to the aminoglycoside antibiotics hygromycin B and kanamycin/G418 respectively. The use of the *hyg B* gene has been developed into a standard selection system for fungal gene transfer permitting transformation of several different fungi (Table 2). Resistance against the systemic fungicide benomyl conferred by *bml*, the structural gene for β-tubulin of a benomyl resistant strain of *N. crassa* (Orbach et al., 1986) has been used several times (Panaccione et al., 1988; Dickman, 1988; Henson et al., 1988). Transformation of the maize pathogen *Cochliobolus heterostrophus* (Turgeon et al., 1985) was at first achieved by using the *amd S* gene of *A. nidulans* (Hynes et al., 1983). The *amd S* gene codes for acetamidase, an enzyme permitting *A. nidulans* to utilize acetamide as sole nitrogen source. Transformation of *Colletotrichum lindemuthianum* (Rodriguez and Yoder, 1987) with the same system demonstrated the utility of *amd S* for gene transfer in pathogens. An additional selection system for *N. crassa* (Avalos et al., 1989) and *C. heterostrophus* (Straubinger and Yoder, pers. comm.) was established using the *bar* gene of *Streptomyces hygroscopicus* (Thompson et al., 1987). The acetyltransferase encoded by the *bar* gene detoxifies the toxic peptide bialaphos and the functional amino acid analogue of the toxin,

Table 1. Antibiotics or fungicides and resistance genes used for DNA-mediated transformation of phytopathogenic fungi

Antibiotic fungicide	Mode of action	Resistance gene	Mechanism or resistance	Origin
Benomyl	depolymerization of the cytoskeleton	*bml*	mutated β-tubulin gene	*N. crassa* (Orbach et al., 1986)
Bialaphos	inhibition of glutamine synthetase	*bar*	acetylation	*S. hygroscopicus* (Thompson et al., 1987)
Hygromycin B	ribosomal function	*hyg B*	phosphorylation	*E. coli* (Gritz and Davies, 1983)
Kanamycin G418	ribosomal function	*APH11*	phosphorylation	*E. coli* (Berg et al., 1975)
Oligomycin	inhibition of ATP synthase	*oli C*	mutated gene for subunit 9 in the ATP synthase	*A. nidulans* (Ward et al., 1986)
Phleomycin, bleomycin	scission of DNA	*ble*	unknown	*E. coli* (Gatignol et al., 1987)

Table 2. Transformed phytopathogenic fungi

Transformed fungus	Selection marker	Reference
Botryotina squamosa	*hyg B*	Huang et al., 1989
Claviceps purpurea	*ble*	van Engelenburg et al., 1989
Cochliobolus heterostrophus	*amd S*	Turgeon et al., 1985
	hyg B	Turgeon et al., 1987
	bar	Yoder, pers. comm.
	uid A	Mönke and Schäfer, unpubl.
Colletotrichum graminicola	*bml*	Panaccione et al., 1988
C. lindemuthianum	*amd S, hyg B*	Rodriguez and Yoder, 1987
	nia D	Daboussi et al., 1989
C. trifolii	*bml, hyg B*	Dickman, 1988
Cryphonectria parasitica	*APH II, bml, hyg B*	Churchill et al., 1990
Fulvia fulva	*hyg B*	Oliver et al., 1987
Fusarium oxysporum	*hyg B*	Kistler and Benny, 1988
	nia D	Malardier et al., 1989
F. sambucinum	*hyg B*	Salch and Beremand, 1988
F. solani f. sp. *phaseoli*	*APH II*	Market et al., 1989
Gaeumannomyces graminis	*bml*	Henson et al., 1988
Leptosphaeria maculans	*hyg B*	Farman and Oliver, 1987
Magnaporthe grisea	*Arg B*	Parsons et al., 1987
	nia D	Daboussi et al., 1989
	hyg B	Leung et al., 1990
Nectria haematococca	*hyg B*	Soliday et al., 1989
	nia D	Daboussi et al., 1990
Pseudocercosporella herpotrichoides	*bml, hyg B*	Blakemore et al., 1989
Septoria nodorum	*hyg B*	Cooley et al., 1988
Ustilago hordei	*hyg B*	Holden et al., 1988
U. maydis	*hyg B*	Wang et al., 1988
U. nigra	*hyg B*	Holden et al., 1988
U. violacea	*hyg B*	Bej and Perlin, 1988

phosphinotricin. The causal agent of ergot of cereals and grasses, *Claviceps purpurea*, can be transformed by using the Tn 5 bleomycin resistance gene (*ble*) (van Engelenburg et al., 1989). The gene product of *ble* inactivates the antibiotic phleomycin, which causes scission of DNA (Gatignol et al., 1987).

Besides the dominant selection marker systems described above, additional markers should be considered for developing gene transfer systems for phytopathogenic fungi. For example, resistance against the inhibitor of mitochondrial ATP synthase, oligomycin, can be inherited both in a nuclear and extranuclear fashion in *A. nidulans* (Rowlands and Turner, 1973). The nuclear resistance locus, *oli C*, coding for the antibiotic resistant subunit 9 of ATP synthase, has been used already for transformation of *A. nidulans* (Ward et al., 1986).

All dominant selection markers based on antibiotic or fungicide resistance have one common disadvantage. The transformation rate using dominant markers is usually lower in comparison to complementation of auxotrophic mutants. On the other hand auxotrophic mutants are often not available or not as pathogenic as the wildtype. The development of a selection system without disturbance of the transformed cells by an antibiotic or fungicide would be helpful. Promising results were obtained using the *uid A* gene of *E. coli* encoding β-glucuronidase (Jefferson et al., 1987) for the transformation of *Cochliobolus heterostrophus* (Mönke and Schäfer, unpubl. results). Colonies transformed with *uid A* were stained blue in the presence of the substrate X-Gluc, thus transformants could be detected visually. The transformation efficiency was at least as high as reported for other selection systems (Turgeon et al., 1985, 1987).

III. Gene Cloning Procedures

In the following we will give a brief description of some methods which were successfully applied in cloning genes from fungi.

Three main approaches have been followed: differential hybridization, functional complementation, and cloning after protein purification. The first two procedures have in common that two different strains have to be available, which differ in the desired gene. This is in many cases not at hand. The third procedure requires a known gene product, e.g., an enzyme, and often an extensive biochemical purification.

A. Cloning by Differential Hybridization

A very sensitive method for cloning by differential hybridization is the cascade hybridization by which Timberlake (1980) showed developmental gene regulation and subsequently cloned developmentally regulated genes in *Aspergillus nidulans* (Zimmermann et al., 1980). In general this method consists in the hybridization of two mRNA populations, one of which was transcribed into cDNA, against each other. This is followed by the isolation of the non-hybridizing, therefore unique (i.e., developmentally specific) mRNAs. These mRNAs can now be cloned, and/or used as a probe to detect the genomic clones in a library of genomic DNA.

In the last ten years this method has been widely used, varied and improved to high sensitivity. For example, single stranded cDNAs of a wild type strain have been biotinylated, hybridized against the single stranded cDNAs of a mutant strain and the unspecific, hybridizing cDNAs removed by binding to avidin (Duguid et al., 1988).

An alternative for the enrichment of cDNA is the differential screening of genomic DNA (John and Davis, 1979). Labelled total RNA (Mangiarotti et al., 1981) or cDNAs (Rinaldy et al., 1988) which contain the desired gene (i.e., mutant or induced strain), are hybridized to a genomic library from the mutant/induced strain. An excess of unlabelled wild type RNA serves as a competitor and allows the detection of the genomic clone of interest.

B. Cloning by Functional Complementation

This procedure requires the transformation of a genomic library into a recipient strain, which lacks a functional version of the gene to be cloned. A main requirement is therefore an efficient transformation system (see above) and the dominance of the desired gene. The resultant transformants are then screened for the new phenotype.

This approach has been successful in several important cases, for example in the cloning of the mating type genes of *Cochliobolus heterostrophus* (Yoder et al., 1989), the *b* locus of *Ustilago maydis* (Kronstad and Leong, 1989; see Banuett and Herskowitz, 1992), and the PDA gene of *Nectria haematococca* (Weltring et al., 1988) (see below).

C. Cloning After Protein Purification

Highly purified proteins can be sequenced or used for raising specific antibodies. If sequence information is available, oligonucleotides can be synthesized and used as radioactive probes to screen genomic or cDNA libraries. For example, the pectin lyase D gene from *Aspergillus niger* was isolated by this method (Gysler et al., 1990).

Alternatively, antibodies can be raised against the purified protein and used to screen an expression library (Young and Davis, 1983). Using this method the pectate lyase A gene of *Aspergillus nidulans* was cloned (Dean and Timberlake, 1989).

IV. Pathogenicity of *Ustilago maydis*

The cloning and characterization of the *b* locus of *Ustilago maydis* is a major break-through in the field of plant pathology. This basidomycete is pathogenic to corn and forms smut galls on leaves, stems and flowering parts of infected plants (for a detailed description of the life cycle of *U. maydis* and related species and the cloning procedure of the *b* locus the reader is referred to Banuett and Herskowitz, 1992). Different alleles at both mating type loci (*a* and *b*) are needed for filamentous growth, but only different b-alleles are necessary for tumor induction and teliospore formation in the galls (Banuett and Herskowitz, 1989). Thus the *b* locus can be considered to control at least these final steps in the life cycle of *U. maydis.*

After establishing an efficient transformation system four different b-alleles which all code for proteins have been cloned and sequenced (Wang et al., 1988; Kronstad and Leong, 1989; Schulz et al., 1990). It has been proposed that interactions between different b-proteins generate regulators for the pathogenic pathway (Schulz et al., 1990). So for the first time it is now possible to address the question of the molecular basis for the mechanism of recognition between different incompatibility loci in a basidiomycete.

Furthermore by using a combination of genetic and molecular approaches one can begin to identify genes which are controlled by the *b* locus and ask for their function in the complex process of gall formation and teliospore production.

The question remains how the *b* locus is involved in pathogenicity. It is long known that haploid strains of *U. maydis* are able to infect their host. But they are never able to induce galls and teliospore formation and the amount of haploid growth in the maize plant depends on the isolate of *U. maydis* involved. The infection with haploid strains results in curling and distortion of colonized leaves, an increased ethylene production, as well as reduced elongation of leaves and shoots of young maize plants (Hanna, 1929; Munnecke, 1948; Rowell and DeVay, 1954; Andrews et al., 1981). If a haploid strain is able to colonize the whole plant or if its growth is more restricted is unknown. But it seems clear that even the haploid form of *U. maydis* can successfully invade its host. There must be a set of genes, yet unknown, which make this basic pathogenicity possible. Now that the b-alleles are cloned even this basic infection process might be addressed in a new light. It should be feasible to evaluate if these haploid infectious strains exhibit a different regulation of the *b* locus compared with less pathogenic lines or if this basic pathogenicity is unlinked to the process regulated by the *b* locus.

V. Fungal Penetration of the Plant Cell

Fungal pathogens have to penetrate their hosts. They can do so by growing through naturally occurring openings such as wounds or stomata. Others can penetrate directly through the intact plant surface. There is an old debate among plant pathologists whether this break-through of the surface barrier is achieved by strictly mechanical forces or by enzymatic degradation. As in many cases it seems now that both sides are right. For each model convincing lines of evidence have accumulated over the years though it seems that the final proof for each of them is still lacking. In the following we will present a well investigated example for mechanical penetration, *Magnaporthe grisea*; and an example for enzymatic degradation, *Nectria haematococca*, where the gene involved has already been cloned.

A. Mechanical Penetration

The fungus *Magnaporthe grisea* causes rice blast disease and is a widespread pathogen of a variety of cereals and grasses. This highly damaging rice pathogen has been investigated in a number of different laboratories, and up to now, several major tasks have been accomplished. Today, it is possible to induce the sexual cycle of *M. grisea* under laboratory conditions, though it has never been found in nature; there are a number of genetic markers available, which make a detailed genetic analysis possible (Leung and Taga, 1988), and to complete the picture, a transformation system has been established (Parsons et al., 1987). All this has made it possible to identify several avirulence genes (see de Wit, 1992).

By studying the pathogenicity of *M. grisea*, a potential role of melanin was suggested (Bell and Wheeler, 1986). The distinctive grey pigment of *M. grisea* is produced by polymerization of 1,8-dihydroxynaphthalene (DHN). Several lines of evidence support the idea that DHN melanin is necessary for pathogenicity of *M. grisea*. First, the rice blast fungicide tricyclazole (5-methyl-1,2,4 triazolo [3,4-b] benzothiazole; Froyd et al., 1976) is not toxic to *M. grisea* but inhibits DHN melanin biosynthesis (Wheeler and Greenblatt, 1988). Second, DHN melanin is produced most abundantly just before penetration when a thick melanin layer is deposited in the inner appressorial cell wall. Third, it mediates the build-up of a high hydrostatic pressure in the appressorium which makes it possible for the fungus to penetrate even artificial surfaces such as polyvinylchloride (Howard and Ferrari, 1989). Fourth, three classes of DHN melanin-deficient mutants of *M. grisea* have been characterized (Chumley and Valent, 1990). The three mutant phenotypes are due to single gene defects at unlinked loci. All mutants are nonpathogenic to the intact host plant, but regain pathogenicity when the plant is wounded. All this strongly indicates that DHN melanin is a pathogenicity factor and that *M. grisea* gains entry into the epidermal plant cell by mechanical force. With the transformation system at hand it is possible now to clone the genes involved by complementation of the DHN melanin mutants and get new insights how the biosynthesis of DHN melanin is regulated and involved in pathogenicity.

Still, some important questions remain. Pigment deficient mutants of *Cochliobolus miyabeanus* are still fully pathogenic to rice plants (Kubo et al., 1989). The same holds true for albino mutants of *Cochliobolus heterostrophus*, which are fully pathogenic under standard conditions in the growth chamber. Furthermore it remains unclear whether other enzymes, e.g., cutinase, are also important for the successful penetration. A cutinase deficient transformant of *M. grisea*, obtained after gene disruption, should answer this.

B. Enzymatic Digestion

The aerial surfaces of higher plants are covered with the cuticle. The structure of the cuticle varies greatly from one plant to the other. The outer layer of epidermal cells consists of a pectic, a cutin and a wax layer. The wax layer is concentrated on the outer side, whereas pectin is deposited at the inner side of the cutin layer. The thickness and composition of the cuticle changes during plant development and depends on environmental factors, such as humidity, light, temperature (Van den Ende and Linskens, 1974).

Though the enzymatic degradation of the epicuticular wax layer by a fungus is far from being understood, much attention has been given to the cutin degrading enzyme, cutinase. Cutinolytic enzyme activity was first described for *Penicillium spinulosum* (Heinen, 1960), an ubiquitous saprophyte of common occurrence in soils, decaying vegetation and foods (Pitt, 1979). Today it is known that cutinase is produced by a variety of fungi (Kolattukudy, 1981) and bacteria (Heinen and de Vries, 1966) but most is known about a cutinase from *Nectria haematococca* mating population VI (anamorph: *Fusarium solani* f. sp. *pisi*). *N. haematococca* causes the much studied root and lower stem disease of pea and, in addition, is pathogenic to at least nine different plant species (VanEtten and Kistler, 1988).

Cutinase activity of *N. haematococca* is repressed by glucose and highly induced by cutin monomers (Lin and Kolattukudy, 1978). Several experiments made the essential role of cutinase during the penetration step of *N. haematococca* evident. First, cutinase is present at the site of penetration (Shaykh et al., 1977). Second, a mutant with a greatly reduced cutinase activity could not infect unwounded pea stems unless exogenous cutinase was added (Dantzig et al., 1986). Similar results were obtained with cutinase deficient laboratory mutants of *Colletotrichum gloeosporioides*. The mutants could not infect intact papaya fruits, but showed full virulence on wounded tissue (Dickman and Patil, 1986). Fourth, antibodies against cutinase in the inoculum prevent infection of intact but not of wounded plants (Maiti and Kolattukudy, 1979). Fifth, specific chemical inhibition of the enzyme prevented infection of pea stems, though the growth rate of the fungus was not effected (Köller et al., 1982a, b).

These results are substantiated by data obtained from transformation experiments with the cutinase gene in a different fungal species. An almost full-length cDNA clone was isolated from an induced culture of *N. haematococca* (Soliday et al., 1984) and subsequently a genomic clone (Soliday et al., 1989) was isolated. Transformation of this cutinase gene from *N. haematococca* into the wound pathogen of the papaya fruit, *Mycosphaerella* spp., enabled the former wound pathogen to infect its host without a wound. In addition the infection could be prevented by antibodies against *N. haematococca* cutinase and the infectivity of the transformants reflected the levels of cutinase transcripts, amount of protein and enzymatic activity (Dickman et al., 1989). This heterologous gene transfer demonstrated unequivocally that cutinase is necessary for *Mycosphaerella* spp. to penetrate through the intact cuticle. These results were made possible by two important factors: First, *Mycosphaerella* spp. possesses the basic set of genes for pathogenicity and readily infects wounded tissue. Second, on the plant side, the major component of the papaya cutin is identical with the most effective inducer of the cutinase gene in *N. haematococca* (Dickman et al., 1989). This allowed the full inducibility of the cutinase gene in the transformants.

The inducibility of the cutinase gene was further elucidated by activation of the gene in isolated *N. haematococca* nuclei by plant cutin monomers. This selective initiation of transcription required in addition to the cutin monomers a protein factor from the fungal extract (Podila et al., 1988). The mode of action of these inducing factors and the DNA sequences regulating the cutinase gene remain up to now unknown.

From these data the following picture emerges: *N. haematococca* spores secrete a low cutinase activity. This low enzyme activity is enough to generate cutin monomers which in turn drastically induce the cutinase gene, thus enabling the fungus to invade the host plant. But of course this picture is far from being complete. The penetration through the epicuticular wax layer remains unexplained as well as the nature of other enzymes necessary for successful infection. A first step in this direction is the cloning of a pectate lyase from *N. haematococca*. Following the approach taken in the cutinase cloning, the fungus was grown on pectin, and a pectate lyase released into the medium, purified to homogeneity. Specific antibodies against this protein were prepared and used to precipitate a precursor polypeptide from in vitro translated mRNA. The pectate lyase antibodies were found to inhibit the ability of the fungus to infect pea stems (Crawford and Kolattukudy, 1987). The gene corresponding to the cDNA was cloned but the plant signals that trigger the expression of the gene and the mechanism of this activation remain obscure (Kolattukudy et al., 1989).

Obviously the test system used for the evaluation of a plant fungus interaction is of central importance for the observed outcome of this relationship. So it is not surprising that with different test systems different results for the same plant-fungus interaction are obtained. For example, *N. haematococca* isolate T30 has been reported to be low in cutinase and weakly virulent on intact, but highly virulent on wounded pea stems. This was shown by infecting sections of epicotyl, 1.5 cm in length, from 6–8 days old etiolated seedlings of pea. They were washed with distilled water and randomly placed on wet filter paper in petri dishes, where the conidial suspension (5 µl; 10^6 conidia) was dropped either on the intact or the needle pricked stem sections. After 72 h incubation in the dark at 22 °C virulence was expressed as the percentage of infected segments showing brown lesions (Köller et al., 1982b). Performing the virulence bioassay with whole plants (Kistler and VanEtten, 1984) instead of pea stem sections gave different results. Fourteen days after the pea stems were inoculated with a plug of mycelium the length of the developing lesions were measured. Lesions produced on needle pricked plants were uniform and reproducible.

Lesion length of unwounded plants were highly variable, and there was no clear difference between the high and low cutinase producer strains, T8 and T30, respectively. Both strains are highly virulent on wounded plants and show a comparable reduced virulence on unwounded pea plants (Stahl and Schäfer, unpubl. results). The variability in virulence on stems not artificially wounded, ranging from no lesions at all to lesions comparable to wounded plants, may be due to naturally occurring wounds. During rapid growth processes, i.e., the elongation of stems, some cells may be proliferating and elongating much faster than their neighbors. This may cause tissue tensions, which in turn may lead to internal wounds by tearing. Taking all of this into consideration, one can say that the high cutinase producing *N. haematococca* strain T8 exhibits a much slower virulence on intact pea stems compared to needle pricked stems and that this pathogenic behavior is comparable to strain T30 with its low cutinase activity.

Looking at the organ specificity of *N. haematococca* the picture gets even more complicated. Many pathogenic fungi show organ specificity to a certain extent. *N. haematococca* infects roots rapidly, stems more slowly and leaves and pods of intact pea plants very poorly (Schäfer et al., unpubl. results). This is in accordance with the observation that *N. haematococca* is a powerful pathogen of the roots and the stem base in the field (Nelson et al., 1981). Keeping in mind that cutinase is regarded as an important pathogenicity factor necessary for penetrating the plant, these results are difficult to explain. The epidermal cutin layer typical for the aerial parts of a plant is not present in pea roots. Instead suberinized wax layers are found in the endodermis and exodermis of roots. Although these layers have been examined histochemically and ultrastructurally, most have not been analysed chemically. Where chemical analyses have been performed, the layers are distinct from, but related to epidermal cuticles (Willison, 1981). If and how cutinase can degrade these layers is unclear. Why *N. haematococca* expressing high cutinase activity can hardly infect the upper stem, leaves and pods of a pea plant awaits further elucidation. It seems that the role of cutinase during the natural infection process of *N. haematococca* still has to be determined.

VI. The PDA Gene, a Gene for Pathogenicity and Host Range

A. The PDA Gene and Nectria haematococca

One hypothesis concerning the cause of plant disease is that the successful pathogen can prevent or obviate the plant defense. The molecular mechanism of the interactions between defender and invader are poorly understood. A plant may defend itself with physical barriers like the cuticle mentioned above, or with preformed toxic chemicals stored in the plant, e.g., in the vacuoles. Among the most important plant defense mechanisms is the synthesis of phytoalexins, i.e., antibiotics actively synthesized in response to a fungal attack.

Sorghum produces a phytoalexin (3-deoxyanthocyanidin flavonoid) which appears to be synthesized in subcellular inclusions within a host epidermal cell that is about to be penetrated by a fungus. This site-restricted synthesis suggests that in sorghum the phytoalexin response occurs initially in the first cells that come under fungal attack and is not simply a response of cells that surround the original infection site (Snyder and Nicholson, 1990). Whether or not 3-deoxyanthocyanidin flavonoid restricts fungal growth and is responsible for the plant's resistance reaction remains to be seen.

The ability of a fungus to detoxify the plant's phytoalexin may be one way for successful plant colonization by fungi. Pisatin is a phytoalexin produced by pea in response to stresses such as microbial invasion. Field isolates of *N. haematococca* vary greatly in their ability to colonize pea. Virulent isolates appear to use a detoxification strategy to circumvent the plant's defense line. All pathogenic isolates are able to detoxify pisatin by demethylation, forming thereby a nontoxic compound, strongly implying that a certain level of demethylating activity is required by *N. haematococca* for pathogenicity on pea (VanEtten et al., 1980). This detoxifying activity is mediated by a pisatin inducible cytochrome P 450 monooxygenase called pisatin demethylase (pda) (VanEtten et al., 1989). A 3.35 kb DNA fragment, harboring a gene (PDA) that controls the production of pda in *N. haematococca* was cloned. This was accomplished by transforming *Aspergillus nidulans*, which does not produce pda, with a cosmid library of DNA fragments from a rapidly demethylating and therefore highly virulent strain of *N. haematococca*. The resulting transformants of *A. nidulans* were tested for their ability to demethylate pisatin. The gene was subcloned and shown to encode a single transcript (Weltring et al., 1988). To assess the role of phytoalexin detoxification during the infection process of *N. haematococca* on pea, the cloned PDA gene was transformed into a nonpathogenic isolate of *N. haematococca* which lacks pda activity. Transformants produced pda and were pathogenic to pea, compared with the untransformed strain (Ciuffetti et al., 1988). This proved that pda is required for pathogenicity of *N. haematococca* on pea. Furthermore this result demonstrated that the production of pisatin in response to fungal attack is indeed a plant defense reaction.

B. The PDA Gene and Cochliobolus heterostrophus

Cochliobolus heterostrophus (anamorph: *Helminthosporium maydis* Nisik. and *Bipolaris maydis* (Nisik.) Shoemaker) is a filamentous ascomycete and pathogenic to corn but not pathogenic to pea (Yoder, 1988). To investigate the possibility that pda can provide pathogenicity to fungi not normally pathogenic to pea, the PDA gene was transformed into *C. heterostrophus* (Schäfer et al., 1989). For this purpose an albino strain of *C. heterostrophus* was chosen because it is nonpathogenic in the field (Fry et al., 1984) and shows reduced epidemiological fitness in the greenhouse (Leonard, 1977) although the strain is normally pathogenic to corn in the growth chamber. Albino strains are very sensitive to ultraviolet radiation, which may explain their inability to survive in nature. In addition, since the sexual stage of *C. heterostrophus* has never been found in nature and there is an abundance of heterokaryon incompatibility alleles in the field population (Leach and Yoder, 1983), it is highly unlikely that recombinant DNA molecules could be

Fig. 1. Symptoms on unexcised stems of pea seedlings caused by *Nectria haematococca* and transformed *Cochliobolus heterostrophus*. Cylinders (4 mm diameter) of agar medium bearing mycelium were placed on pin-prick wounds, and the plants were incubated 6 days in a moist chamber. From left to right: *C. heterostrophus* (control), transformed *C. heterostrophus* (high pda), *N. haematococca* (high pda). Arrowheads delimit the extent of each lesion

passed on to other organisms in the event a strain escaped and survived briefly in the field. The albino mutant of *C. heterostrophus* provides therefore an excellent biological containment for recombinant DNA molecules. After the PDA gene was transformed into *C. heterostrophus*, a transformant was selected which produced twice the pda activity as the highly virulent strain of *N. haematococca* where the gene originated. The untransformed albino strain of *C. heterostrophus* showed little or no pda activity. Though the PDA gene was highly expressed in *C. heterostrophus* it was differently regulated than in wild type *N. haematococca*. The PDA gene in *N. haematococca* was inducible by pisatin and repressible by glucose but was constitutively expressed in *C. heterostrophus*. This suggests that either the 3.35 kb DNA fragment carrying the PDA gene lacks some cis acting regulatory sequences or that *C. heterostrophus* lacks the corresponding trans acting factors. To evaluate the significance of high pda activity for pathogenicity virulence on pea plants was assessed using an assay that was developed to measure virulence of different *N. haematococca* isolates (Kistler and VanEtten, 1984). The pda producing *C. heterostrophus* caused a stem lesion clearly larger than those produced by the control strain (Fig. 1). Thus high pda activity alone allowed *C. heterostrophus* to cause significant damage to pea stems although the lesions were only about 30% of the size of *N. haematococca* lesions (Fig. 1). It is clear that *N. haematococca* has additional factors besides pda activity important for virulence on pea that are lacking in *C. heterostrophus*.

C. The PDA Gene and Aspergillus nidulans

As it became clear that the expression of a single gene can expand the host range of a transgenic pathogenic fungus the effect of the PDA gene on the saprophyte *Aspergillus nidulans* was investigated. A mutant strain was chosen that provided a biological containment for recombinant DNA (Schäfer et al., 1989). A transformant of *A. nidulans* was selected that produced as much pda as wild type *N. haematococca*. In the transformant the pda activity was constitutively expressed and not inducible by pisatin. Glucose repressed pda in the transformant, as it does in *N. haematococca*. Assessed for pathogenicity on pea plants, a control strain and the pda producing *A. nidulans* caused no detectable symptoms on pea (Fig. 2). This result indicated that the high production of pda was not sufficient to alter the phenotype of the saprophyte *A. nidulans* on pea. As pda alone can not confer pathogenicity to this saprophyte, the pathogenic fungus *C. heterostrophus* seems to have a basic set of pathogenicity genes, at present unknown, that *A. nidulans* lacks. These results and others (VanEtten et al., 1989) demonstrated that for high virulence of a fungal pathogen on pea several gene products are required only one of which is pda.

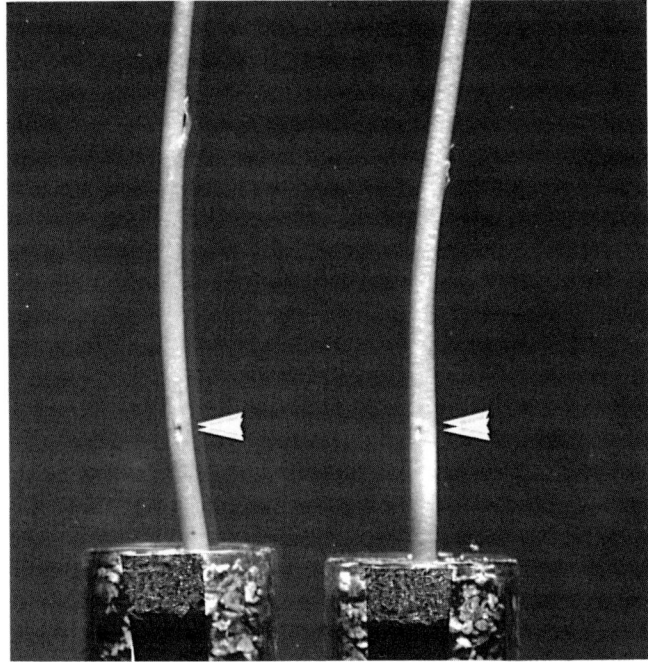

Fig. 2. Symptoms on unexcised stems of pea seedlings caused by untransformed and transformed *Aspergillus nidulans*. Cylinders (4 mm diameter) of agar medium bearing mycelium were placed on pin-prick wounds, and the plants were incubated 6 days in a moist chamber. From left to right: *A. nidulans* (control), transformed *A. nidulans* (high pda). Arrowheads delimit the extent of each lesion

D. The PDA Gene and Organ Specificity

Most fungal pathogens display a specificity for the plant organs they invade successfully. *N. haematococca* and *C. heterostrophus* display different organ specificities in their respective hosts. *N. haematococca* is a pathogen of the roots and the basal stem of pea whereas *C. heterostrophus* readily colonizes the leaves but not the roots of maize. A pda producing *C. heterostrophus* transformant was tested for its capacity to infect pea roots and leaves. When *N. haematococca* and *C. heterostrophus* strains producing similarly high amounts of pda were compared on pea plants, *N. haematococca* caused root rot but was not infectious on leaves. The pda producing *C. heterostrophus* transformant caused no lesions at all on roots but clearly produced lesions on pea leaves. Thus, *C. heterostrophus* retained its organ specificity even when expressing a foreign pathogenicity gene and colonizing a pea plant not normally its host (Schäfer and Yoder, unpubl. results).

VII. Some Aspects of Further Development in Plant Pathology

A. Studying Fungal Avirulence

There is convincing evidence that the first fungal gene responsible for avirulence has been cloned (see de Wit, 1992). The cloning and further characterization of the A9 gene of *Fulvia fulva* (anamorph: *Cladosporium fulvum*) makes it now possible to study the molecular basis for fungal avirulence in a gene for gene relationship.

B. Transposable Elements in Pathogenic Fungi

It is well known that fungal pathogens can rapidly evolve new virulent races and overcome the resistance of previously unsusceptible cultivars. High mutation rates to virulence have been observed (Dinoor et al., 1988). A possible explanation is the existence of transposable elements, known to play a role in pathogenicity of *Xanthomonas campestris* vs. *vesicatoria* (Kearney et al., 1988). Until recently there was no evidence for transposable elements in pathogenic fungi. Now it could be shown that *Fulvia fulva*, a fungal pathogen of tomato, expresses genes for a fungal reverse transcriptase (McHale et al., 1989). This enzyme is required for the replication of retroviruses and retroviral-like transposable elements. The sequence of a 225 bp clone revealed a striking similarity to reverse transcriptase genes of LTR retrotransposons of the Gypsy class. The sequence was present in 50–100 copies in the genome. Several full length copies have now been isolated and analysed. Long terminal repeats, a putative target site duplication, a terminal inverted duplication, and putative first and second strand primer binding sites were identified. Sucrose gradient fractionation of fungal homogenates showed copurification of reverse transcriptase activity, RNA homologous to the retroelements, and virus-like particles. These results suggest that the transposable element is actively expressed and packaged into virus-like particles. The occurrence of retroviruses could explain the variability of fungal pathogens. Transposons may inactivate genes or may provide dispersed regions of homology in the genome, promoting rearrangements. Knowing their structure and mode of action could help to overcome the difficulties working with genetically uncharacterized organism (Oliver, pers. comm.).

C. Establishing a Model System

To understand the nature of plant–pathogen interactions it is necessary to isolate and characterize the involved genes from the pathogen as well as from the infected plant. Most systems where a plant–pathogen relation is studied have the major disadvantage that molecular studies on one side of the interaction are difficult or even impossible. Obligate parasites and many host plants are difficult to transform and have a complex genome, hindering the molecular approach. The crucifer *Arabidopsis thaliana* has become a model organism in plant molecular biology (Meyerowitz, 1989). Classical genetic studies are facilitated by its short generation time and high fertility. A molecular approach is feasible because of its small genome, the availability of defined mutants, a genetic map, and a transformation system. Hence this plant appears to be well suited for studies of pathogenicity of both bacteria (Simpson and Johnson, 1990) and fungi (Koch and Slusarenko, 1990). *Xanthomonas campestris* pv. *campestris* and *Peronospora parasitica* were recognized as pathogens of *A. thaliana*. Though *P. parasitica*, a downy mildew fungus, is an obligate parasite and will certainly be difficult to study, for the plant side things look more promising. Besides the highly susceptible *Arabidopsis* strain Weiningen, a resistant strain RLD was found (Koch and Slusarenko, 1990). This opens up the possibility of cloning resistance determinants from a host plant against a fungal pathogen. As other fungi are already described as being pathogens of *Arabidopsis* (Koch and Slusarenko, 1990), it is likely that fungal pathogens more amenable to molecular biology can be studied and a model system to study the basic molecular events underlying the plant–pathogen interactions will be established.

VIII. References

Andrews EC, Hanowski C, Beiderbeck R (1981) Wachtumsveränderungen an Maiskeimlingen nach Befall mit haploiden Linien des Maisbrandes (*Ustilago maydis*). Phytopathol Z 102: 10–20

Armaleo D, Ye G-N, Klein TM, Shark KB, Sanford JC, Johnston SA (1990) Biolistic nuclear transformation of *Saccharomyces cerevisiae* and other fungi. Curr Genet 17: 97–103

Avalos J, Geever RF, Case ME (1989) Bialaphos resistance as a dominant selectable marker in *Neurospora crassa*. Curr Genet 16: 369–372

Banuett F, Herskowitz L (1989) Different a-alleles of *Ustilago maydis* are necessary for maintenance of filamentous growth but not for meiosis. Proc Natl Acad Sci USA 86: 5878–5882

Bannuett F, Herskowitz I (1992) Pathogenicity determinants in the smut fungi of cereals. In: Boller T, Meins F (eds) Genes involved in plant defense. Springer, Wien New York, pp 115–128 [Dennis ES et al (eds) Plant gene research. Basic knowledge and application]

Bej AK, Perlin M (1988) Apparent transformation and maintenance in basidiomycete mitochondria of a plasmid bearing the hygromycinB gene. Genome 19 [Suppl 1]: 300

Bell AA, Wheeler MH (1986) Biosynthesis and functions of fungal melanins. Annu Rev Phytopathol 24: 411–451

Berg DE, Davies J, Allet B, Rochaix J-D (1975) Transposition of R factor genes to bacteriophage lambda. Proc Natl Acad Sci USA 72: 3628–3632

Blakemore EJA, Dobson MJ, Hocart MJ, Lucas JA, Peberdy JF (1989) Transformation of *Pseudocercosporella herpotrichoides* using two heterologous genes. Curr Genet 16: 177–180

Case ME, Schweizwer M, Kushner SR, Giles NH (1979) Efficient tranformation of *Neurospora crassa* by utilizing hybrid plasmid DNA. Proc Natl Acad Sci USA 76: 5259–5263

Chumley FG, Valent B (1990) Genetic analysis of melanin deficient, nonpathogenic mutants of *Magnaporthe grisea*. Mol Plant Microbe Interact 3: 135–143

Churchill ACL, Ciuffetti LM, Hansen DR, Van Etten HD, Van Alfen NK (1990) Transformation of the fungal pathogen *Cryphonectria parasitica* with a variety of heterologous plasmids. Curr Genet 17: 25–31

Ciuffetti LM, Weltring KM, Turgeon BG, Yoder OC, VanEtten HD (1988) Tranformation of *Nectria haematococca* with a gene for pisatin demethylating activity, and the role of pisatin detoxification in virulence. J Cell Biochem 12C: 278

Cooley RN, Shaw RK, Franklin FCH, Caten CE (1988) Transformation of the phytopathogenic fungus *Septoria nodorum* to hygromycinB resistance. Curr Genet 13: 383–389

Crawford MS, Kolattukudy PE (1987) Pectate lyase from *Fusarium solani* f. sp. *pisi*: purification, characterization, in vitro translation of the mRNA, and involvement in pathogenicity. Arch Biochem Biophys 258: 196–205

Daboussi MJ, Djeballi A, Gerlinger A, Blaiseau PL, Bouvier I, Cassan M, Lebrun MH, Parisot D, Brygoo Y (1989) Transformation of seven species of filamentous fungi using the nitrate reductase gene of *Aspergillus nidulans*. Curr Genet 15: 453–456

Dantzig AH, Zuckerman SH, Andonov-Roland MM (1986) Isolation of a *Fusarium solani* mutant reduced in cutinase activity and virulence. J Bacteriol 168: 911–916

Dean AR, Timberlake WE (1989) Regulation of the *Aspergillus nidulans* pectate lyase gene (pelA). Plant Cell 1: 275–284

De Wit PJGM (1992) Functional models to explain gene-for-gene relationships in plant–pathogen interactions. In: Boller T, Meins F (eds) Genes involved in plant defense. Springer, Wien New York, pp 25–47 [Dennis ES et al (eds) Plant gene research. Basic knowledge and application]

Dickman MB (1988) Whole cell transformation of the alfalfa pathogen *Colletotrichum trifolii*. Curr Genet 14: 241–246

Dickman MB, Patil SS (1986) Cutinase deficient mutants of *Colletotricum gloeosporioides* are nonpathogenic to papaya fruit. Physiol Mol Plant Pathol 28: 235–242

Dickman MB, Podila GK, Kolattukudy PE (1989) Insertion of cutinase gene into a wound pathogen enables it to infect intact host. Nature 342: 446–448

Dinoor A, Eshed N, Nof E (1988) *Puccinia coronata*, crown rust of oats and grasses. In: Sidhu GS (ed) Genetics of plant pathogenic fungi. Academic Press, London, pp 333–344 (Advances in plant pathology, vol 6)

Duguid JR, Rohwer RG, Seed B (1988) Isolation of cDNAs of scrapie-modulated RNAs by subtractive hybridisation of a cDNA library. Proc Natl Acad Sci USA 85: 5738–5742

Farman ML, Oliver RP (1987) The transformation of protoplasts of *Leptoshaeria maculans* to hygromycin B resistance. Curr Genet 13: 327–330

Fincham JR (1989) Transformation in fungi. Microbiol Rev 53: 148–170

Froyd JD, Paget CJ, Guse LR, Dreikorn BA, Pafford JL (1976) A new systemic fungicide for control of *Pyricularia oryzae* on rice. Phytopathology 66: 1135–1139

Fry WE, Yoder OC, Apple AE (1984) Influence of naturally occurring marker genes on the ability of *Cochliobolus heterostrophus* to induce field epidemics of southern corn leaf blight. Phytopathology 74: 175–178

Gatignol A, Baron M, Tiraby G (1987) Phleomycin resistance encoded by the *ble* gene from transposon Tn5 as a dominant selectable marker in *Saccharomyces cerevisiae*. Mol Gen Genet 207: 342–348

Goldmann GH, Van Montagu M, Herrera-Estrella A (1990) Transformation of *Trichoderma harzianum* by high voltage electric pulse. Curr Genet 17: 169–174

Gritz L, Davies J (1983) Plasmid-encoded hygromycin B resistance: the sequence of hygromycinB phosphotransferase gene and its expression in *Escherichia coli* and *Saccharomyces cerevisiae*. Gene 25: 179–188

Gysler C, Harmsen JAM, Kester HCM, Visser J, Heim J (1990) Isolation and structure of the pectin lyase D-encoding gene from *Aspergillus niger*. Gene 89: 101–108

Hanna WF (1929) Studies in the physiology and cytology of *Ustilago zeae* and *Sorosphorium reilianum*. Phytopathology 19: 415–442

Heinen W (1960) Über den enzymatischen Cutinabbau I. Mitteilung: Nachweis eines Cutinase-Systems. Acta Bot Neerl 9: 167–190

Heinen W, de Vries H (1966) Stages during the breakdown of plant cutin by soil micro-organisms. Arch Microbiol 54: 331–338

Henson JM, Blake NK, Pilgeram A (1988) Transformation of *Gaeumannomyces graminis* to benomyl resistance. Curr Genet 14: 113–117

Hinnen A, Hicks JB, Fink GR (1978) Transformation of yeast chimaeric ColE1 plasmid carrying LEU 2. Proc Natl Acad Sci USA 75: 1929–1933

Holden DW, Wang J, Leong SA (1988) DNA-mediated transformation of *Ustilago hordei* and *Ustilago nigra*. Physiol Mol Plant Pathol 33: 235–239

Howard RJ, Ferrari MA (1989) The role of melanin in appressorium function. Exp Mycol 13: 403–418

Huang D, Bhairi D, Staples RC (1989) A transformation procedure for *Botryotinia squamosa*. Curr Genet 15: 411–414

Hynes MJ, Corrick CM, King JA (1983) Isolation of genomic clones containing the amdS gene of *Aspergillus nidulans* and their use in the analysis of structural and regulatory mutations. Mol Cell Biol 3: 1430–1439

Ito H, Fukuda Y, Murata K, Kimura A (1983) Transformation of yeast cells treated with alkali cations. J Bacteriol 153: 163–168

Jefferson RA, Kavanagh TA, Bevan MW (1987) GUS fusions: β-glucuronidase as a sensitive and versatile gene fusion marker in higher plants. EMBO J 6: 3901–3907

John TP St, Davis RW (1979) Isolation of galactose-inducible DNA sequences from *Saccharomyces cerevisiae* by differential plaque filter hybridisation. Cell 16: 443–452

Kearney B, Ronald PC, Dahlbeck D, Staskawicz B (1988) Molecular basis for evasion of plant host defence in bacterial spot disease of pepper. Nature 322: 541–543

Kistler HC, Benny UK (1988) Genetic transformation of the fungal plant wilt pathogen, *Fusarium oxysporum*. Curr Genet 13: 145–149

Kistler HC, Van Etten HD (1984) Regulation of pisatin demethylation in *Nectria haematococca* and its influence on pisatin tolerance and virulence. J Gen Microbiol 130: 2605–2613

Koch E, Slusarenko A (1990) *Arabidopsis* is susceptible to infection by a downy mildew fungus. Plant Cell 2: 437–445

Köller W, Allan CR, Kolattukudy PE (1982a) Protection of *Pisum sativum* from *Fusarium solani* f. *pisi* by inhibition of cutinase with organophosphorus pesticides. Phytopathology 72: 1425–1430

Köller W, Allan CR, Kolattukudy PE (1982b) Role of cutinase and cell wall degrading enzymes in infection of *Pisum sativum* by *Fusarium solani* f. sp. *pisi*. Physiol Plant Pathol 20: 47–60

Kolattukudy PE (1981) Structure, biosynthesis and biodegradation of cutin and suberin. Annu Rev Plant Physiol 32: 539–567

Kolattukudy PE, Podila GK, Mohan R (1989) Molecular basis of the early events in plant-fungus interaction. Genome 31: 342–349

Kronstad JW, Leong SA (1989) Isolation of two alleles of the b locus of *Ustilago maydis*. Proc Natl Acad Sci USA 86: 978–982

Kubo Y, Tsuda M, Furasawa I, Shishiyama J (1989) Genetic analysis of genes involved of melanin biosynthesis of *Cochliobolus miyabeanus*. Exp Mycol 13: 77–84

Leach J, Yoder OC (1983) Heterokaryon incompatibility in the plant-pathogenic fungus *Cochliobols heterostrophus* J Hered 74: 149–152

Leonhard K (1977) Virulence, temperature optima, and competitive abilities of isolines of races T and O of *Bipolaris maydis*. Phytopathology 67: 1273–1279

Leung H, Taga M (1988) *Magnaporthe grisea* the blast fungus. In: Sidhu GS (ed) Genetics of plant pathogenic fungi. Academic Press, London, pp 175–188 (Advances in plant pathology, vol 6)

Leung H, Lehtinen U, Karjalainen R, Skinner D, Tooley P, Leong H, Ellingboe A (1990) Transformation of the rice blast fungus *Magnaporthe grisea* to hygromycinB tolerance. Curr Genet 17: 409–411

Lin TS, Kolattukudy PE (1978) Induction of a biopolyester hydrolase (cutinase) by low levels of cutin monomers in *Fusarium solani* f. *pisi*. J Bacteriol 133: 942–951

Malardier L, Daboussi MJ, Julien J, Roussel F, Scazzocchio C, Brygoo Y (1989) Cloning of the nitrate reductase gene (*niaD*) of *Aspergillus nidulans* and its use for transformation of *Fusarium oxysporum*. Gene 78: 147–156

Mangiarotti G, Chung S, Zuker C, Lodish HF (1981) Selection and analysis of cloned developmentally-regulated *Dictyostelium discoideum* genes by hybridization-competition. Nucleic Acids Res 9: 947–963

Marek ET, Schardl CL, Smith DA (1989) Molecular transformation of *Fusarium solani* with an antibiotic resistance marker having no fungal homology. Curr Genet 15: 421–428.

Maiti IB, Kolattukudy PE (1979) Prevention of fungal infection of plants by specific inhibition of cutinase. Science 205: 507–508

McHale MT, Roberts IN, Talbot NJ, Oliver RP (1989) Expression of reverse transcriptase genes in *Fulvia fulva*. Mol Plant Microbe Interact 2: 165–168

Meyerowitz EM (1989) *Arabidopsis*, a useful weed. Cell 56: 263–269

Munnecke DE (1948) An unusual host response to certain paired monosporidial cultures of *Ustilago zeae*. Phytopathology 38: 19

Nelson PE, Toussoun TA, Cook RJ (1981) *Fusarium* diseases, biology, and taxonomy. The Pennsylvania State University Press, University Park

Oliver RP, Roberts IN, Harling R, Kenyon L, Punt PJ, Dingemanse MA, van den Hondel CAMJJ (1987) Transformation of *Fulvia fulva*, a fungal pathogen of tomato, to hygromycin resistance. Curr Genet 12: 231–233

Orbach MJ, Porro EB, Yanofsky C (1986) Cloning and characterization of the gene for β-tubulin from a benomyl-resistant mutant of *Neurospora crassa* and its use as a dominant selection marker. Mol Cell Biol 6: 2452–2461

Panaccione DG, McKiernan M, Hanau RM (1988) *Colletotrichum graminicola* transformed with homologous and heterologous benomyl-resistance genes retains expected pathogenicity to corn. Mol Plant Microbe Interact 1: 113–120

Parsons KA, Chumley FG, Valent B (1987) Genetic transformation of the fungal pathogen responsible for rice blast disease. Proc Natl Acad Sci USA 84: 4161–4165

Pitt JI (1979) The genus *Penicillium* and its telemorphic states *Eupenicillium* and *Talaromyces*. Academic Press, London

Podila GK, Dickman MB, Kolattukudy PE (1988) Transcriptional activation of a cutinase gene in isolated fungal nuclei by plant cutin monomers. Science 242: 922–925

Punt PJ, Oliver RP, Dingemanse MA, Pouwels PH, Van den Hondel CAMJJ (1987) Transformation of *Aspergillus* based on the hygromycinB resistance marker from *Escherichia coli*. Gene 56: 117–124

Rambosek J, Leach J (1987) Recombinant DNA in filamentous fungi: progress and prospects. CRC Crit Rev Biotechnol 6: 357–393

Rinaldy AR, Dodson ML, Darling TL, Lloyd RS (1988) Gene cloning using cDNA libraries in a differential competition hybridization strategy: application to cloning XP-A related genes. DNA 7: 563–570

Rodriguez RJ, Yoder OC (1987) Selectable genes for transformation of the fungal pathogen *Gomerella cingulata* f. sp. *phaseoli* (*Colletotrichum lindemuthianum*). Gene 54: 73–81

Rowell JB, DeVay JE (1954) Genetics of *Ustilago zeae* in relation to basic problems of its pathogenicity. Phytopathology 44: 356–362

Rowlands RT, Turner G (1973) Nuclear and extranuclear inheritance of oligomycin resistance in *Aspergillus nidulans*. Mol Gen Genet 126: 201–216

Salch YP, Beremand MN (1988) Development of a transformation system for *Fusarium sambucinum*. J Cell Biochem [Suppl] 12C: 290

Schäfer W, Straney D, Ciuffetti L, VanEtten HD, Yoder OC (1989) One enzyme makes a fungal pathogen, but not a saprophyte, virulent on a new host plant. Science 246: 247–249

Schulz B, Banuett F, Dahl M, Schlesinger R, Schäfer W, Martin T, Herskowitz I, Kahmann R (1990) The b alleles of *Ustilago maydis*, whose combinations program pathogenic development, code for polypeptides containing a homeodomain related motif. Cell 60: 295–306

Shaykh M, Soliday C, Kolattukudy PE (1977) Proof of the production of cutinase by *Fusarium solni* f. *pisi* during penetration into its host, *Pisum sativum*. Plant Physiol 60: 170–172

Simpson RB, Johnson LJ (1990) *Arabidopsis thaliana* as a host for *Xanthomonas campestris* pv. *campestris*. Mol Plant Microbe Interact 3: 233–237

Snyder BA, Nicholson RL (1990) Synthesis of phytoalexins in sorghum as a site-specific response to fungal ingress. Science 248: 1637–1639

Soliday CL, Flurkey WH, Okita TW, Kolattukudy PE (1984) Cloning and structure determination of cDNA for cutinase, an enzyme involved in fungal penetration of plants. Proc Natl Acad Sci USA 81: 3939–3943

Soliday CL, Dickman MB, Kolattukudy PE (1989) Structure of the cutinase gene and detection of promotor activity in the 5′-flanking region by fungal transformation. J Bacteriol 171: 1942–1951

Thompson CJ, Movva NR, Tizard R, Crameri R, Davies JE, Lauwereys M, Botterman J (1987) Characterization of the herbicide-resistance gene bar from *Streptomyces hygroscopicus*. EMBO J 6: 2519–2523

Tilburn J, Scazzocchio G, Taylor GG, Zabicky-Zissima JH, Lockington RA, Davis RW (1983) Transformation by integration in *Aspergillus nidulans*. Gene 26: 205–221

Timberlake WE (1980) Developmental gene regulation in *Aspergillus nidulans*. Dev Biol 78: 497–510

Turgeon GB, Garber RC, Yoder OC (1985) Transformation of the fungal maize pathogen *Cochliobolus heterostrophus* using the *Aspergillus nidulans* amdS gene. Mol Gen Genet 201: 450–453

Turgeon GB, Garber RC, Yoder OC (1987) Development of a fungal transformation system based on selection of sequences with promoter activity. Mol Cell Biol 7: 3297–3305

Van den Ende G, Linskens HF (1974) Cutinolytic enzymes in relation to pathogenesis. Annu Rev Phytopathol 12: 247–258

Van Engelenburg F, Smit R, Goosen T, van den Broek H, Tudzynski P (1989) Transformation of *Claviceps purpurea* using a bleomycin resistance gene. Appl Microbiol Biotechnol 30: 364–370

VanEtten HD, Kistler HC (1988) *Nectria haematococca* mating populations I and VI. In: Sidhu GS (ed) Genetics of plant pathogenic fungi. Academic Press, London, pp 189–206 (Advances in plant pathology, vol 6)

VanEtten HD, Mathews PS, Tegtmeier KJ, Dietert MF, Stein JI (1980) The association of pisatin tolerance and demethylation with virulence on pea in *Nectria haematococca*. Physiol Plant Pathol 16: 257–268

VanEtten HD, Mathews DE, Mathews PS (1989) Phytoalexin detoxification: importance for pathogenicity and practical implications. Annu Rev Phytopathol 27: 143–164

Wang J, Holden DW, Leong SA (1988) Gene transfer system for the phytopathogenic fungus *Ustilago maydis*. Proc Natl Acad Sci USA 85: 865–869

Ward M, Wilkinson B, Turner G (1986) Transformation of *Aspergillus nidulans* with a cloned, oligomycin-resistant ATP synthase subunit 9 gene. Mol Gen Genet 202: 265–270

Weltring KM, Turgeon BG, Yoder OC, VanEtten HD (1988) Isolation of a phytoalexin-detoxifying gene from the plant pathogenic fungus *Nectria haematococca* by detecting its expression in *Aspergillus nidulans*. Gene 68: 335–344

Wheeler MH, Greenblatt GA (1988) The inhibition of melanin biosynthetic reactions in *Pyricularia oryzae* by compounds that prevent rice blast disease. Exp Mycol 12: 151–160

Willison JHM (1981) Secretion of cell wall material in higher plants. In: Tanner W, Loewus FA (eds) Plant carbohydrates II. Springer, Berlin Heidelberg New York, pp 513–541 [Pierson A, Zimmermann MH (eds) Encyclopedia of plant physiology, vol 13B]

Yelton MM, Hamer JE, Timberlake WE (1984) Transformation of *Aspergillus nidulans* by using a *tryC* plasmid. Proc Natl Acad Sci USA 81: 1470–1474

Yoder OC, (1988) *Cochliobolus heterostrophus*, cause of southern corn leaf blight. In: Sidhu GS (ed) Genetics of plant pathogenic fungi. Academic Press, London, pp 93–112 (Advances in plant pathology, vol 6)

Yoder OC, Turgeon BG, Schäfer W, Ciuffetti L, Bohlmann H, VanEtten HD (1989) Molecular analysis of mating type and expression of a foreign pathogenicity gene in *Cochliobolus heterostrophus*. In: Nevalainen H, Penttilä M (eds) Proceedings of the EMBO-alko Workshop on Molecular Biology of Filamentous Fungi. Found Biotech Indust Ferment Res 6: 189–196

Young RA, Davis RW (1983) Efficient isolation of genes by using antibody probes. Proc Natl Acad Sci USA 80: 1194–1198

Zimmermann CR, Orr WC, Leclerc RF, Barnard EC, Timberlake WE (1980) Molecular cloning and selection of genes regulated in *Aspergillus* development. Cell 21: 709–715

Section III
Perception of Pathogens and Signal Transduction

Chapter 7

Interactions Between *Agrobacterium tumefaciens* and Its Host Plant Cells

Stephen C. Winans

Section of Microbiology, Cornell University, Ithaca, New York 14853, U.S.A.

With 5 Figures

Contents

I. Introduction

Agrobacterium tumefaciens is the causative agent of the crown gall disease of dicotyledonous plants (Smith and Townsend, 1907; reviewed in Braun, 1982). The finding 13 years ago that the bacterium can transfer a discrete segment of tumorigenic DNA (T-DNA) to the genome of the plant host (Chilton et al., 1977, 1980; Willmitzer, 1980) attracted the interest of a large number of laboratories around the world. This was largely due to the possibility of exploiting this unprecedented interkingdom DNA transfer to create transgenic plants. There have been at least two consequences of this research. The first is that *Agrobacterium* has indeed been used to create transgenic plants of several dozen species containing genes of scientific or commercial importance. The second consequence, perhaps somewhat accidental, is that *Agrobacterium* now provides the best available model for studying the molecular interactions between plants and their bacterial pathogens.

This chapter will attempt to review recent findings concerning the molecular mechanisms underlying these interactions. This rapidly moving field of research has been the subject of many other recent reviews. Some of these have treated general aspects of crown gall tumorigenesis (Binns and Thomashow, 1988; Nester et al., 1984; Sinkar et al., 1987; Hooykaas, 1989; Gelvin, 1990; White and Sinkar, 1987; Zambryski, et al., 1989; Zambryski, 1988). Other reviews have emphasized either signal exchange in plant–microbe interactions (De Cleene, 1988; Halverson and Stacey, 1986; Morris, 1986; Morris and Powell, 1987; Stachel and Zambryski, 1986b; Weiler and Schroder, 1987), or mechanisms of T-DNA transfer (Koukolikova-Nicola et al., 1987; Ream, 1989; Stachel and Zambryski, 1986b, 1989). Still other reviews have discussed the use of *Agrobacterium* in creating transgenic plants (Fraley et al., 1986; Klee et al., 1987; Lichtenstein and Fuller, 1987; Weising et al., 1988; White, 1990).

II. An Overview of Crown Gall Tumorigenesis

The genus *Agrobacterium* contains two widely studied pathogenic species: *A. tumefaciens* and *A. rhizogenes*. Both infect a broad variety of dicotyledonous plants, and infect only at wound sites. They infect individual cells at the site of infection, and do not kill these cells, but rather cause these cells to proliferate, causing in the case of *A. tumefaciens*, both galls of disorganized callus tissue and teratomas containing stunted shoots, and in the case of *A. rhizogenes*, proliferations of morphologically distinctive roots (Nester et al., 1984). In each case, this is achieved by the transfer of plasmid-encoded DNA to the nuclei of plant cells, where it is integrated into genomic DNA.

The transferred DNA carries genes which are specifically devoted to causing proliferation. The *A. tumefaciens* T-DNA contains the *iaa M* and *iaa H* genes, which encode enzymes that catalyze the conversion of tryptophan to auxin. Also transferred is the *ipt* gene, which encodes the enzyme that catalyzes the synthesis of isopentenyl-AMP, a key intermediate in the synthesis of *trans*-ribosylzeatin and *trans*-zeatin, the two naturally occurring cytokinins (Binns and Thomashow, 1988). The overproduction of these compounds causes the transformed cell to undergo multiple rounds of cell division. At a gross level, the transformation of many wounded cells causes them as well as untransformed neighboring cells to form a chimeric, neoplastic tissue, which is capable of unlimited rounds of cell division in planta. These tissues can also be propagated indefinitely in vitro, even in the absence of exogenous phytohormones. In the case of *A. rhizogenes*, the oncogenes of central importance are known as *rol* (root locus) genes. Each of these genes (*rol A, B, C*, and *D*) contributes to the rooty morphology of transformed cells (White and Nester, 1980). Collectively, they do not appear to alter the concentration of phytohormones, but rather confer increased sensitivity towards the transformed cells' endogenous auxins (Shen et al., 1988).

Also transferred to the plant cell by both types of bacteria are genes directing the synthesis of a wide range of derivatized amino acids and sugars collectively known as opines. These compounds are synthesized in and secreted by transformed cells, and can be catabolized by *Agrobacterium* in or near the tumor. The "opine hypothesis" holds that the benefits derived by the bacterium from infection are attributable to the production and excretion of opines by infected plant cells (Petit and Tempe, 1985).

The mechanism of transfer of the T-DNA is probably the most widely studied aspect of *Agrobacterium* research. A great deal is known about the metabolism of T-DNA in bacteria, while little is known about events that occur in the plant cells. The T-DNA is flanked on both sides by 25 base pair imperfect direct repeats known as borders (Yadav et al., 1982). These and nearby sequences known as "overdrive" are the only *cis* acting sites required for transfer. After stimulation of plasmid-encoded genes required for transfer (see below), the bottom strands of these borders contain nicks at identical positions. About 50% of these bottom strands are found in a single-stranded, linear form. Furthermore, one molecule of the protein which catalyzed the nick remains tightly attached to the 5' ends of the T-DNA. These molecules are believed to represent transfer intermediates which are transferred to the plant in this form, where they acquire a complementary strand and are integrated into the host cell genome.

The genes encoding proteins which catalyze the transfer process are not themselves transferred, but most are found on non-transferred portions of the Ti-plasmid. There are seven operons of Ti-plasmid encoded genes (*vir* genes), encoding about 23 proteins, which mediate or facilitate transfer. They include proteins which catalyze border nicks (*vir D1* and *vir D2*), proteins which play an ancillary role in nicking or formation of single-strand DNA (*vir C1* and *vir C2*), a protein which binds to single stranded DNA (*vir E2*), a number of proteins which are probably secreted or membrane spanning and therefore thought to form a pore (encoded by the *vir B* operon), and two proteins which control the transcription of the regulon during infection (*vir A* and *vir G*). In addition to the Ti-plasmid genes, there are a number of chromosomal genes required for transfer. The genes *chv A*, *chv B*, *cel*, and *att* are thought to play a role in binding of *Agrobacterium* to the plant cell, while *chv D*, *chv E* and *ros* play some role in transcriptional regulation of *vir* genes.

III. A Chronology of Tumorigenesis

A. Chemotaxis

Agrobacterium is a highly motile organism, and there are several reports indicating that motility and chemotaxis can play a role in the early events of infection. An assay was developed in which strains of *Agrobacterium* could be tested for the ability to translocate toward root cap cells (Hawes et al.,

1988). Wild type strains containing or lacking the Ti plasmid do indeed exhibit chemotaxis toward these cells. This phenomenon is reminiscent of observations that *Rhizobium* exhibits chemotaxis toward root exudates (Gaworzewska and Carlile, 1982). Transposon insertion mutants of *Agrobacterium* were obtained which are deficient in chemotaxis. Among the strains isolated were (*i*) those completely non-motile, (*ii*) those motile but non-chemotactic to any tested chemoattractant, and (*iii*) strains which are deficient in chemotaxis toward plant cells but not toward several other compounds (Hawes et al., 1988). These data indicate that *Agrobacterium* may have a chemoreceptor devoted specifically to sensing compounds released by plant cells. In a subsequent study, these mutants were assayed for the ability to form tumors (Hawes and Smith, 1989). Under most conditions, they were virulent when assayed using pea seedlings as a host, but when soil was infested with *Agrobacterium*, dried, and then used to grow the plants, the non-chemotactic mutants showed highly attenuated virulence.

Other studies indicate that *Agrobacterium* is chemotactic toward a group of phenolic compounds previously identified as *vir* gene inducers (see below). Acetosyringone and related compounds were highly chemotactic in two different assays. Chemotaxis required the Ti plasmid, and specifically, the *vir* gene regulatory genes *vir A* and *vir G* (Shaw et al., 1988). Three properties of chemotaxis toward acetosyringone suggest that *vir* gene induction is *not* required: (*i*) the concentration most effective for chemotaxis (50 nM) is far below that required for efficient *vir* gene induction, (*ii*) the assays were performed over an interval of time too short for *vir* gene induction, and (*iii*) the pH used (7.0) is too alkaline for efficient induction. These results indicate that Vir A and Vir G proteins carry out two functions: chemotaxis, and *vir* gene induction. A third laboratory could not detect chemotaxis toward acetosyringone, and found that chemotaxis toward related compounds did not require the Ti-plasmid (Parke et al., 1987).

B. Attachment

A second early step in infection is the binding of bacteria to target plant cells. A number of bacterial strains have been isolated having mutations affecting this step. All such mutations were found to be chromosomal, and no known *vir* mutation affects binding (Douglas et al., 1985; Krens et al., 1985). *Agrobacterium* binding to plant cells has been described as occurring in two steps. The first step is a rapid and reversible binding of the bacteria to a specific plant cell receptor (Lippincott and Lippincott, 1969). The second step is a slower, non-reversible binding and aggregation that is mediated by bacterial polysaccharides including cellulose (Matthysse et al., 1981). A series of mutations in the *att* gene have been isolated which appear to block the first step (Matthysse, 1987). These mutants are avirulent and lack a number of outer membrane proteins. A mutation in the *cel* gene encoding a protein important in cellulose biosynthesis prevents the second step (Matthysse, 1983). The *cel* mutants still bind to plant cells, but do so individually rather than forming aggregates. These mutants are still virulent, but are far more susceptible to being washed from carrot cells than wild type bacteria.

Three other genes, *chv* A, *chv* B, and *exo* C, are required for the synthesis of a cyclic β-1,2 glucan which has also been implicated in plant cell binding. Mutations in *chv* A, *chv* B, or *exo* C, abolish virulence (Douglas et al., 1985; Cangelosi et al., 1987; Thomashow et al., 1987; Kamoun et al., 1989) and cause a roughly 10-fold decrease in binding of bacteria to tobacco mesophyll cells (Douglas et al., 1985). *Chv* B is involved in biosynthesis of the glucan (Zorreguieta et al., 1986), while *chv* A appears to be required for the export of this polysaccharide from the cytoplasm to the periplasm and extracellular fluid (Cangelosi et al., 1989; De Iannino et al., 1989). *Chv* A is homologous to a family of membrane bound ATPases involved in active transport (Cangelosi et al., 1989). The cyclic β-1,2 glucan has been implicated in resistance of the bacteria to high osmotic pressure (Miller et al., 1986), and indeed, *chv* A or *chv* B mutants are osmotically sensitive. However, the associated avirulence is found in conditions of either high or low osmotic pressure (Cangelosi et al., 1990). Genes homologous to *chv* A, *chv* B, and *exo* C, have been found in *Rhizobium*; strains containing mutations in these genes are unable to colonize root nodules (Dylan et al., 1986; Cangelosi et al., 1987)

C. Vir Gene Induction

Approximately 25 Ti-plasmid encoded *vir* genes are required for tumorigenesis. These genes are arranged in 7 operons, as shown in Fig. 1. All Ti-plasmid encoded *vir* operons are transcriptionally induced during infection by a family of related phenolic compounds, some of which are released from plant wounds (Stachel and Nester, 1986; Stachel et al., 1986a; Winans et al., 1987). Transcriptional induction can be observed by cocultivating *Agrobacterium* with isolated plant tissues, cultured tissues and protoplasts, or by incubating the bacteria in the presence of synthetic phenolics such as acetosyringone and hydroxyacetosyringone, vanillin, sinapinic acid, guaiacol, and many other structurally similar compounds (Stachel et al., 1985; Spencer and Towers, 1988; Melchers et al., 1989; Boulton et al., 1986) (see Fig. 2). The fact that these compounds include conferyl alcohol and sinapyl alcohol (Spencer and Towers, 1988), is especially intriguing, since these compounds are precursors of lignin, and lignin is generally present in scar tissue (Kahl, 1982). It is plausible that lignin precursors are a ubiquitous feature of plant wounds, and therefore are useful compounds to use as chemical cues. Recent reports indicate that the response to these compounds is potentiated by low levels of opines, while higher levels inhibit

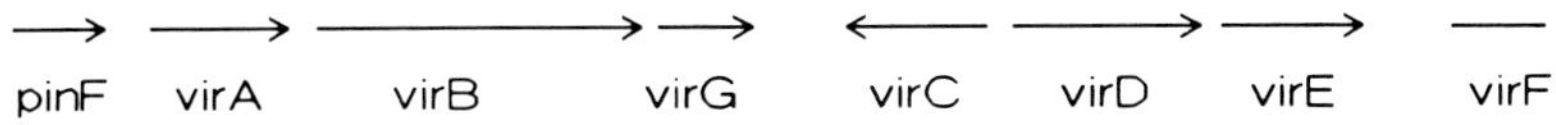

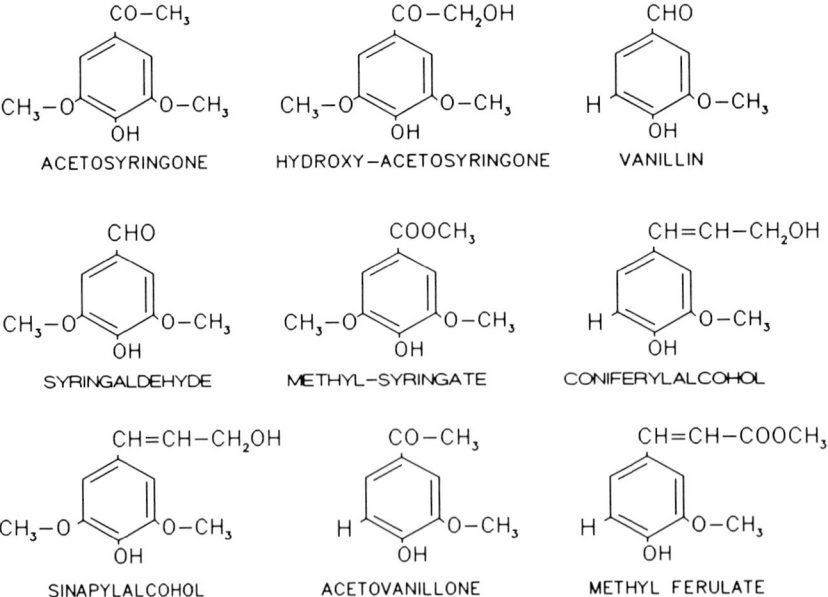

pinF	virA	virB	virG	virC	virD	virE	virF

LOCUS	SIZE	ORFS	FUNCTION
virA	2.8 kb	1	Environmental Sensor
virG	1.0 kb	1	Transcriptional Activator
virC	1.5 kb	2	T–DNA Processing
virD	4.5 kb	4	Border Endonuclease
virE	2.2 kb	2	ssDNA Binding Protein
virB	9.5 kb	11	Membrane Proteins (pore?)
virF	?	?	?
pinF	2.8 kb	2	Cytochrome P450's

Fig. 1. Genetic organization of the *vir* region of the octopine Ti plasmid pTiA6. Arrows indicate transcriptional units. ORFS, Number of genes in each operon

Fig. 2. Chemical structures of a representative group of phenolic compounds which are active in induction of the *vir* regulon

induction (Veluthambi et al., 1989; Krishnan and Gelvin, manuscript submitted).

Induction of *vir* gene requires two plasmid-encoded proteins, Vir A and Vir G; mutations in either gene encoding these proteins completely block the response to plant phenolics (Stachel and Zambryski, 1986a; Winans et al., 1987). Vir A is 829 amino acids in length, and has a typical leader sequence, suggesting that at least a portion of the protein crosses the cytoplasmic membrane. About 250 amino acids toward the C terminus is a second hydrophobic region, followed by a strongly positively charged region (Leroux et al., 1987; Melchers et al., 1987). This resembles a number of stop-transfer sequences found in membrane-spanning proteins. The hypothesis that Vir A spans the cytoplasmic membrane was confirmed by use of transposon mutagenesis using Tn*pho A* and by exposing spheroplasts of cells expressing Vir A to proteolysis followed by analysis of remaining proteins by Western blotting (Winans et al., 1988; Melchers et al., 1989b). Protease treatment removed about 270 amino acid residues from the N terminus. Western blot analysis of fractionated cells also has shown that Vir G is cytoplasmically localized.

Vir A and Vir G are members of a gene family of "two component" regulatory systems (Winans et al., 1986; Leroux et al., 1987). Vir A is a member of the "sensor-transmitter" class, many of whose members are known to be protein kinases. Vir G is a member of the "receiver-effector" class of proteins, whose N terminal halves are the targets of phosphorylation, and whose C-terminal halves are generally promoter-binding proteins (reviewed in Stock et al., 1989; Albright et al., 1990; Miller et al., 1989). Recently, data from several groups have shown that Vir A and Vir G do indeed have the predicted properties. Vir A is autophosphorylated at a histidine residue which is essential for protein function (Jin et al., 1990a; Y. Huang et al., 1990). It is presumed to transfer this phosphate to Vir G and hence, to modulate its promoter activating properties. A family of similar sequences to which Vir G protein might bind is found upstream of each *vir* promoter (Winans et al., 1987). Vir G binds to at least some of these sequences, called *vir* boxes (Jin et al., 1990b; Pazour et al., 1990).

An unusual feature of Vir A is that it contains at its C terminus a domain homologous to the receiver domain of Vir G. Similar receiver domains are found in a small number of sensor protein in this gene family (Stock et al., 1989). The function of this domain is unknown, but deletions removing part of it render the protein non-functional (Melchers et al., 1989b). Another surprising feature of Vir A was found during the construction of Vir A-Tar hybrid proteins (Melchers et al., 1989b). The periplasmic domain of Vir A was found not to be required for responsiveness to acetosyringone, as deletions of this domain did not abolish acetosyringone responsiveness. Assuming that Vir A itself does bind acetosyringone at all, the binding site must lie in the cytoplasmic portion of the protein. Deletion of this domain did however, alter the normal acidic pH (less than 5.5) and low temperature (29 °C) optima of the wild type protein. Taken together, these data suggest

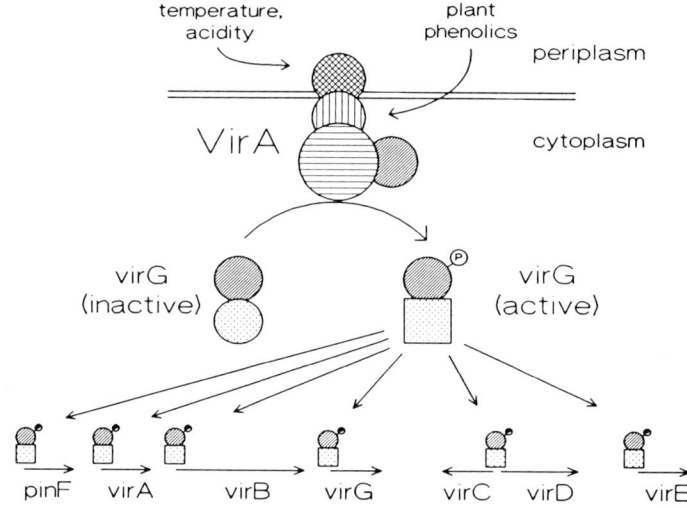

Fig. 3. A model describing the proposed function of the regulatory proteins Vir A and Vir G. Vir A is a transmembrane protein kinase that may directly bind phenolic inducers, while Vir G is the target of the Vir A kinase and binds to *vir* promoters to activate their transcription

that Vir A must contain a least four domains. A model for Vir A and Vir G function is shown in Fig. 3.

Efficient induction requires a third protein: that encoded by the chromosomal *chv E* gene (M.-L.W. Huang et al., 1990). The Chv E protein is homologous to a family of periplasmically localized sugar binding proteins involved in chemotaxis and active transport. These data suggest that Chv E may also be periplasmically localized and may have a role in binding of the phenolic inducers. However, the fact that a residual level of induction is observed in Chv E mutants indicates that it does not contain the only binding site. Alternatively, Chv E could be required for efficient carbon utilization. It would be interesting to test *chv E* mutant strains for induction in media containing different carbon sources.

A subset of *vir* genes appear to be responsive to additional regulatory systems. *Vir G* is one example. *Vir G* has two promoters, and is responsive to three environmental stimuli (Stachel and Zambryski, 1986a; Winans et al., 1997; Veluthambi et al., 1986). One promoter is induced by acetosyringone in strains containing Vir A and Vir G proteins (Winans, 1990). In this sense, *vir G* expression is identical to all other *vir* genes. However, this same promoter is also induced by phosphate starvation (Winans, 1990). This induction does not require Vir A or Vir G or any other Ti plasmid encoded genes. Finally, a second promoter, 50 bp downstream from the first, is induced by a family of environmental stresses, including extremes of pH, heat, ethanol, DNA-damaging agents and heavy metals (Mantis and Winans, unpubl. obs.). This induction also does not require any Ti-plasmid-encoded genes. This promoter has a strong sequence similarity to the family of heat-shock promoters of *E. coli*. It is believed that activation of *vir G* by phosphate starvation or other environmental stresses may play an important role in the initial stages of induction, in order to increase the concentration of Vir G protein to sufficient levels to allow the upstream promoter to function (Winans, 1990). This "pump-priming" expression of Vir G would occur only in conditions of environmental stress, which would suggest that stress should potentiate *vir* gene induction. This hypothesis is currently being tested.

The regulation of *vir C* and *vir D* also seem to be distinguishable from other *vir* genes, in that they alone are transcribed at elevated levels in strains containing a mutation at the chromosomal *ros* locus (Close et al., 1985, 1987). This mutant was recovered by selecting elevated expression of a *vir D :: cat* fusion. The mutant also results in elevated expression of *vir C*, whose promoter is close to and divergent from that of *vir D*, but does not affect transcription of *vir B, vir E, vir B*, or *pin F*.

D. Metabolism of T-DNA

The junctions between the transferred and non-transferred DNA contain imperfect direct repeats 25 base pairs long (Yadav et al., 1982; Zambryski et al., 1982). The right border is essential for efficient tumorigenesis and acts in a polar fashion, directing transfer of sequences to its left (Wang et al., 1984; Shaw et al., 1984). In contrast to the right border, the left border is dispensable for tumorigenesis (Joos et al., 1983).

In addition to the right border, a second sequence called overdrive (or *ode*) is needed for efficient tumorigenesis (Peralta and Ream, 1985; Peralta et al., 1986). The *ode* sequence can be moved away from the right border at least 5 kb in either direction, or in inverted orientation without disrupting its function. For these reasons, it has been compared to eukaryotic transcriptional enhancers. This sequence has been extensively characterized in the octopine strains. Although no sequence similar to overdrive has been found in nopaline plasmids, sequences flanking the right border enhance transfer efficiency (Wang et al., 1987a), suggesting that some site functionally similar to overdrive may be present. In contrast, sequences flanking the left border inhibited transfer efficiency.

Upon induction of the *vir* regulon, a number of striking alterations of the T-DNA have been observed. Single-stranded scissions occur at identical positions between the third and fourth base pairs from the left end of each border (Albright et al., 1987; Wang et al., 1987b; Durrenberger et al., 1989; Jayaswal et al., 1987). In the octopine strain pTiA6, which has two T-DNAs and four borders, nicking was found at each border (Stachel et al., 1987). This nicking requires the products of the *vir A, vir G, vir D1*, and *vir D2* (Yanofsky et al., 1986; Stachel et al., 1987; Veluthambi et al., 1987). The first two of these proteins were thought to be required merely to induce the transcription of the last two. This was proven to be the case by the finding that *vir D1* and *vir D2*, when expressed in *E. coli* from a foreign promoter, are sufficient to catalyze site-specific nicking at a border supplied in *trans* (Yanofsky et al., 1986; Jayaswal et al., 1987).

A large portion of the DNA between the single-stranded scissions was found in a single-stranded linear form. This was most elegantly demonstrated by the finding that this DNA could be size-fractionated and transferred to nitrocellulose without any prior denaturation (Stachel et al., 1986b). Since only single-stranded DNA binds nitrocellulose under these conditions, this finding suggested that the DNA was single-stranded. These sequences could be hybridized using strand-specific probes complementary to the bottom strand but not the top strand, indicating that only the bottom strand is recovered in single-stranded form. The top strand was not found in single-stranded form, indicating either that this strand is selectively degraded, or, more likely, that it is made double-stranded by synthesis of complementary DNA sufficient to replace the bottom strand.

In early studies, purification of the "T-strands" involved use of protease. In later studies, experiments were conducted in which protease was omitted. Under these conditions, a protein was observed to bind tightly to the T-strands. A number of groups have concluded that this binding occurs at the 5' end of the DNA and that the bound protein is encoded by *vir D2* (Herrera-Estrella et al., 1988; Howard, et al., 1989; Ward and Barnes, 1988; Young and Nester, 1988). Perhaps the most elegant proof was obtained by digesting the DNA with DNase I, which degrades the entire T strand except for a small fragment attached to the protein, and hybridizing the digestion products to an oligonucleotide. It was found that a probe complementary to the 5' end did hybridize, while a probe complementary to the 3' end failed to hybridize (Ward and Barnes, 1988). After size-fractionation of the protein-DNA complex by SDS-PAGE and autoradiography, a radioactive protein the size of Vir D2 was identified. Furthermore, when an endonuclease proficient Vir D2-β-galactosidase fusion protein was used instead of the wild type protein, a radioactive band was detected having the molecular weight of the fusion protein (Ward and Barnes, 1988). The Vir D2-T-strand complex survives boiling in 1% SDS or treatment with reducing agents or 6 M urea, and is therefore thought to be caused by a covalent linkage, perhaps created at the same time as endonucleolytic cleavage. This protein has been described as a pilot protein to guide the T-strands to the nucleus. The carboxy terminus of Vir D2 contains a sequence similar to a family of nuclear targeted proteins of animals and yeast (Stachel and Zambryski, 1989; Chelsky et al., 1989). Deletion of this sequence abolishes virulence (T. Steck, pers. comm.), indicating that it plays some essential role in T-DNA transfer.

Two other proteins may interact with Vir D1 and Vir D2 during nicking. The products of Vir C1 and Vir C2 are thought to bind to *ode* and to somehow contribute to T-strand formation. Indirect evidence comes from the observation that a strain containing both a *vir C* mutation and an *ode* mutation is as deficient in tumorigenesis as a strain containing only one or the other mutation (Ji et al., 1988). More direct evidence was provided by the observation that Vir C1 protein binds specifically to *ode* in a gel retardation assay and can protect it from nuclease digestion in a footprint assay (Toro et al., 1989). It has also been reported that Vir C proteins stimulate production of T-strands in strains of *E. coli* that express low amounts of the Vir D1 and Vir D2, while in strains synthesizing high levels of Vir D, Vir C proteins are not needed for efficient T-strand production (De Vos and Zambryski, 1989). Although one group reported that *vir C* mutants are slightly deficient in nicking (Toro et al., 1988), others reported no deficiency (Stachel et al., 1987). If it is true that *vir C* products do not play a role in nicking, but do play a role in T-strand formation, then it appears possible that they might aid in initiation of the displacement synthesis of the bottom strand which is thought to be required to replace the T strand.

At least one other protein appears to bind to the T-strands. A number of groups have independently found that the product of *vir E2* binds to single-stranded DNA (Gietl et al., 1987; Christie et al., 1988; Das, 1988; Citovsky et al., 1988). Binding has no apparent sequence-specificity. Vir E2 does not bind duplex DNA or single-stranded RNA. Binding is highly cooperative (Sen et al., 1989). Sufficient Vir E2 is found in induced *Agrobacterium* to bind all the intracellular T-strands. It has therefore been proposed that Vir E2 may bind along the whole length of the T-strand, perhaps to protect it from endonucleases it may encounter during its sojourn from bacterial cytoplasm to nucleoplasm of the plant cell. A model for the formation of T-strands is shown in Fig. 4.

Mutations in either *vir E2* and *vir E1* genes are peculiar in that their avirulent phenotype is rescuable by coinfection with a strain having an intact *vir* region but lacking T-DNA (Otten et al., 1984; Christie et al., 1989). This suggests either (*i*) that the Vir E proteins can be donated to the mutant strain in such a way that the mutant strain can reuse them during infection, or (*ii*) that T-DNA can be transferred from the *vir E* mutant to the rescuing strain and from there to the plant. Mutations in any *vir* gene of the rescuing strain except *vir C1* and *vir C2* abolish rescuing ability. The *vir* mutations tested were due to transposon insertion, and most of these are strongly polar, so it remains unknown how many Vir proteins play a role in rescuing. It is surprising, however, that at least the last protein of the *vir B* and *vir D* operons are required for rescuing. The data suggest that these proteins, and possibly other Vir B and Vir D proteins, play a role in export of Vir E proteins. Furthermore, an endonuclease-proficient, non-polar insertion of Vir D2 also failed to rescue (Christie et al., 1989), indicating that the carboxy terminal portion of Vir D2 plays a role in rescuing. However, the rescuing strains do not need to contain T-DNA, as strains containing the disarmed plasmid pLA4404 are proficient in rescuing VirE mutants. The fact that Vir E1 and Vir E2 mutants are both rescuable suggests that they may play similar roles in virulence. Further evidence that Vir E1 may

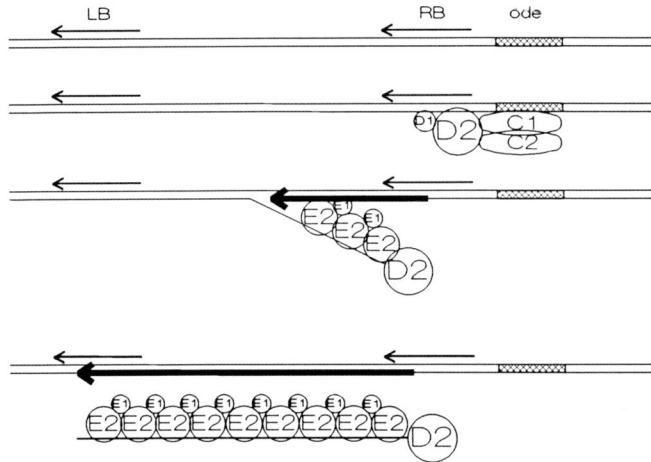

Fig. 4. Proposed mechanism of synthesis of single-stranded T-DNA. LB, RB, and *ode*, Left border, right border and overdrive sites, respectively. Thick arrows indicate newly synthesized DNA which could displace bottom strand of T-DNA. C1, C2, D1, and D2, E1, E2, Products of the *vir C1*, *vir C2*, *vir D1*, *vir D2*, *vir E1*, and *vir E2* genes, respectively.

interact with VirE2 comes from the observation that VirE1 appears to increase the stability of VirE2 (McBride et al., 1988). This would mean that VirE1, like VirE2, might bind to T-strands.

Complementation of a mutation in one bacterium by a gene in a different bacterium, although somewhat unusual, is not without precedent. For example, a gene required for conjugal transfer of *inc N* plasmids could be rescued by extracellular complementation using any strain that expressed all genes required for pilus synthesis (Winans and Walker, 1985). A second example is rescue of the fix⁻ phenotype of *exo* mutants of *Rhizobium* spp. by exo⁺ strains (Djordjevic et al., 1987). *Myxococcus xanthus* provides several well documented examples. First, no fewer than four proteins required for fruiting body formation are rescuable in this way and are thought to provide extracellular signals between these social bacteria (reviewed in Kaiser, 1989). Secondly, mutations in as many a seven genes required for gliding mobility can be rescued (Hodgkin and Kaiser, 1977). In most or all of these examples, the products of these genes, or other macromolecules synthesized by those proteins, are found on the outer surface of bacterial cells or in the extracellular medium.

How might this hypothetical T-strand-protein complex traverse the bacterial membranes and peptidoglycan? There is some evidence that *vir B* products could create a pore through which DNA could pass. This 11 kilobase operon has been sequenced in three different laboratories (Ward et al., 1988; Thompson et al., 1988; Kuldau et al., 1990), and a small number of reading frame discrepancies exists in the data described. The open reading frame assignments reported by Ward et al. (1988) for the pTiA6 plasmid, and modified by their published erratum (Ward et al., 1990a), correspond closely to assignments of Kuldau et al. (1990) for the pTiC58 plasmid; therefore, these two DNA sequences are likely to be correct. All three groups report 11 open reading frames, and significantly, 10 of these have a hydropathy profile which suggests an extracellular or membrane-spanning topology. Recently, avirulent mutants were obtained using Tn*pho*A, a transposon which identifies secreted proteins (Manoil and Beckwith, 1986). Many of the mutants obtained by this method were in the *vir*B operon, indicating that these proteins are indeed partly or fully secreted (G. Cangelosi, pers. comm.). Ward has looked more closely at the Vir B8, Vir B9, and Vir B10 proteins, and finds that each of these are also either secreted or membrane-spanning (Ward et al., 1990b). In contrast, Vir B11, a hydrophilic protein which binds and hydrolyses ATP, is capable of autophosphorylation and shows protein homology to the *Bacillus subtilis* Com G ORF 1 protein (Christie et al., 1989). The *Bacillus* protein is required for transformation competence, and may play a role in the transcriptional regulation of other competence genes. However, there is no evidence that Vir B11 is a regulatory protein.

1. The Conjugation Model for T-DNA Transfer

A striking similarity between the metabolism of T-DNA and the metabolism of plasmid DNA during bacterial conjugal was first noted by Stachel and Zambryski (1986b), and later expanded upon by Zambryski (1987), Ream (1989), and others. In both systems, the transferred DNA appears to be a single-stranded molecule with a molecule of the nicking enzyme covalently attached to the 5′ end. In both cases, a single stranded DNA binding protein has been implicated (Willetts and Skurray, 1987). Two kinds of evidence support the model that *vir* genes evolved from *tra* genes. The first is the finding that the *ori T* site of broad host range plasmid RK2 is similar to the T-DNA borders and is nicked at precisely the same position (D. Guiney, pers. comm.). Furthermore, the protein encoded by the *virD4* gene is homologous to with the RK2 Tra G protein (E. Lanka, pers. comm.).

The second kind of evidence is the finding that derivatives of the broad host range, mobilizable plasmid RSF1010 containing a plant-selectible gene but not containing a T-DNA border could be transferred to plant nuclei and integrated into nuclear DNA (Buchanan-Wollaston et al., 1987). Transfer required the RSF1010 *ori T*, at least one of the three *mob* gene products, and *Agrobacterium vir* genes. RSF1010 is an example of a class of bacterial plasmids that are not self-transmissible, but are mobilizable by coresident transmissible plasmids. Mobilization requires an *ori T* in *cis*, and the products of *mob* genes supplied in *cis* or in *trans*. These *mob* gene products constitute a site specific endonuclease which causes nicks at *ori T*. All other transfer proteins are provided by the coresident transmissible plasmid. Mobilizable plasmids tend to be promiscuous in that they can be mobilized by a variety of transmissible plasmids having quite diverse conjugation systems. The finding that RSF1010 can also be mobilized by *vir* genes to plant cells suggests that it is able to recognize Vir proteins as though they were Tra proteins, indicating a very strong similarity between the two. Very recently, it has been shown that two different conjugal plasmids are able to transfer to the yeast *S. cerevisiae* (Heinemann and Sprague, 1989), indicating that perhaps all conjugal transfer systems are more promiscuous than previously thought.

As the similarities between T-DNA transfer and conjugal transfer have been noted in other recent reviews, it may be useful here to emphasize the differences between the two systems. One major difference is the presence of single-stranded, linear T-DNA in the donor before transfer. In the bacterial conjugation systems, the transferred DNA is single-stranded only transiently during transfer, and a complementary strand is quickly (though not obligately) synthesized in both donor and recipient (Willetts and Skurray, 1987). Another major difference concerns the fate of the transferred DNA. Conjugally transferred DNA recircularizes in the recipient to create an autonomously replicating plasmid indistinguishable from that found in the donor. This DNA may also be integrated into recipient DNA by homologous recombination. In contrast, *Agrobacterium* T-DNA is found covalently integrated by illegitimate recombination into plant genomic DNA.

2. An Alternative Model for T-DNA Mobilization

In addition to single-stranded scission, T-DNA has been reported to undergo other kinds of events. Double-stranded breaks at each border have been reported by several groups (Veluthambi et al., 1987; Durrenberger et al., 1989; Steck et al., 1989). Joining of these borders by Campbell recombination has been inferred from indirect genetic observations but never observed by physical methods (Koukolikova-Nicola et al., 1985). Covalently joined left and right borders, forming "hybrid" or "joint" borders have been recovered both in *Agrobacterium* and *E. coli* (Alt-Moerbe et al., 1986; Koukolikova-Nicola et al., 1985; Machida et al., 1986). Furthermore, the frequency with which these molecules are recovered increases dramatically in cells that express *vir D1* and *vir D2*. It cannot be denied that these circular molecules must exist, but their existence does not mean that they represent transfer intermediates. An alternative explanation for their occurrence is that the nicks catalyzed by Vir D1 and Vir D2 provide substrates for host-mediated homologous recombination. According to this model, such circular molecules result from a side reaction and are not true transfer intermediates. Bakkeren et al. (1989) attempted to recover circular molecules introduced into plants by the process of "agroinfection." One genome length of CaMV was inserted between T-DNA borders, transferred to plants, and allowed to form circular molecules able to cause symptoms of viral infection. Viruses recovered from plant tissues did *not* contain hybrid borders, suggesting that double-stranded circles are not transfer intermediates.

3. Integration

In contrast to the many recent advances in our understanding of T-DNA metabolism in *Agrobacterium*, very little is known about the fate of this DNA in plant cytoplasm and nucleoplasm. Although one study has been published in which this DNA was characterized only a few hours after transfer (Virts and Gelvin, 1985), the overwhelming number of studies concentrate on characterization of integrated T-DNA in the descendants of infected cells. A large number of studies have been undertaken to determine the number and location of T-DNA inserts. It can be concluded that one or more than one T-DNA can be transferred and integrated (Spielman and Simpson, 1986; Thomashow et al., 1980; Chyi et al., 1986; Ursic et al., 1983; Jorgensen et al., 1987). In octopine strains of *Agrobacterium*, which contain two T-DNAs, T_R, which contains genes encoding opines but no oncogenes, may or may not be transferred to a given transformed plant cell. In cells containing more than one T-DNA, the extra copies have been found, in some cases, to be tightly linked to each other while in other studies they have been found dispersed throughout the plant genome. Those copies that are linked can be in tandem or inverted orientation (Chyi et al., 1986; Jorgensen et al., 1987).

The DNA sequence of several junctions between transferred and host DNA has been determined (Holsters et al., 1983; Simpson et al., 1982; Yadav et al., 1982; Zambryski et al., 1982; Kwok et al., 1985). Gheysen et al., (1987), have pointed out that these junctions, in general, appear more variable than the junctions created by transposons, retroviruses, or retro-transposons. Rather, they are more similar to the junctions created by insertions of non-retro transforming viruses such as SV40 or adenovirus. In general terms, it seems that the right junction is somewhat less variable than the left. Recall that the right end of the T-strand created in Agrobacteria contains the 3 base pairs of the right border, while the left end contains 22 base pairs of the left border, and that Vir D2 protein is attached to the right end. In most cases the right junction includes all but 1 to 3 base pairs of the T-strand. In contrast, the left end usually lacks some of the sequences of the left end of the T-strand. The number of missing bases ranges from a few base pairs up to 100 base pairs. It is tempting to speculate that T-strand integration, like its formation, could be initiated at its right end, and could even be mediated in part by Vir D2.

The target site of T-DNA integration has been sequenced in only one study (Gheysen et al., 1987). That integration caused a duplication of 158 base pairs of host DNA, a small deletion, a translocation, and several single base pair transitions. These data hint at the generally variable nature of integration. On the other hand, the 158 base pairs duplication may not be a general feature of integration, as such long direct repeats probably would have been detected in some of the earlier studies. Clearly, there is a need for more such studies, and these could be greatly facilitated by the technique of PCR (polymerase chain reaction), particularly inverse PCR (Ochman et al., 1988).

E. Expression of Transferred Genes

Although there is evidence that the genes found in T-DNA were *not* evolutionarily derived from plants, they have never-the-less evolved eukaryotic-like regulatory sites. Transcription of all T-DNA genes is inhibited by α-amanitin, indicating that their expression, like other protein-encoding genes, depends upon RNA polymerase II (Willmitzer et al., 1981). Inspection of the DNA sequence of these promoters reveals CAAT and TATA boxes typical of plant promoters (reviewed in Nester et al., 1984). Furthermore, the mRNA from these genes contain eukaryotic poly-A tails, and polyadenylation sites have been found in several of these genes. A final similarity between T-DNA genes and plant genes is that at least some of transferred genes contain upstream activating sites that are similar to transcriptional enhancers (Ellis et al., 1987; Singh et al., 1989). On the other hand, unlike most plant genes, no intron has been described for any T-DNA gene.

Most or all of the genes found in T-DNA can be categorized as being involved either in upsetting the host cell's normal balance of phyto-hormones, or in the production of opines. Two genes have been implicated in the overproduction of auxin. *Iaa M* and *iaa H* (also known as transcripts 1 and 2, or as *tms* 1 and *tms* 2) encode proteins that act in concert to convert tryptophan to IAA via indolacetamide (Thomashow et al., 1984; Schroder et al., 1984). The *ipt* gene (also known as transcript 3 or *tmr*) catalyzes the condensation of isopentenyl pyrophosphate and adenosine monophosphate to form isopentenyl-AMP, which is converted by host enzymes to *trans*-zeatin and *trans*-ribosylzeatin (Akiyoshi et al., 1984; Buchmann et al., 1985). In addition, two other genes play a poorly characterized ancillary role in tumor formation. Transcript 6b (also known as *tml*) and transcript 5, while not oncogenic alone, appear to modulate tumorigenicity when present in combination with other oncogenes (Spanier et al., 1989; Tinland et al., 1989; Leemans et al., 1982).

In contrast to *A. tumefaciens*, *A. rhizogenes* appears to use a different strategy for oncogenesis. No strain of *A. rhizogenes* transfers an *ipt* gene (though at least one strain does have a homologous non-transferred gene, see Regier et al., 1989). Some strains do transfer the *iaa* genes, while others do not (White et al., 1980). The genes most important for tumorigenicity appear to be the four *rol* genes (root locus; see White et al., 1980; Cardarelli et al., 1985). These genes do not change the levels of endogenous phyto-hormones, but rather, appear to increase the sensitivity of transformed roots to exogenously applied auxins (Shen et al., 1988). They presumably also sensitize host cells to endogenous auxins, and this is probably the underlying mechanism of hairy root formation. Certain features of the hairy root syndrome appear to be attributable to individual *rol* genes, suggesting that they may each independently confer auxin hypersensitivity (Spena et al., 1987; Sinkar et al., 1988; Schmulling et al., 1988). The mechanisms by which they do this are completely unknown, and will certainly be the source of exciting research in the next few years.

All known tumorigenic strains of *Agrobacterium* are able to direct the synthesis of derivatized amino acids or sugars known as opines (Tempe and Goldman, 1982). Most strains transfer genes directing the synthesis of at least two of these compounds. Almost without exception, a strain which directs the synthesis of a particular opine has a corresponding non-transferred gene able to direct the catabolism of the same opine (Tempe and Petit, 1982). At least 20 of these compounds have been described and doubtless many more remain to be discovered (Petit and Tempe, 1985). Figure 5 shows the structure of some representative examples of opines.

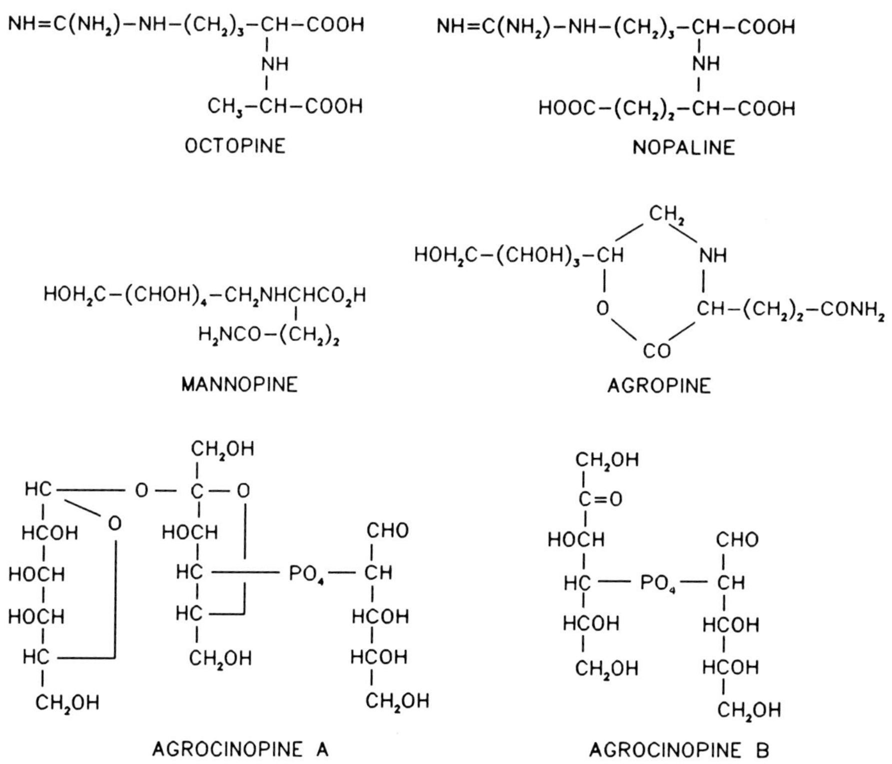

$$NH=C(NH_2)-NH-(CH_2)_3-CH-COOH$$
$$|$$
$$NH$$
$$|$$
$$CH_3-CH-COOH$$

OCTOPINE

$$NH=C(NH_2)-NH-(CH_2)_3-CH-COOH$$
$$|$$
$$NH$$
$$|$$
$$HOOC-(CH_2)_2-CH-COOH$$

NOPALINE

$$HOH_2C-(CHOH)_4-CH_2NHCHCO_2H$$
$$|$$
$$H_2NCO-(CH_2)_2$$

MANNOPINE

$$HOH_2C-(CHOH)_3-CH \quad CH_2 \quad NH$$
$$| \qquad \qquad |$$
$$O \qquad \qquad CH-(CH_2)_2-CONH_2$$
$$CO$$

AGROPINE

AGROCINOPINE A

AGROCINOPINE B

Fig. 5. Chemical structure of representative opines. Synthesis of octopine, mannopine and agropine is directed by the A6 and ACH5 Ti plasmids, while synthesis of nopaline and agrocinopines A and B is directed by the C58 and T37 Ti plasmids, among others

Octopine is made by the reductive condensation of arginine and pyruvate, a reaction catalyzed by the *ocs* gene. Nopaline is made the product of the *nos* gene, and is synthesized by a similar mechanism utilizing α-ketoglutarate in place of pyruvate. A separate transferred gene (transcript 6a) appears to mediate the excretion of these opines from plant cells (Messens et al., 1985).

The octopine-type plasmids pTiA6 and pTi15955 direct the synthesis of additional opines. In addition to making octopine, the *ocs* product can synthesize lysopine, histopine, and octopinic acid by reductive condensation of pyruvate with lysine, histidine, and ornithine, respectively. In addition to the octopine family of opines, these plasmids transfer genes directing the synthesis of mannopine, mannopinic acid, agropine, and agropinic acid (Ellis et al., 1984). Genes 2' and 1' of the T_R DNA direct the reductive condensation of glucose with glutamine or glutamic acid to form mannopine or mannopinic acid, respectively. Mannopine can be enzymatically lactonized by the product of the O' gene to form agropine. Mannopine can also undergo spontaneous lactonization to form agropinic acid. The "nopaline" plasmids also direct the synthesis of more than just nopaline. The *nos* product can synthesize nopalinic acid by reductive condensation of lysine with α-ketoglutarate. These plasmids also direct the condensation of arabinose with fructose or sucrose, resulting in agrocinopines A or B (Ellis and Murphy, 1981). Other strains of *Agrobacterium* make still other opines, some of whose structures are shown in Fig. 5. In addition to serving as a source of carbon, nitrogen, and sometimes phosphorus, a subset of opines, known as conjugal opines, serve as transcriptional inducers of Ti plasmid conjugal transfer (Ellis et al., 1982; Klapwijk and Schilperoort, 1979). Although the opine hypothesis predicts that opine utilization provides a selective advantage over other bacteria in the rhizosphere (Petit and Tempe, 1985), the ability to catabolize opines is not in fact limited to *Agrobacterium*, as various isolates of *Pseudomonas* are also able to do so (Bergeron et al., 1990; Bouzar and Moore, 1987).

Compounds similar to opines have recently been described in *Rhizobium meliloti*. This organism fixes nitrogen in a mutualistic association with leguminous plants. Bacteria colonize roots of these plants, and terminally differentiate into forms called "bacteroids". These bacteroids synthesize and secrete derivatized sugars called rhizopines which are catabolized by their soil-dwelling relatives. In this way, bacteria inhabiting a nodule can feed free-living bacteria. These findings may suggest a possible evolution of the opines in *Agrobacterium*.

IV. References

Akiyoshi DE, Klee H, Amasino R, Nester EW, Gordon MP (1984) T-DNA of *Agrobacterium tumefaciens* encodes an enzyme of cytokinin biosynthesis. Proc Natl Acad Sci USA 81: 5994–5998

Albright LM, Yanofsky MF, Leroux B, Ma D, Nester EW (1987) Processing of the T-DNA of

Agrobacterium tumefaciens generates border nicks and linear, single-stranded T-DNA. J Bacteriol 169: 1046–1055

Albright LM, Huala E, Ausubel FM (1989) Prokaryotic signal transduction mediated by sensor and regulator protein pairs. Annu Rev Genet 23: 311–336

Alt-Moerbe J, Rak B, Schroder J (1986) A 3.6-kbp segment from the *vir* region of Ti plasmids contains genes responsible for border sequence-directed production of T region circles in *E. coli*. EMBO J 5: 1129–1135

Ashby AM, Watson MD, Loake GJ, Shaw CH (1988) Ti Plasmid-specified chemotaxis of *Agrobacterium tumefaciens* C58C[1] toward *vir*-inducing phenolic compounds and soluble factors from monocotyledonous and dicotyledonous plants. J Bacteriol 170: 4181–4187

Bakkeren G, Koukolikova-Nicola Z, Grimsley N, Hohn B (1989) Recovery of *Agrobacterium tumefaciens* T-DNA molecules from whole plants early after transfer. Cell 57: 847–857

Bergeron J, MacLeod RA, Dion P (1990) Specificity of octopine uptake by *Rhizobium* and *Pseudomonas* strains. Appl Environ Microbiol 56: 1453–1458

Binns AN, Thomashow MF (1988) Cell biology of *Agrobacterium* infection and transformation of plants. Annu Rev Microbiol 42: 575–606

Bouzar H, Moore LW (1987) Isolation of different *Agrobacterium* biovars from a natural oak savanna and tallgrass prairie. Appl Environ Microbiol 53: 717–721

Braun A (1982) A history of the crown gall problem In: Kahl G, Schell JS (eds) Molecular biology of plant tumors. Academic Press, New York, pp 155–210

Buchanan-Wollaston V, Passiatore JE, Cannon F (1987) The *mob* and *oriT* mobilization functions of a bacterial plasmid promote its transfer to plants. Nature 328: 172–175

Buchmann I, Marner FJ, Schroder G, Waffenschmidt S, Schroder J (1985) Tumor genes in plants: T-DNA encoded cytokinin biosynthesis. EMBO J 4: 853–859

Cangelosi GA, Hung L, Puvanesarajah V, Stacey G, Ozga DA, Leigh JA, Nester EW (1987) Common loci for *Agrobacterium tumefaciens* and *Rhizobium meliloti* exopolysaccharide synthesis and their roles in plant interactions. J Bacteriol 169: 2086–2091

Cangelosi, GA, Martinetti G, Leigh JA, Lee CC, Theines C, Nester EW (1989) Role of *Agrobacterium tumefaciens* Chv A protein in export of β-1,2-glucan. J Bacteriol 171: 1609–1615

Cangelosi GA, Martinetti G, Nester EW (1990) Osmosensitivity phenotypes of *Agrobacterium tumefaciens* mutants that lack periplasmic β-1,2-glucan. J Bacteriol 172: 2172–2174

Cardarelli M, Spano L, DePaolis A, Mauro ML, Vitali G (1985) Identification of the genetic locus responsible for non-polar root induction by *Agrobacterium rhizogenes*. 1855. Plant Mol Biol 5: 385–391

Chelsky D, Ralph P, Jonak G (1989) Sequence requirements for synthetic peptide-mediated translocation to the nucleus. Mol Cell Biol 9: 2487–2492

Chilton M-D, Drummond MH, Merlo DJ, Sciaky D, Montoya AL, Gordon MP, Nester EW (1977) Stable incorporation of plasmid DNA into higher plant cells: the molecular basis of crown gall tumorigenesis. Cell 11: 263–271

Chilton M-D, Saiki RK, Yadav N, Gordon MP, Quetier F (1980) T-DNA from *Agrobacterium* Ti plasmid is in the nuclear DNA fraction of crown gall tumor cells. Proc Natl Acad Sci USA 77: 4060–4064

Christie PJ, Ward JE, Winans SC, Nester EW (1988) The *Agrobacterium tumefaciens vir E2* gene product is a single-stranded-DNA-binding protein that associates with T-DNA. J Bacteriol 170: 2659–2667

Christie PJ, Ward JE, Gordon MP, Nester EW (1989) A gene required for transfer of T-DNA to plants encodes an ATP-ase with autophosphorylating activity. Proc Natl Acad Sci USA 86: 9677–9681

Chyi YS, Jorgensen RA, Goldstein D, Tanksley SD, Loaiza-Figueroa F (1986) Locations and stability of *Agrobacterium*-mediated T-DNA insertions in the *Lycopersicon* genome. Mol Gen Genet 204: 64–69

Citovsky V, DeVos G, Zambryski P (1988) Single-stranded DNA binding protein encoded by the *virE* locus of *Agrobacterium tumefaciens*. Science 240: 501–504

Close TJ, Tait RC, Kado CI (1985) Regulation of Ti plasmid virulence genes by a chromosomal locus of *Agrobacterium tumefaciens*. J Bacteriol 164: 774–781

Close TJ, Rogowsky PM, Kado CI, Winans SC, Yanofsky MF, Nester EW (1987) Dual control of the *Agrobacterium tumefaciens* Ti plasmid virulence genes. J Bacteriol 169: 5113–5118

Das A (1988) The *A. tumefaciens virE* operon encodes a single stranded DNA binding protein Proc Natl Acad Sci USA 85: 2609–2913

De Cleene M (1988) The susceptibility of plants of *Agrobacterium*: a discussion of the role of phenolic compounds. FEMS Microbiol Rev 54: 1–8

De Iannino NI, Ugalde RA (1989) Biochemical characterization of avirulent *Agrobacterium tumefaciens chvA* mutants: synthesis and excretion of β-(1-2)glucan. J Bacteriol 171: 2842–2849

De Vos G, Zambryski P (1989) Expression of *Agrobacterium* nopaline-specific VirD1, VirD2, and VirC1 proteins and their requirement for T-strand formation in *E. coil*. Mol Plant Microbe Interact 2: 43–52

Djordjevic SP, Chen H, Batley M, Redmond JW, Rolfe BG (1987) Nitrogen fixation ability of exopolysaccharide synthesis mutants of *Rhizobium* sp. strain NGR234 and *Rhizobium trifolii* is restored by the addition of homologous exopolysaccharides. J Bacteriol 169: 53–60

Douglas CJ, Staneloni RJ, Rubin RA, Nester EW (1985) Identification and genetic analysis of an *Agrobacterium tumefaciens* chromosomal virulence gene. J Bacteriol 161: 850–860

Durrenberger F, Crameri A, Hohn B, Koukolikova-Nicola Z (1989) Covalently bound VirD2 protein of *Agrobacterium tumefaciens* protects the T-DNA from exonucleolytic degradation. Proc Natl Acad Sci USA 86: 9154–9158

Dylan T, Ielpi L, Stanfield S, Kashyap L, Douglas C, Yanofsky M, Nester E, Helinski DR, Ditta G (1986) *Rhizobium meliloti* genes required for nodule development are related to chromosomal virulence genes in *Agrobacterium tumefaciens*. Proc Natl Acad Sci USA 83: 4403–4407

Ellis JG, Murphy PJ (1981) Four new opines from crown gall tumors—their detection and properties. Mol Gen Genet 181: 36–43

Ellis JG, Kerr A, Petit A, Tempe J (1982) Conjugal transfer of nopaline and agropine Ti-plasmids – the role of agrocinopines. Mol Gen Genet 186: 269–273

Ellis JG, Ryder MH, Tate ME (1984) *Agrobacterium tumefaciens* T$_R$-DNA encodes a pathway for agropine biosynthesis. Mol Gen Genet 195: 466–473

Ellis JG, Llewellyn DJ, Walker JC, Dennis ES, Peacock WJ (1987) The *ocs* element: a 16 base pair palindrome essential for activity of the octopine synthase enhancer. EMBO J 6: 3203–3208

Fraley RT, Rogers SG, Horsch RB (1986) Genetic transformation in higher plants. Crit Rev Plant Sci 4: 1–46

Gaworzewska RT, Carlile MJ (1982) Positive chemotaxis of *Rhizobium leguminosarum* and other bacteria towards root exudates from legumes and other plants. J Gen Microbiol 128: 1179–1188

Gietl C, Koukolikova-Nicola Z, Hohn B (1987) The mobilization of the T-DNA from *Agrobacterium* to the plant cells involves a single-stranded DNA binding protein. Proc Natl Acad Sci USA 84: 9006–9010

Gelvin SB (1990) Crown gall disease and hairy root disease: a sledgehammer and a tackhammer. Plant Physiol 92: 281–285

Grimsley N, Hohn B, Ramos C, Kado C, Rogowsky P (1989) DNA transfer from *Agrobacterium* to *Zea mays* or *Brassica* by agroinfection is dependent on bacterial virulence functions. Mol Gen Genet 217: 309–316

Halverson LJ, Stacey G (1986) Signal exchange in plant–microbe interactions. Microbiol Rev 50: 193–225

Hawes MC, Smith LY (1989) Requirement for chemotaxis in pathogenicity of *Agrobacterium tumefaciens* on roots of soil-grown pea plants. J Bacteriol 171: 5668–5671

Hawes MC, Smith LY, Howarth AJ (1988) *Agrobacterium tumefaciens* mutants deficient in chemotaxis to root exudates. Mol Plant Microbe Interact 1: 182–186

Heinemann JA, Sprague GF (1989) Bacterial conjugative plasmids mobilize DNA transfer between bacterial and yeast. Nature 340: 205–209

Herrera-Estrella A, Chen Z, Van Montagu M, Wang K (1988) Vir D proteins of *Agrobacterium tumefaciens* are required for the formation of a covalent DNA-protein complex at the 5' terminus of T-strand molecules. EMBO J 7: 4055–4062

Hodgkin J, Kaiser D (1977) Cell-to-cell stimulation of movements in nonmotile mutants of *Myxococcus*. Proc Natl Acad Sci USA 74: 2938–2942

Holsters M, Villarroel R, Gielen J, Seurinck J, De Greve H, Van Montagu M, Schell J (1983) An analysis of the boundaries of the octopine T_L-DNA in tumors induced by *Agrobacterium tumefaciens*. Mol Gen Genet 190: 35–41

Hooykaas PJJ (1989) Transformation of plant cells via *Agrobacterium*. Plant Mol Biol 13: 327–336

Howard EA, Winsor BA, De Vos G, Zambryski P (1989) Activation of the T-DNA transfer process in *Agrobacterium* results in the generation of a T-strand-protein complex: tight association of Vir D2 with the 5' ends of T-strands. Proc Natl Acad Sci USA 86: 4017–4021

Huang M-LW, Cangelosi GA, Halperin W, Nester EW (1990) A chromosomal *Agrobacterium tumefaciens* gene required for effective plant signal transduction. J Bacteriol 172: 1814–1822

Huang Y, Morel P, Powell B, Kado CI (1990) Vir A, a coregulator of Ti-specified virulence genes, is phosphorylated in vitro. J Bacteriol 172: 1142–1144

Jayaswal RK, Veluthambi K, Gelvin SB, Slightom JL (1987) Double stranded T-DNA cleavage and the generation of single stranded T-DNA molecules in *E. coli* by a *vir D* encoded border specific endonuclease from *A. tumefaciens* J Bacteriol 169: 5035–5045

Ji JM, Martinez A, Dabrowski M, Veluthambi K, Gelvin SB, Ream W (1988) The overdrive enhancer sequence stimulates production of T-strands from the *Agrobacterium tumefaciens* tumor-inducing plasmid. In: Staskawicz B, Ahlquist P, Yoder P (eds) Molecular biology of plant–pathogen interactions. AR Liss, New York, pp 19–31 (UCLA Symp Mol Cell Biol)

Jin S, Roitsch T, Ankenbauer RG, Gordon MP, Nester EW (1990a) The Vir A protein of *Agrobacterium tumefaciens* is autophosphorylated and is essential for *vir* gene induction. J Bacteriol 172: 525–530

Jin S, Roitsch T, Christie PJ, Nester EW (1990b) The regulatory Vir G protein specifically binds to a *cis*-acting regulatory sequence involved in transcriptional activation of *Agrobacterium tumefaciens* virulence genes. J Bacteriol 172: 531–537

Jorgensen RA, Snyder C, Jones JDG (1987) T-DNA is organized predominantly in inverted repeat structures in plants transformed in *A. tumefaciens* C58 derivatives. Mol Gen Genet 207: 471–477

Joos H, Inze D, Caplan A, Sormann M, Van Montagu M, Schell J (1983) Genetic analysis of T-DNA transcripts in nopaline crown gall. Cell 32: 1057–1067

Kahl G (1982) Molecular biology of wound healing: the conditioning phenomenon. In: Kahl G, Schell JS (eds) Molecular biology of plant tumors. Academic Press, London, pp 211–268

Kaiser D (1989) Multicellular development in *Myxobacteria*. In: Hopwood DA, Chater KF (eds) Genetics of bacterial diversity. Academic Press, London, pp 243–263

Kamoun S, Cooley MB, Rogowsky UM, Kado CI (1989) Two chromosomal loci involved in production of exopolysaccharide in *Agrobacterium tumefaciens*. J Bacteriol 171: 1755–1759

Klapwijk PM, Schilperoort RA (1979) Negative control of octopine degradation and transfer genes of octopine Ti plasmids in *Agrobacterium tumefaciens*. J Bacteriol 141: 424–431

Klee H, Horsch R, Rogers S (1987) *Agrobacterium*-mediated plant transformation and its further applications to plant biology. Annu Rev Plant Physiol 38: 467–486

Koukolikova-Nicola A, Shillita RD, Hohn B, Wang K, Van Montagu M, Zambryski P (1985) Involvement of circular intermediates in the transfer of T-DNA from *Agrobacterium tumefaciens* to plant cells. Nature 313: 191–196

Koukolikova-Nicola Z, Albright L, Hohn B (1987) The mechanism of T-DNA transfer from *Agrobacterium tumefaciens* to the plant cell. In: Hohn T, Schell J (eds) Plant DNA infectious agents. Springer, Wien New York, pp 110–138 [Dennis ES et al (eds) Plant gene research. Basic knowledge and application]

Krens FA, Molendijk L, Wullems GJ, Schilperoort RA (1985) The role of bacterial attachment in the transformation of cell-wall-regenerating tobacco protoplasts by *Agrobacterium tumefaciens*. Planta 166: 300–308

Kuldau GA, DeVos G, Owen J, McCaffrey G, Zambryski P (1990) The *virB* operon of *Agrobacterium tumefaciens* pTiC58 encodes 11 open reading frames. Mol Gen Genet 221: 256–266

Kwok WW, Nester EW, Gordon MP (1985) Unusual plasmid DNA organization in an octopine crown gall tumor. Nucleic Acid Res 13: 459–471

Leemans J, Deblaere R, Willmitzer L, DeGreve H, Hernalsteens JP, Van Montagu M, Schell J (1982) Genetic identification of functional of T_L-DNA transcripts in octopine crown galls. EMBO J 1: 147–152

Leroux B, Yanofsky MF, Winans SC, Ward JE, Zeigler SF, Nester EW (1987) Characterization of the *virA* locus of *Agrobacterium tumefaciens*: a transcriptional regulator and host range determinant. EMBO J 6: 849–856

Lichtenstein DP, Fuller SL (1987) Vectors for the genetic engineering of plants. In: Rigby PWJ (eds) Genetic engineering, vol 6. Academic Press, New York, pp 103–183

Lippincott BB, Lippincott JA (1969) Bacterial attachment to a specific wound site as essential stage in tumor initiation by *Agrobacterium tumefaciens*. J Bacteriol 97: 620–628

McBride KE, Knauf VC (1988) Genetic analysis of the *virE* operon of the *Agrobacterium* Ti plasmid pTiA6. J Bacteriol 170: 1430–1437

Machida Y, Usami S, Yamamoto A, Takebe I (1986) Plant-inducible recombination between the 25-base-pair border sequence of T-DNA in *Agrobacterium tumefaciens*. Mol Gen Genet 204: 374–382

Manoil C, Beckwith J (1986) A genetic approach to analyzing membrane protein topology. Science 233: 1403–1408

Matthysse AG (1983) Role of bacterial cellulose fibrils in *Agrobacterium tumefaciens* infection. J Bacteriol 154: 906–915

Matthysse AG (1987) Characterization of nonattaching mutants of *Agrobacterium tumefaciens*. J Bacteriol 169: 313–323

Matthysse AG, Holmes KV, Gurlitz RHG (1981) Elaboration of cellulose fibrils by *Agrobacterium tumefaciens* during attachment to carrot cells. J Bacteriol 145: 583–595

Melchers LS, Thompson DV, Idler KB, Neuteboom TC, deMaagd RA, Schilperoort RA, Hooykaas PJJ (1987) Molecular characterization of the virulence gene *vir A* of the *Agrobacterium tumefaciens* Ti plasmid. Plant Mol Biol 9: 635–645

Melchers LS, Regensburg-Tuink AJG, Schilperoort RA, Hooykaas PJJ (1989a) Specificity of signal molecules on the activation of *Agrobacterium* virulence gene expression. Mol Microbiol 3: 969–977

Melchers LS, Regensburg-Tuink TJF, Bourret RB, Sedee NJA, Schilperoort RA, Hooykaas PJJ (1989b) Membrane topology and functional analysis of the sensory protein Vir A of *Agrobacterium tumefaciens*. EMBO J 8: 1919–1925

Messens E, Lenaerts EA, Van Montagu M, Hedges RW (1985) Genetic basis for opine secretion from crown gall tumor cells. Mol Gen Genet 199: 344–348

Miller JF, Mekalanos JJ, Falkow S (1989) Coordinate regulation and sensory transduction in the control of bacterial virulence. Science 243: 916–922

Miller KJ, Kennedy EP, Reinhold VN (1986) Osmotic adaptation by gram-negative bacteria: possible role for periplasmic oligosaccharides. Science 213: 48–51

Morris RO (1986) Genes specifying auxin and cytokinin biosynthesis in phytopathogens. Annu Rev Plant Physiol 37: 509–538

Morris RO, Powell GK (1987) Genes specifying cytokinin biosynthesis in prokaryotes. Bioessays 6: 23–28

Nester EW, Gordon MP, Amasino RM, Yanofsky MF (1984) Crown gall: a molecular and physiological analysis. Annu Rev Plant Physiol 35: 387–413

Nixon BT, Ronson CW, Ausubel FM (1986) Two-component regulatory systems responsive to environmental stimuli share strongly conserved domains with the nitrogen assimilation regulatory genes *ntr B* and *ntr C*. Proc Natl Acad Sci USA 83: 7850–7854

Ochman H, Gerber AS, Hartl DL (1988) Genetic application of an inverse polymerase chain reaction. Genetics 120: 621–623

Otten L, De Greve H, Leemans J, Hain R, Hooykaas P, Schell J (1984) Restoration of virulence of *vir* region mutants of *Agrobacterium tumefaciens* strain B6S3 by coinfection with normal and mutant *Agrobacterium* strains. Mol Gen Genet 195: 159–163

Parke D, Ornston LN, Nester EW (1987) Chemotaxis to plant phenolic inducers of virulence genes is constitutively expressed in the absence of the Ti plasmid in *Agrobacterium tumefaciens*. J Bacteriol 169: 5336–5338

Pazour GJ, Das A (1990) *vir G*, an *Agrobacterium tumefaciens* transcriptional activator, initiates translation at a UUG codon and in a sequence-specific DNA-binding protein. J Bacteriol 172: 1241–1249

Peralta EG, Ream LW (1985) T-DNA border sequence required for crown gall tumorigenesis. Proc Natl Acad Sci USA 82: 5112–5116

Peralta EG, Hellmiss R, Ream LW (1986) Overdrive, a T-DNA transmission enhancer on the *A. tumefaciens* tumor-inducing plasmid. EMBO J 5: 1137–1142

Petit A, Tempe J (1985) The function of T-DNA in nature. In: van Vloten-Doting L, Groot G, Hall T (eds) Molecular form and function of the plant genome. Plenum, New York, pp 625–636

Ream W (1989) *Agrobacterium tumefaciens* and interkingdom genetic exchange. Annu Rev Phytopathol 27: 583–618

Regier DA, Akiyoshi DE, Gordon MP (1989) Nucleotide sequence of the *tzs* gene from *Agrobacterium rhizogenes* strain A4. Nucleic Acids Res 17: 88–85

Ronson CW, Nixon BT, Ausubel FM (1987) Conserved domains in bacterial regulatory proteins that respond to environmental stimuli. Cell 49: 579–581

Schmülling T, Schell J, Spena A (1988) Single genes from *Agrobacterium rhizogenes* influence plant development. EMBO J 7: 2621–2629

Schroder G, Waffenschmidt S, Weiler EW, Schroder J (1984) The T-region of Ti plasmids codes for an enzyme synthesizing indole-3-acetic acid. Eur J Biochem 138: 387–391

Sen P, Pazour GJ, Anderson D, Das A (1989) Cooperative binding of *Agrobacterium tumefaciens* Vir E2 protein to single-stranded DNA. J Bacteriol 171: 2573–2580

Shaw CH, Watson MD, Carter GH, Shaw CH (1984) The right hand copy of the nopaline Ti plasmid 25 bp repeat is required for tumor formation. Nucleic Acids Res 12: 6031–6041

Shaw CHJ, Ashby AM, Brown A, Royal C, Loake GJ, Shaw CH (1989) Vir A and G are necessary for acetosyringone chemotaxis in *Agrobacterium tumefaciens*. Mol Microbiol 2: 413–417

Shen WH, Petit A, Guern J, Tempe J (1988) Hairy roots are more sensitive to auxin than normal roots. Proc Natl Acad Sci USA 85: 3417–3421

Simpson RB, O'Hara PJ, Kwok W, Montoya AL, Licktenstein C, Gordon MP, Nester EW (1982) DNA from the A6S/2 crown gall tumor contains scrambled Ti plasmid sequences near its junctions with the plant DNA. Cell 29: 1005–1014

Singh K, Tokuhis JG, Dennis ES, Peacock WJ (1989) Saturation mutagenesis of the octopine synthase enhancer: correlation of mutant phenotypes with binding of a nuclear protein factor. Proc Natl Acad Sci USA 86: 3733–3737

Sinkar VP, White FF, Gordon MP (1987) Molecular biology of Ri plasmid: a review. J Biosci 11: 47–58

Sinkar VP, Pythoud F, White FF, Nester EW, Gordon MP (1988) *rol A* locus of the Ri plasmid directs developmental abnormalities in transgenic tobacco plants. Genes Dev 2: 688–697

Smith EF, Townsend CO (1907) A plant-tumour of bacterial origin. Science 25: 671–673

Spanier K, Schell J, Schreier PH (1989) A functional analysis of T-DNA gene 6b: the fine tuning of cytokinin effects on shoot development. Mol Gen Genet 219: 209–216

Spena A, Schmülling T, Koncz C, Schell JS (1987) Independent and synergistic activity of *rol A, B*, and *C* loci in stimulating abnormal growth in plants. EMBO J 6: 3891–3899

Spencer PA, Towers GHN (1988) Specificity of signal compounds detected by *Agrobacterium tumefaciens*. Phytochemistry 27: 2781–2785

Spielman A, Simpson RB (1986) T-DNA structure in transgenic tobacco plants with multiple independent integration sites. Mol Gen Genet 205: 34–41

Stachel SE, Nester EW (1986) The genetic and transcriptional organization of the *vir* region of the A6 Ti plasmid of *Agrobacterium tumefaciens*. EMBO J 5: 1445–1454

Stachel SE, Zambryski PC (1986a) *vir A* and *vir G* control the plant-induced activation of the T-DNA transfer process of *A. tumefaciens*. Cell 46: 325–333

Stachel SE, Zambryski PC (1986b) *Agrobacterium tumefaciens* and the susceptible plant cell: a novel adaptation of extracellular recognition and DNA conjugation. Cell 47: 155–157

Stachel SE, Zambryski PC (1989) Generic trans-kingdom sex? Nature 340: 190–191

Stachel SE, Messens E, Van Montagu M, Zambryski P (1985) Identification of the signal molecules produced by wounded plant cells that activate T-DNA transfer in *Agrobacterium tumefaciens*. Nature 318: 624–629

Stachel SE, Nester EW, Zambryski PC (1986a) A plant cell factor induces *Agrobacterium tumefaciens vir* gene expression. Proc Natl Acad Sci USA 83: 379–383

Stachel SE, Timmerman B, Zambryski P (1986b) Generation of single-stranded T-DNA molecules during the initial stages of T-DNA transfer from *Agrobacterium tumefaciens* to plant cells. Nature 322: 706–712

Stachel SE, Timmerman B, Zambryski P (1987) Activation of *Agrobacterium tumefaciens vir* gene expression generates multiple single-stranded T-strand molecules from the pTiA6 region: requirement for 5' *vir D* products. EMBO J 6: 857–863

Steck TR, Close TJ, Kado CI (1989) High levels of double-stranded transferred DNA (T-DNA) processing from an intact nopaline Ti plasmid. Proc Natl Acad Sci USA 86: 2133–2137

Stock JB, Ninfa AJ, Stock AM (1989) Protein phosphorylation and regulation of adaptive responses in bacteria. Microbiol Rev 53: 450–490

Tempe J, Goldmann A (1982) Occurrence and biosynthesis of opines. In: Kahl G, Schell J (eds) Molecular biology of plant tumors. Academic Press, New York, pp 427–449

Tempe J, Petit A (1982) Opine utilization by *Agrobacterium*. In: Kahl G, Schell J (eds) Molecular biology of plant tumors. Academic Press, New York, pp 451–459

Thomashow MF, Nutter R, Montoya AL, Gordon MP, Nester EW (1980) Integration and organization of Ti plasmid sequences in crown gall tumors. Cell 19: 729–739

Thomashow LS, Reeves S, Thomashow MF (1984) Crown gall oncogenesis: evidence that a T-DNA gene from the *Agrobacterium* Ti plasmid pTiA6 encodes an enzyme that catalyzes synthesis of indoleacetic acid. Proc Natl Acad Sci USA 81: 5071–5075

Thomashow MF, Karlinsey JE, Marks JR, Hurlbert RE (1987) Identification of a new virulence locus in *Agrobacterium tumefaciens* that affects polysaccharide composition and plant cell attachment. J Bacteriol 169: 3209–3216

Thompson DV, Melchers LS, Idler KB, Schilperoort RA, Hooykaas PJJ (1988) Analysis of the complete nucleotide sequence of the *Agrobacterium tumefaciens vir B* operon. Nucleic Acids Res 16: 4621–4636

Tinland B, Huss B, Paulus F, Bonnard G, Otten L (1989) *Agrobacterium tumefaciens 6b* genes are strain-specific and affect the activity of auxin as well as cytokinin genes. Mol Gen Genet 219: 217–224

Toro N, Datta A, Yanofsky M, Nester E (1988) Role of the overdrive sequence in T-DNA border cleavage in *Agrobacterium*. Proc Natl Acad Sci USA 85: 8558–8562

Toro N, Datta A, Carmi OA, Young C, Prusti RK, Nester EW (1989) The *Agrobacterium tumefaciens vir C1* gene product binds to overdrive, a T-DNA transfer enhancer. J Bacteriol 171: 6845–6849

Ursic D, Slightom JL, Kemp JD (1983) *A. tumefaciens* T-DNA integrates into multiple sites of the sunflower crown gall genome. Mol Gen Genet 190: 494–503

Veluthambi K, Jayaswal RK, Gelvin SB (1987) Virulence genes A, G, and D mediate the double stranded border cleavage of T-DNA from the *Agrobacterium* Ti plasmid. Proc Natl Acad Sci USA 84: 1881–1885

Veluthambi K, Krishnan M, Gould JH, Smith RH, Gelvin SB (1989) Opines stimulate induction of the *vir* genes of *Agrobacterium tumefaciens* Ti plasmid. J Bacteriol 171: 3696–3703

Virts EL, Gelvin SB (1985) Analysis of transfer of tumor-inducing plasmids from *Agrobacterium tumefaciens* to *Petunia* protoplasts. J Bacteriol 162: 1030–1038

Wang K, Herrera-Estrella L, Van Montagu M, Zambryski P (1984) Right 25 bp terminus sequences of the nopaline T-DNA is essential for and determines direction of DNA transfer from *Agrobacterium* to the plant genome. Cell 38: 35–41

Wang K, Genetello C, Van Montagu M, Zambryski P (1987a) Sequence context of the T-DNA border repeat element determines its relative activity during T-DNA transfer to plant cells. Mol Gen Genet 210: 338–346

Wang K, Stachel SE, Timmerman B, Van Montagu M, Zambryski P (1987b) Site-specific nick in the T-DNA border sequence as a result of *Agrobacterium vir* gene expression. Science 235: 587–591

Ward ER, Barnes WM (1988) Vir D2 protein of *Agrobacterium tumefaciens* very tightly linked to the 5′ end of T-strand DNA. Science 242: 927–930

Ward JE, Akiyoski DE, Regier D, Datta A, Gordon MP, Nester EW (1988) Characterization of the *virB* operon from an *Agrobacterium tumefaciens* Ti plasmid. J Biol Chem 263: 5804–5814

Ward JE, Akiyoski DE, Regier D, Datta A, Gordon MP, Nester EW (1990a) Correction: characterization of the *virB* operon from an *Agrobacterium tumefaciens* Ti plasmid. J Biol Chem 265: 4768

Ward JE, Dale EM, Christie PJ, Nester EW, Binns AN (1990b) Complementation analysis of *Agrobacterium tumefaciens* Ti plasmid *virB* genes by use of a *vir* promoter expression vector; *virB9*, *virB10*, and *virB11* are essential virulence genes. J Bacteriol 172: 5187–5199

Weiler EW, Schroder J (1987) Hormone genes and crown gall disease. Trends Biochem Sci 12: 271–275

Weising K, Schell J, Kahl G (1989) Foreign genes in plants: transfer, structure, expression and applications. Annu Rev Genet 22: 421–477

White FF (1990) Vectors for gene transfer in higher plants. In: Kung S-D, Arntzen CJ (eds) Plant biotechnology. Butterworths, Boston, pp 3–34

White FF, Nester EW (1980) Hairy root plasmid encodes virulence traits in *Agrobacterium rhizogenes*. J Bacteriol 141: 1134–1141

White FF, Sinkar VP (1987) *Agrobacterium rhizogenes*. In: Hohn T, Schell J (eds) Plant DNA infectious agents. Springer, Wien New York, pp 149–178 [Dennis ES et al (eds) Plant gene research. Basic knowledge and application]

Willetts N, Skurray R (1987) Structure and junction of the F factor and mechanism of conjugation. In: Neidhardt FC, Ingraham JL, Low KB, Magasanik B, Schaechter M, Umbarger HE (eds) *Escherichia coli* and *Salmonella typhimurium*: cellular and molecular biology, vol 2. American Society for Microbiology, Washington, DC, pp 1110–1133

Willmitzer L, Schmalenbach W, Schell J (1981) Transcription of T-DNA in octopine and nopaline crown gall tumours is inhibited by low concentrations of α-aminitin. Nucleic Acids Res 9: 4801–4812

Winans SC (1990) Transcriptional induction of an *Agrobacterium* regulatory gene at tandem promoters by plant-released phenolic compounds, phosphate starvation and acidic growth media. J Bacteriol 172: 2433–2438

Winans SC, Walker GC (1985) Conjugal transfer system of the N incompatibility plasmid pKM101. J Bacteriol 161: 402–410

Winans SC, Ebert PR, Stachel SE, Gordon MP, Nester EW (1986) A gene essential for *Agrobacterium* virulence is homologous to a family of positive regulatory loci. Proc Natl Acad Sci 83: 8278–8282

Winans SC, Kerstetter RA, Nester EW (1988) Transcriptional regulation of the *virA* and *virG* genes of *Agrobacterium tumefaciens*. J Bacteriol 170: 4047–4054

Winans SC, Kerstetter RA, Ward JE, Nester EW (1989) A protein required for transcriptional regulation of *Agrobacterium* virulence genes spans the cytoplasmic membrane. J Bacteriol 171: 1616–1622

Willmitzer L, Debeuckeleer M, Lemmers M, Van Montagu M, Schell J (1980) DNA from Ti plasmid present in nucleus and absent from plastids of crown gall plant cells. Nature 287: 359–361

Yadav NS, Vanderlayden J, Bennett DR, Barnes WM, Mary-Dell Chilton (1982) Short direct repeats flank the T-DNA on a nopaline Ti plasmid. Proc Natl Acad Sci USA 79: 6322–6326

Yanofsky MF, Porter SG, Young C, Albright LA, Gordon MP, Nester EW (1986) The *virD* operon of *Agrobacterium tumefaciens* encodes a site-specific endonuclease. Cell 47: 471–477

Young C, Nester EW (1988) Association of the Vir D2 protein with the 5' end of T strands in *Agrobacterium tumefaciens*. J Bacteriol 170: 3367–3374

Zambryski P (1988) Basic processes underlying *Agrobacterium*-mediated DNA transfer to plant cells. Annu Rev Genet 22: 1–30

Zambryski P, Depicker A, Kruger K, Goodman H (1982) Tumor induction by *Agrobacterium tumefaciens*: analysis of the boundaries of T-DNA. J Mol Appl Genet 1: 361–370

Zambryski P, Tempe J, Schell J (1989) Transfer and function of T-DNA genes from *Agrobacterium* Ti and Ri plasmids in plants. Cell 56: 193–201

Zorreguieta A, Ugalde RA (1986) Formation in *Rhizobium* and *Agrobacterium* spp. of a 235-kilodalton protein intermediate in β-D(1-2)glucan synthesis. J Bacteriol 167: 947–951

Chapter 8

Elicitor Recognition and Signal Transduction

Jürgen Ebel[1] and Dierk Scheel[2]

[1]Biologisches Institut II, Universität Freiburg, D-W-7800 Freiburg, and
[2]Abteilung Biochemie, Max-Planck-Institut für Züchtungsforschung, D-W-5000
Köln 30, Federal Republic of Germany

With 2 Figures

Contents

I. Introduction

The majority of plants are highly resistant to attack by most micro-organisms (non-host or species resistance). The number of true host/pathogen combinations is comparatively small, but these interactions cause significant losses in agriculture. Within a susceptible plant species, resistance may be expressed in a number of host cultivars to certain races of a pathogen (cultivar resistance). The biochemical mechanisms of non-host resistance and cultivar resistance are similar and include a wide range of inducible defense responses. Many of these responses involve gene activation (e.g., the phytoalexin response), whereas others do not (e.g., callose formation). Compatible plant/pathogen combinations are characterized by a delay or apparent absence of the typical defense response, which may be caused by pathogen-mediated disturbances at any level of the reaction sequence(s) leading to successful resistance.

Despite the multiplicity of induced resistance reactions, it is postulated that plant defense mechanisms share certain features which may be arranged in four stages: (1) recognition of the pathogen by the plant as non-self; (2) perception of pathogen-derived signal(s) and (3) intracellular signal transduction leading to (4) the initiation of defense responses. Recognition as non-self (step 1) has not been well defined in any of the plant/pathogen systems studied thus far. It is, therefore, difficult to judge to what degree any of the factors that are involved in this level of recognition are also involved in the activation of the plant's defenses (steps 2 to 4).

Inducible plant defenses can be activated, not only upon challenge of plant tissues by microbes, but also upon exposure to elicitors. Elicitors are now widely used in experimental systems of reduced complexity (e.g., plant cell cultures) to study the mechanisms underlying steps 2 to 4. Selected plant/elicitor systems are described in this chapter to illustrate recent progress in the characterization of the molecular details of the induction process.

II. Elicitors and Plant Responses

The term "elicitor", used originally to refer to compounds that stimulate phytoalexin synthesis in plants (Keen and Bruegger, 1977; West, 1981; Darvill and Albersheim, 1984), has more recently been applied, in addition, to substances that induce other typical defense mechanisms, such as the synthesis of cell wall-associated phenylpropanoid compounds, the deposition of callose (1,3-β-glucan), the accumulation of hydroxyproline-rich glycoproteins, and the synthesis of certain hydrolytic enzymes (i.e., β-glucanases and chitinases) (Ebel, 1986; Hahlbrock and Scheel, 1987). It has also been shown that at least some elicitors stimulate more than one defense response (Ebel, 1986) and that certain elicitors interact synergistically (Darvill and Albersheim, 1984; Preisig and Kuc, 1985; Davis et al., 1986a; Davis and Hahlbrock, 1987).

Several attempts have been made to identify so-called "specific elicitors" that reflect the specificity of certain pathogen races towards a number of host cultivars in interactions exhibiting gene-for-gene relationships (Flor, 1971). Despite a number of reports of "specific elicitors" (Keen et al., 1989), very few have been isolated and characterized. A peptide consisting of 28 amino acids from *Cladosporium fulvum* has been isolated from intercellular washing fluids of infected leaves and shown to be a race/cultivar-specific elicitor of necrosis in tomato (*Lycopersicon esculentum*) cultivars carrying the *Cf9* resistance gene (Schottens-Toma and De Wit, 1988). A necrosis-inducing peptide of M_r 7,200 isolated from the culture medium of *Rhynchosporium secalis* appeared to be a race/cultivar-specific elicitor of accumulation of RNA encoding a thaumatin-like pathogenesis-related protein in barley (*Hordeum vulgare* L.) (Knogge et al., 1991). Other examples of specific elicitors are discussed by Keen and Dawson (1992). It should be noted that, at least in some cases, non-specific elicitors are complemented by additional factors (Barz et al., 1989; Yamada et al., 1989), that mediate the race/cultivar specificity observed in vivo.

Many other types of elicitor have been described, none of which reflects the race/cultivar specificity observed in the plant–pathogen interaction. These "general elicitors" may be involved in general resistance. Biotic elicitors include carbohydrates from fungal and plant cell walls, lipids,

Table 1. Selected elicitors of plant defense responses

Source	Chemical structure	Plant response	Ref[a]
Fungi			
Cladosporium fulvum (apoplastic space of infected tomato leaves)	polypeptide ($M_r = 3,186$)	chlorosis and necrosis in *Lycopersicon esculentum*	1
Fusarium solani f. sp. *phaseoli* (mycelium)	oligo-1,4-β-glucosamine (chitosan)	pisatin in *Pisum sativum*	2
Monilinia fructicola (mycelium)	polypeptide ($M_r = 8,000$)	phaseollin in *Phaseolus vulgaris*	3
Phytophthora capsici (culture filtrate)	protein ($M_r = 10,155$)	systemic necrosis and acquired resistance in *N. tabacum* cv. *xanthi*	4
P. cryptogea (culture filtrate)	protein ($M_r = 10,323$)	systemic necrosis and acquired resistance in *N. tabacum* cv. *xanthi*	4
P. infestans (mycelium)	eicosapentaenoic and arachidonic acids	sesquiterpenoids in *Solanum tuberosum*	5
P. megasperma f. sp. *glycinea* (culture filtrate and mycelial cell walls)	branched β-glucan with 3-, 6-, and 3,6-linked glucosyl residues	glyceollins in *Glycine max*	6, 7
P. megasperma f. sp. *glycinea* (culture filtrate)	glycoprotein ($M_r = 42,000$)	furanocoumarins in *Petroselinum crispum*	8
P. parasitica var. *nicotianae* (culture filtrate)	glycoprotein ($M_r = 46,000$)	capsidiol in *N. tabacum*	9
Puccinia graminis f. sp. *tritici* (germ tube walls)	peptidoglycan ($M_r = 67,000$)	lignin-like compounds in *Triticum aestivum*	10
Rhynchosporium secalis	polypeptide ($M_r = 7,200$)	thaumatin-like protein in *Hordeum vulgare*	11
Plants			
Glycine max (cell walls and citrus pectin)	oligo-1,4-α-galacturonide	glyceollins in *Glycine max*	12, 13
Ricinus communis, (citrus pectin and polygalacturonic acid)	oligo-1,4-α-galacturonide	casbene synthase activity and lignin-like compounds in *Ricinus communis*	14, 15

[a] 1, Schottens-Toma and De Wit, 1988; 2, Hadwiger and Beckman, 1980; 3, Cruickshank and Perrin, 1968; 4, Ricci et al., 1989; 5, Preisig and Kuc, 1985; 6, Darvill and Albersheim, 1984; 7, Sharp et al., 1984; 8, Parker et al., 1991; 9, Farmer and Helgeson, 1987; 10, Kogel et al., 1988; 11, Knogge et al., 1991; 12, Davis et al., 1986b; 13, Nothnagel et al., 1983; 14, Bruce and West, 1989; 15, Jin and West, 1984

microbial enzymes, and polypeptides or glycoproteins, representative examples of which are given in Table 1. Abiotic elicitors constitute an equally diverse group of compounds (Grisebach and Ebel, 1978). This discussion will be limited to purified biotic elicitors of the carbohydrate, polypeptide and glycoprotein type.

Three major types of carbohydrate-based elicitors have been identified: β-glucans and chitin or chitosan fragments (oligo-1,4-β-N-acetyl glucosamine or oligo-1,4-β-glucosamine) from fungal cell walls and oligo-1,4-α-galacturonides from plant cell walls. Several authors have suggested, therefore, that plants possess an oligosaccharide-dependent communication system that regulates the expression of defense responses (Darvill and Albersheim, 1984; Ryan, 1988). This model must now be extended to include proteinaceous compounds as well, to accommodate more recent findings (Farmer and Helgeson, 1987; Ricci et al., 1989; Scheel et al., 1989; Parker et al., 1991).

Chitosan has been shown to elicit various defense responses in a number of plants, such as phytoalexin accumulation in pea (*Pisum sativum*) pods (Hadwiger and Beckman, 1980), cultured soybean (*Glycine max*) (Köhle et al., 1984) and parsley (*Petroselinum crispum*) cells (Conrath et al., 1989); callose formation in soybean (Köhle et al., 1984, 1985), parsley (Conrath et al., 1989) and *Catharanthus* (Kauss et al., 1989a); and synthesis of proteinase inhibitor in tomato (Walker-Simmons et al., 1983, 1984; Walker-Simmons and Ryan, 1984). Increases in chitinase activity were stimulated in melon (*Cucumis melo*) by chitin fragments, which were released from chitin by partial acid hydrolysis (Roby et al., 1987). Since this treatment could result in partial deacetylation, it is possible that oligo-1,4-β-glucosamine structures were the active components in these experiments, too.

Fragments of pectic polysaccharides of plant cell walls released either by enzymatic or mild acid hydrolysis can also elicit phytoalexin accumulation (Nothnagel et al., 1983; Jin and West, 1984; Davis et al., 1986b; Ryan, 1988; Bruce and West, 1989). In addition, synergistic effects have been observed between these oligogalacturonides and phytoalexin elicitors of fungal origin in soybean and parsley (Darvill and Albersheim, 1984; Davis et al., 1986a; Davis and Hahlbrock, 1987). However, it is not known at which stage in the activation of the defense response the two stimuli interact.

The phytoalexin defense response, elicitor recognition and elicitor-mediated signal transduction have been studied in some detail in soybean. Various soybean tissues, including cotyledons, hypocotyls, roots and suspension-cultured cells, respond to treatment with a β-glucan elicitor derived from the cell walls of *Phytophthora megasperma* f. sp. *glycinea*, the causal agent of stem and root rot of soybean, by synthesizing a group of pterocarpan phytoalexins, the glyceollins (Ebel, 1986; Ebel and Grisebach, 1988). Although soybean cultivars are differentially resistant to several

P. megasperma races (Sinclair and Shurtleff, 1975), elicitor preparations from both avirulent and virulent races of the fungus are very similar in structure and identical in their abilities to stimulate glyceollin accumulation in the same soybean cultivar (Ayers et al., 1976). These elicitors alone, therefore, are insufficient to account for the known race/cultivar specificity of the *P. megasperma*–soybean interaction.

The phytoalexin elicitors of *P. megasperma* active in soybean have been characterized as β-glucans (Darvill and Albersheim, 1984), which are prominent structural polysaccharides of the cell walls of oomycetes. Investigations of the structure, distribution, and biological activity of these glucans have shown that the configuration required for elicitor activity is contained in oligosaccharides with 3-, 6-, and 3,6-linked glucosyl residues (Darvill and Albersheim, 1984; Sharp et al., 1984). Structure–activity relationships have also been analyzed with a series of chemically synthesized oligoglucosides (Davis et al., 1986a; Hahn et al., 1989; Cheong et al., 1991). A hexa-β-glucosyl glucitol and the corresponding hepta-β-glucoside were the most potent elicitors of phytoalexin accumulation in a bioassay, being active at concentrations of about 10 nM. It remains to be established how the structural motif of the oligoglucoside is integrated within the β-glucan of the *P. megasperma* cell wall. In addition, to what extent this cell wall component is processed in the plant–pathogen interaction to release elicitor-active glucan fragments in situ, also remains to be determined.

Crude elicitor preparations from *P. megasperma* also activate several defense responses in suspension-cultured parsley cells and protoplasts freshly prepared from these cells. Among these responses are the synthesis and secretion of furanocoumarin phytoalexins (Tietjen et al., 1983; Hauffe et al., 1986; Scheel et al., 1986) and the transcriptional activation of a characteristic set of additional defense-related genes (Somssich et al., 1989). The same responses have been observed in parsley leaves infected by *P. megasperma* (Scheel et al., 1986; Jahnen and Hahlbrock, 1988; Schmelzer et al., 1989). The parsley cell and protoplast systems have, therefore, been useful tools for detailed analyses of several aspects of the induction process.

Significant progress in the understanding of the parsley cell–elicitor interaction resulted from systematic analysis of an elicitor-active component of crude *P. megasperma* fractions. Although crude cell-wall preparations of this fungus were potent elicitors in both soybean and parsley, pure glucan fractions isolated from this material were active only in soybean and completely inactive in parsley (Parker et al., 1988). Treatment of crude preparations, either by proteolytic digestion or by chemical deglycosylation using trifluoro-methanesulphonic acid demonstrated that, in parsley, proteinaceous components are the active eliciting compounds. These are inactive in soybean. On the other hand, with similar preparations from *P. infestans*, both parsley and potato (*Salanum tuberosum*) respond to carbohydrate constituents (Scheel et al., 1989). There seems, therefore, to be no predictable consensus in the structural determinants of elicitor activity.

It appears that soybean, parsley and potato not only perceive different signals from *P. megasperma* and *P. infestans*, but are also unable to respond to certain fungal components recognized by the other plants.

Purification from *P. megasperma* culture filtrates of a component exhibiting elicitor activity in parsley correlated with enrichment of a glycoprotein of M_r 42,000 (Scheel et al., 1989; Parker et al., 1991). The purified glycoprotein has been demonstrated to elicit phytoalexin synthesis in parsley cells and protoplasts at concentrations as low as 1 nM. Deglycosylation and proteinase treatments confirmed that the protein, rather than the carbohydrate component, is critical for activity.

Use of a polyclonal antiserum raised against the deglycosylated elicitor demonstrated that the proteinaceous elicitor is present in higher quantities in the cell wall than in the culture filtrate of *P. megasperma*, although it represents a lower proportion of the total extractable protein in the former. A protein of slightly lower molecular weight than that of *P. megasperma* was detected with this antiserum in cell walls from each of *P. nicotianae* var. *parasitica* and *P. parasitica* (Parker et al., 1991). No crossreaction with the antiserum was found in extracts of mycelial walls from *P. infestans*, *Sclerotinia sclerotiorum*, *Fusarium solani*, *Rhynchosporium secalis*, and *Alternaria carthami*.

A number of other proteinaceous elicitors have also been identified. The first was isolated by Cruickshank and Perrin (1968) from *Monilinia fructicola* and called monilicolin A. It was found to have a molecular weight of approximately 8,000 and stimulated the accumulation of phaseollin in french bean (*Phaseolus vulgaris*), which is not a host plant for this fungus.

An extracellular glycoprotein of M_r 46,000 from *P. parasitica* var. *nicotianae* was found to elicit the accumulation of the sesquiterpenoid phytoalexin, capsidiol, in tobacco (*Nicotiana tabacum*) callus (Farmer and Helgeson, 1987). Boiling or digestion of the protein with pronase destroyed the elicitor activity, whereas periodate treatment did not, indicating that the activity resided in the protein component. A polyclonal antiserum against this glycoprotein detected crossreacting mycelial proteins in *P. cinnamomi*, *P. parasitica* var. *nicotianae*, *P. infestans*, *P. megasperma* f. sp. *glycinea*, and *P. cactorum*, but not in *Pythium ultimum* and *Fusarium oxysporum* f. sp. *pisi*.

Two other proteins isolated from *P. cryptogea* and *P. capsici*, respectively, elicited systemic leaf necrosis in tobacco, which is not a host plant for these fungi (Ricci et al., 1989). Treatment of tobacco with the pure proteins has been reported to protect the plant from the pathogen *Phytophthora nicotianae*, which does not produce such an elicitor (Ricci et al., 1989). A phenomenon such as this could be of potential use in plant protection. Development of appropriate inducers which are effective under field conditions may improve control of infectious plant diseases.

A peptidoglycan (M_r 67,000) in which elicitor activity resided in the carbohydrate moiety has been purified from germ tube walls of *Puccinia graminis* f. sp. *tritici* (Kogel et al., 1988), the causal agent of wheat-stem rust disease. The pure compound induced the accumulation of lignin-like compounds in wheat (*Triticum aestivum*), which was preceded by transient increases in the levels of phenylalanine ammonia-lyase, 4-coumarate:CoA ligase, and cinnamyl alcohol dehydrogenase activities (Moerschbacher et al., 1986). The carbohydrate portion of this elicitor consisted of mannose (22%), galactose (77%) and glucose (< 1%) (Beissmann and Reisener, 1990). The elicitor was equally active in two wheat cultivars whether or not they carried the *Sr5* gene for rust resistance.

III. Perception of Elicitor Stimulus

The characterization of primary events in the perception of elicitor signals by plant cells requires the measurement of at least one typical defense reaction in response to treatment with a pure elicitor. Experiments of this type have been performed with elicitors of callose formation and phytoalexin accumulation.

Chitosan, in stimulating callose synthesis, likely acts through allosteric activation of 1,3-β-glucan synthase (Kauss et al., 1989b). Its ability to stimulate callose formation in protoplasts of *Catharanthus roseus* increases with its chain length, but decreases with the degree of N-acetylation (Kauss et al., 1989a). It is believed that this polycation binds electrostatically to the negatively charged moieties of phospholipids, thereby severely disturbing plasma membrane integrity. Consistent with this hypothesis, fluorescein-labeled chitosan was distributed evenly over the surface of intact protoplasts from *Catharanthus roseus* (Kauss et al., 1990).

For elicitors other than chitosan the initial event in the activation of plant defense genes is thought to be the interaction with a specific primary target site(s) or receptor on the host cell surface. Experimental results which support such a hypothesis are the short apparent lag phase in the induction of the response, the sensitivity of the various target cells to elicitors, parameters of the dose-response relationships, the species-specificity of some purified elicitors, and the stringent structural requirements for the elicitor activity of certain compounds (e.g., the glucan-based fragments from *P. megasperma* (Darvill and Albersheim, 1984; Sharp et al., 1984). The most likely location of such target sites is the plasma membrane, although the involvement of soluble internal or cell wall-bound sites cannot be ruled out. The role of the plant cell wall is not clear, but could be related more to signal modulation than to perception.

Initial identification of binding sites for a fungal β-glucan in soybean cell membranes was obtained with [^{14}C]mycolaminaran from *Phytophthora* spp. (Yoshikawa et al., 1983). Two major disadvantages in using this ligand, however, were its low elicitor activity and the low specific radioactivity achieved by labeling in vivo. More recently, ^{3}H- and ^{125}I-labeled *P. megasperma* branched 1,3- and 1,6-β-glucan fragments with high elicitor activity and a narrow size range (18 to 22 glucose units) were used to demonstrate high-affinity binding to soybean membranes (Schmidt and Ebel, 1987; Cosio et al., 1988). The β-glucan fragments used in these studies were prepared by partial acid hydrolysis of the fungal cell wall and were shown to possess a glycosyl linkage composition typical of the *P. megasperma* branched β-glucan (Schmidt and Ebel, 1987). Modification of the glucans at the reducing end with phenylalkylamine reagents had no effect on binding affinity. This was exploited to synthesize an oligoglucosyl tyramine derivative suitable for radioiodination (Cosio et al., 1988).

The [^{125}I]glucan (> 100 Ci/mmol) thus obtained provided higher sensitivity and lower detection limits for binding assays while behaving in a manner identical to the [^{3}H]glucan labeled by reduction of the glucan fraction with boro[^{3}H]hydride (Schmidt and Ebel, 1987). The ^{3}H- and ^{125}I-labeled β-glucan ligands exhibited high-affinity binding (K_ds between 10 and 40 nM) which was both saturable and reversible (Cosio et al., 1988; Ebel et al., 1989). Specific glucan binding was detected in membrane preparations from soybean roots, hypocotyls, cotyledons, and cell-suspension cultures, but not in soluble protein extracts from these tissues. The binding followed the distribution of a plasma-membrane marker, 1,3-β-glucan synthase, in sucrose density gradients (Schmidt and Ebel, 1987). This finding agrees well with the observation that intact protoplasts isolated from soybean cell cultures show competable β-glucan binding with an affinity identical to that found in membrane preparations (Cosio et al., 1988).

Competition studies with a variety of poly- and oligosaccharides have been carried out to investigate the specificity of β-glucan binding to these sites (Schmidt and Ebel, 1987; Cosio et al., 1988). A strong positive correlation was found between the ability of the different ligands to displace the labeled glucan ligand and to elicit glyceollin accumulation in a bioassay. Oligoglucosides of varying degrees of polymerisation were obtained by acid hydrolysis of fungal polysaccharides. The ability of these compounds to compete with the ^{125}I-labeled glucan for binding to soybean membranes increased with their degree of polymerization (Fig. 1) (Cosio et al., 1990b). Ligand concentrations required for 50% inhibition of binding (IC_{50}) are presented in Fig. 1. Assuming a hyperbolic relationship, a minimum IC_{50} value of approximately 3 nM can be predicted for an "optimal" glucan competitor. Most interestingly, the β-glucan-binding sites displayed the highest affinity of all ligands tested to date (IC_{50} = 8 nM) (Fig. 1) for a heptaglucoside which was synthesized according to a published method (Fügedi et al., 1987). Competition with this ligand was complete and followed a uniform sigmoidal pattern, indicating access of the heptaglucoside to all sites available to the fungal ^{125}I-glucan fraction (Cosio et al., 1990b). This is the first demonstrated binding of a ligand clearly defined in terms of structure and biological activity to the soybean β-glucan-binding sites, which were originally identified by using a ligand fraction of structural isomers (Schmidt and Ebel, 1987). The high affinity of the β-glucan-binding sites for the hepta-β-glucoside was confirmed in ligand-saturation ex-

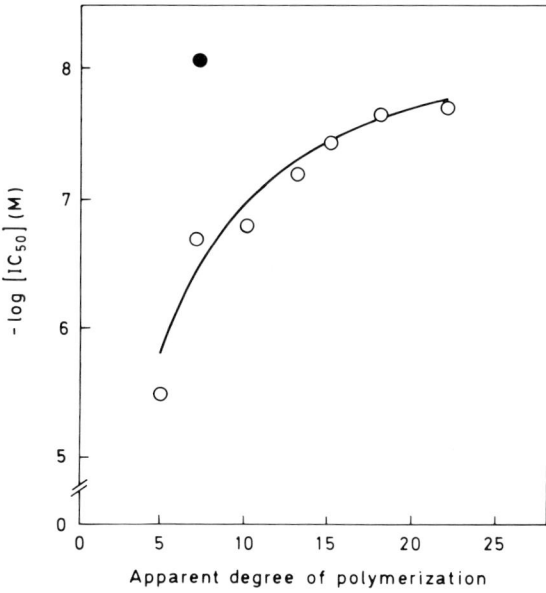

Fig. 1. Ability of oligoglucosides of varying apparent degrees of polymerization (DP), which were obtained by acid hydrolysis of *Phytophthora megasperma* β-glucans, to compete with an ^{125}I-labeled β-glucan (DP approximately 20) for binding to soybean (*Glycine max*) membranes (○). Strongest competition (IC$_{50}$ = 8 nM) was obtained with a synthetic heptaglucoside of defined structure (●) (IC$_{50}$: concentration giving 50% inhibition of binding) (Cosio et al., 1990b)

periments giving K_d values of 0.75 to 3 mM (Cosio et al., 1990b; Cheong and Hahn, 1991).

Further studies are aimed at elucidating the role of the β-glucan-binding sites in mediating the elicitor stimulus in soybean cells. To achieve this goal, the binding sites must be isolated and the putative protein(s) characterized. To this end, glucan-binding activity has been solubilized using detergents that contain C_{12} alkyl chains such as zwittergent 3-12, 1-dodecanoyl propanediol-3-phosphorylcholine (ES12H), and dodecyl maltoside (Cosio et al., 1990a). The solubilized binding site(s) displayed a binding affinity for β-glucans similar to that reported for the membrane-bound site(s). In gel-permeation chromatography, detergent-protein micelles with glucan-binding activity eluted with approximate M_r-values of 300,000 (zwittergent) and 380,000 (ES12H). On a chromatofocusing column, elution occurred between pH 6.2 and 6.6 (zwittergent and ES12H, respectively), suggesting that the apparent pI is nearly neutral. Glucan-binding in membranes was inhibited by concanavalin A but not by other lectins and the solubilized binding site(s) was retained by ConA-agarose, suggesting that the binding protein(s) are glycoproteins (E. Cosio and J. Ebel, unpubl. results). Ligand-affinity techniques, which have been indispensable in the isolation of less abundant receptor proteins, are presently being developed to achieve an efficient purification of the β-glucan-binding site.

The peptidoglycan from germ tube walls of *Puccinia graminis* f. sp. *tritici*, described above, was recently radioactively labeled to a specific activity of approximately 8×10^5 cpm/µg and used for binding studies with purified wheat plasma membranes (Kogel et al., 1991). Saturable and reversible binding was reported to be higher on plasma membranes than on intracellular membranes. An apparent K_d-value of 2 µM and a maximum number of binding sites of 250 pmol/mg of plasma membrane protein were calculated from these data. The results obtained with the radioactively labeled ligand were confirmed by ELISA experiments using polyclonal antisera against the elicitor. Elicitor binding sites were present in plasma membranes from wheat cultivars with and without the *Sr5* gene for rust resistance. This result is in agreement with the observation that the elicitor stimulates lignification in a cultivar non-specific manner. Treatment of protein blots of wheat plasma membranes after separation on an SDS gel with the radioactively labeled elicitor identified two proteins as candidates for elicitor binding sites.

IV. Transduction of Elicitor Signal

Plant defense gene activation, as involved in the phytoalexin response, requires the transduction of elicitor signals from the site of primary perception at the cell surface to the nucleus where transcription of specific genes is initiated. Knowledge about transmembrane signaling in response to environmental stimuli in plants is fragmentary and existing models rely on results obtained with animal systems. Well-established second messengers in animals include cAMP, cGMP, inositol 1,4,5-trisphosphate, Ca^{2+}, and diacylglycerol. In plants, a reasonable body of evidence exists for a role of Ca^{2+} as a second messenger as well as some indirect evidence for that of inositol 1,4,5-trisphosphate. Little evidence, however, exists that cAMP, cGMP or diacylglycerol act as second messengers (West et al., 1989).

In several plant/elicitor systems, convincing evidence has been obtained for the involvement of Ca^{2+} in defense gene activation. The reduction of extracellular Ca^{2+} concentrations in cell cultures of soybean (Stäb and Ebel, 1987), carrot (*Daucus carota*) (Kurosaki et al., 1987a), and parsley (Scheel et al., 1989) significantly lowered the levels of phytoalexins accumulated in response to elicitor. In cultured potato cells, the elicitor-stimulated increase in the activities of phenylalanine ammonia-lyase and tyrosine decarboxylase, two enzymes involved in transcriptionally regulated defense responses, was drastically reduced upon Ca^{2+} withdrawal (Scheel et al., 1989). Omission of Ca^{2+} from the medium of parsley protoplasts resulted in a corresponding reduction in run-off transcription rates of elicitor-responsive genes, but did not affect transcription of constitutively expressed or UV-inducible genes (Table 2) (C. Colling and D. Scheel, unpubl. results). The

Table 2. Effect of Ca^{2+} in the culture medium on the run-off transcription rate of elicitor-responsive genes in parsley (*Petroselinum crispum*) protoplasts treated with a crude cell wall-derived elicitor (Ayers et al., 1976) from *Phytophthora megasperma* f. sp. *glycinea* (C. Colling and D. Scheel, unpubl. results)

Gene[a]	Run-off transcription rate [% of control][b]	
	$+ Ca^{2+}$	$- Ca^{2+}$
CON2	100	98 ± 9
PAL	100	53 ± 11
4CL	100	63 ± 24
BMT	100	63 ± 12
ELI5	100	37 ± 20
ELI12	100	20 ± 4
PR1	100	56 ± 15

[a] Labelled transcripts were hybridized to filter-bound cDNAs specifying the following genes: phenylalanine ammonia-lyase (PAL) (Schulz et al., 1989) and 4-coumarate:CoA ligase (4CL) (Douglas et al., 1987) from the general phenylpropanoid pathway, SAM; bergaptol O-methyltransferase (BMT) (K.-D. Hauffe and D. Scheel, unpubl. results), specifically involved in phytoalexin biosynthesis, three genes (ELI5, ELI12, PR1) of unknown function, all of which are elicitor-responsive, and one gene (CON2) that is expressed constitutively (Somssich et al., 1989)

[b] Run-off transcription was performed as described by Chappell and Hahlbrock (1984) in the presence or absence of Ca^{2+} in the protoplast media. Mean values and standard deviations of five independent experiments are shown

Ca^{2+} ionophore, A23187, stimulated phytoalexin synthesis in soybean (Stäb and Ebel, 1987) and carrot (Kurosaki et al., 1987a), but not in parsley (C. Colling and D. Scheel, unpubl. results). Consistent with these results, the Ca^{2+} channel blocker, verapamil, inhibited elicitor-stimulated phytoalexin accumulation in carrot (Kurosaki et al.,1987a) and soybean cells (Stäb and Ebel, 1987), but had no effect on furanocoumarin accumulation in elicitor-treated parsley cells (C. Colling and D. Scheel, unpubl. results). In potato tuber disks, addition of Ca^{2+} or Sr^{2+}, but not Mg^{2+}, enhanced arachidonic acid-stimulated phytoalexin accumulation, which was, however, inhibited by EGTA and La^{3+} (Zook et al., 1987). In both cases, this inhibition was overcome by Ca^{2+} addition.

In elicitor-treated parsley cells, direct measurements of external Ca^{2+}, Cl^-, K^+ and protons with ion-selective electrodes and of Ca^{2+} fluxes using $^{45}Ca^{2+}$ revealed alkalinization of the culture medium and rapid uptake of

Ca^{2+} as well as massive efflux of Cl^- and K^+. These changes were detectable within 2 min of elicitor addition to cultured cells (Scheel et al., 1991). Taken together, it appears that more complex changes of plasma membrane ion fluxes are involved in defense gene activation in parsley than in soybean, carrot or potato. It is not yet known if these ion fluxes are primary events in transduction of the elicitor signal or an indirect response to depletion of mobilized internal Ca^{2+} stores. Patch-clamp analysis on single parsley protoplasts, which retain their responsiveness to elicitor (Dangl et al., 1987), may provide the means to detect and characterize elicitor-responsive ion channels, thereby resolving this question (Hedrich and Schroeder, 1989).

Very similar fluxes of Ca^{2+}, K^+, and protons were observed upon treatment of plant cells with elicitors of callose formation, such as chitosan (Kauss et al., 1989b). Since chitosan also triggered phytoalexin accumulation in soybean (Köhle et al., 1984) and parsley cells (Conrath et al., 1989), it was postulated that the two processes, activation of 1,3-β-glucan synthase and induction of defense gene transcription, share these initial steps (Kauss et al., 1989b). However, the pure glycoprotein elicitor of phytoalexin accumulation in parsley described previously did not stimulate callose formation in parsley protoplasts, and chitosan failed to induce furanocoumarin synthesis under conditions optimal for stimulation of callose formation in this system (Scheel et al., 1991). Precise determination of ion fluxes in response to these different types of elicitor may, therefore, reveal whether quantitative differences in release or uptake of specific ions lead to qualitative differences in defense responses. Moreover, the ability of chitosan to stimulate phytoalexin accumulation in cultured cells appears to be limited to a narrow range of relatively high concentrations (Köhle et al., 1984; Conrath et al., 1989). Under these conditions, chitosan may act by cell toxification, in a manner similar to abiotic elicitors.

In order to fulfill its function as a second messenger, Ca^{2+} is present in the cytosol at low concentrations which are tightly controlled (Marmé, 1989). In animals phosphoinositides play an important role in the regulation of intracellular Ca^{2+} levels (Berridge and Irvine, 1989). All elements necessary for the operation of this second messenger system appear to be present in plants as well (Boss, 1989). In addition, inositol 1,4,5-trisphosphate treatment released Ca^{2+} from zucchini (*Cucurbita pepo*) microsomes (Drøbak and Ferguson, 1985) and opened Ca^{2+} channels in red beet (*Beta vulgaris*) root vacuole membranes (Alexandre et al., 1990). Elicitor treatment of parsley or soybean cells for 0.5 to 20 min, however, did not result in significant changes in the concentrations of phosphatidylinositol 4-phosphate and phosphatidylinositol 4,5-bisphosphate (Strasser et al., 1986). In addition, Li^+, which inhibits the enzymatic hydrolysis of inositol phosphates, thereby reducing the supply of inositol (Berridge et al., 1989), does not affect elicitor-stimulated phytoalexin accumulation in parsley (C. Colling and D. Scheel, unpubl. results). In carrot cells on the other hand, an elicitor-mediated increase in D-*myo*-inositol 1,4,5-trisphosphate was reported to precede 6-methoxymellein synthesis and to indicate the involvement of phosphoinositides in transduction of the elicitor signal in this system (Kurosaki et al., 1987b). However, the extremely complex biochemistry of phosphoinositides in plants, which was recognized only recently (Boss, 1989), calls for more detailed analyses to confirm these results.

Results from animal systems suggest a direct coupling of cytosolic Ca^{2+} and pH in signal transduction (Dickens et al., 1990). The elicitor-mediated alkalinization of the parsley cell culture medium, mentioned above, is probably due to proton uptake and may, therefore, result in corresponding changes in the cytosolic pH. In several non-plant systems, cytosolic pH shifts precede changes in cellular metabolism and development (Busa and Nuccitelli, 1984). It has been postulated that these pH changes act as "synergistic messengers" in concert with second messengers, such as Ca^{2+}. At present only indirect evidence from ^{31}P NMR studies exists for intracellular pH changes in plant cells as part of the elicitor response. Transient acidification of the vacuole and the cytosol was observed in these experiments within a few minutes of elicitor treatment of cultured parsley and french bean cells (Strasser et al., 1983; Ojalvo et al., 1987).

In contrast to Ca^{2+}, which appears to play an important role both in animals and plants, it is still controversial whether cyclic nucleotides act as second messengers in plants as they do in animals and microorganisms. Cyclic AMP is present in plants at levels of a few pmoles per gram fresh weight (Colling et al., 1988). This concentration did not change in soybean hypocotyls, cultured soybean cells, parsley cells or parsley protoplasts in response to elicitor treatment, when measured from 0.5 to 18 h after elicitor application in soybean (Hahn and Grisebach, 1983), or 1 min to 2 h in parsley (Scheel et al., 1989). In contrast, transient increases in cAMP levels were reported for elicitor-treated carrot cells with maximum concentrations at 20 to 40 min after elicitor addition (Kurosaki et al., 1987a). Cultured carrot cells and sweet potato (*Ipomoea batatas*) roots accumulated phytoalexins in response to exogenously applied dibutyryl cAMP and cAMP, respectively, whereas parsley cells and protoplasts did not react to similar treatments (Oguni et al., 1976; Kurosaki et al., 1987a; C. Colling and D. Scheel, unpubl. results).

Several other processes, namely elicitor-mediated lipid peroxydation (Rogers et al., 1988), NADPH-dependent generation of superoxide anions (Doke, 1985) and H_2O_2 production (Lindner et al., 1988; Apostol et al., 1989) have recently been proposed to play a role in signal transduction leading to plant defense gene activation (Boller, 1989; Dixon and Lamb, 1990). Although these processes offer interesting alternative models, evidence for their involvement in transmembrane signaling is still scant.

Little is known at present about the integration of intracellular Ca^{2+} in the transduction pathway linking elicitor perception to defense gene activation. Ca^{2+}-dependent phosphorylation of proteins, such as specific transcription factors, may represent one such mechanism. Protein kinases are key elements in signal transduction and amplification in animal cells (Hunter, 1987). This is likely to be the case in plant cells as well (Ranjeva and Boudet, 1987), where the modulation of a number of physiological processes, in particular by Ca^{2+}, appears to involve protein phosphorylation (Veluthambi and Poovaiah, 1986; Ranjeva and Boudet, 1987). Elicitor treatment of soybean cells, prelabeled with radioactive phosphate, resulted in transient increases, as well as decreases, in the labeling of several proteins when compared with those of untreated control cells (Grab et al., 1989). An endogenous factor was isolated from soybean cells that activated the in vitro phosphorylation of a protein with a molecular weight of 69,000, which was dephosphorylated in vivo in response to elicitor. Further work is required to characterize the target proteins as well as this factor and to elucidate their function in vivo. Approximately sixteen different proteins were transiently labeled by in vivo phosphorylation in elicitor-treated parsley cells (Dietrich et al., 1990). Most of these were phosphorylated specifically in response to elicitor and not other stress conditions, such as UV irradiation or heat shock. Two of these proteins were found to be located in the nucleus, thirteen were soluble and one was associated with microsomal as well as cytoplasmic fractions. Elicitor-mediated phosphorylation of these proteins was greatly reduced in the absence of external Ca^{2+}, as were mRNA accumulation (Dietrich et al., 1990) and transcription rates of defense responsive genes (Table 2). These latter results support the idea that the signal transduction chain leading to plant defense gene activation involves Ca^{2+}-dependent protein kinases, although conclusive evidence for a causal relationship is still lacking. Similar results were obtained with cultured tomato cells treated with an elicitor from yeast extract. The protein kinase inhibitors K-252a and staurosporine suppressed elicitor responses, such as alkalinization of the culture medium, ethylene synthesis, and increases in activity of phenylalanine ammonia-lyase, and completely prevented elicitor-stimulated changes in protein phosphorylation (Grosskopf et al., 1990; Felix et al., 1991).

V. Regulation of Plant Responses

The elicitors of plant defenses described in this chapter were characterized and in some cases purified, by their ability to cause necrosis, callose formation or phytoalexin accumulation in distinct experimental systems. In most cases, the same systems have also been used successfully to study the regulation of the particular response. While the molecular basis of elicitor-induced necrosis is still poorly defined, considerable progress has been made in understanding the regulation of callose synthesis and defense gene activation.

The biosynthesis of the 1,3-β-glucan, callose, from UDP-glucose is catalyzed by a plasma membrane-located 1,3-β-glucan synthase, which is allosterically activated by Ca^{2+} (Fink et al., 1987; Kauss, 1987). The binding sites for both UDP-glucose and Ca^{2+} appear to be located on the inside of the plasma membrane. An increase in the levels of cytosolic Ca^{2+}, which may be caused by chitosan treatment of plant cells, was recently concluded to be essential, but not sufficient, for the onset of callose formation in vivo (Kauss et al., 1989b). Additional factors required for in vivo regulation are still unknown.

Phytoalexin accumulation is stimulated by the specific activation of genes encoding the appropriate biosynthetic enzymes (Chappell and Hahlbrock, 1984; Ebel, 1986; Hahlbrock and Scheel, 1989; Dixon and Lamb, 1990). Rapid increases in the transcription rates of these genes are generally accompanied by the activation of an entire set of additional plant defense genes (Hahlbrock and Scheel, 1987; Somssich et al., 1989; Dixon and Lamb, 1990). The promoters of a number of these genes have been sequenced and some of them analyzed for the presence of specific regulatory elements (Dixon and Lamb, 1990; Dixon and Harrison, 1990). In vivo footprinting experiments with elicitor-treated and UV-irradiated parsley cells defined putative *cis*-elements in the promoter of the phenylalanine ammonia-lyase-1 gene (*Pc*PAL-1) (Fig. 2) (Lois et al., 1989). Although conclusive evidence for the functional involvement of these motifs in the induction process is still lacking, sequences with striking homology are present in the promoter regions of a number of stress-inducible genes and

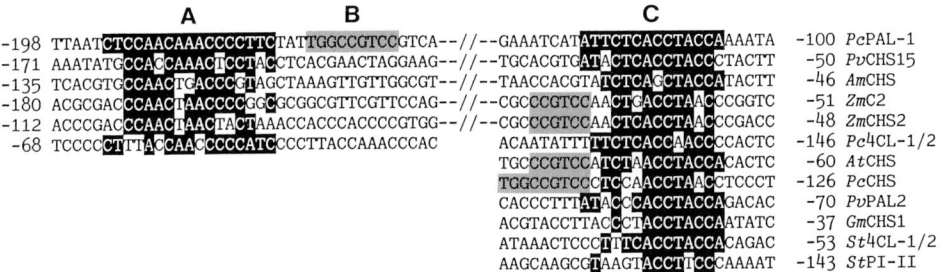

Fig. 2. UV light- (A, C) and elicitor-inducible (A–C) in vivo footprints in the parsley phenylalanine ammonia-lyase promoter in comparison to conserved motifs in the promoters of several genes involved in phenylpropanoid biosynthesis and/or responses to stress. Distances from the transcription start sites are indicated on either side of the sequences. *Pc*PAL-1, *Petroselinum crispum* phenylalanine ammonia-lyase-1 (Lois et al., 1989); *Pv*CHS15, *Phaseolus vulgaris* chalcone synthase 15 (Dron et al., 1988); *Am*CHS, *Antirrhinum majus* chalcone synthase (Sommer and Saedler, 1986); *Zm*C2 and *Zm*CHS2, *Zea mays* chalcone synthases C2 and 2, respectively (Niesbach-Klösgen, 1987); *Pc*4CL-1/2, *Petroselinum crispum* 4-coumarate:CoA ligases 1 and 2, respectively (Douglas et al., 1987); *At*CHS, *Arabidopsis thaliana* chalcone synthase (Feinbaum and Ausubel, 1988); *Pc*CHS, *Petroselinum crispum* chalcone synthase (Herrmann et al., 1988); *Pv*PAL2, *Phaseolus vulgaris* phenylalanine ammonia-lyase 2 (Cramer et al., 1989); *Gm*CHS1, *Glycine max* chalcone synthase 1 (Wingender et al., 1989); *St*4CL-1/2, *Solanum tuberosum* 4-coumarate:CoA ligases 1 and 2, respectively (Becker-André et al., 1991); *St*PI-II, *Solanum tuberosum* proteinase inhibitor II (Palm et al., 1990)

genes involved in the synthesis of phenylpropanoid compounds (Fig. 2). Within the families of chalcone synthase genes of soybean (Wingender et al., 1989) and phenylalanine ammonia-lyase genes of french bean (Cramer et al., 1989), only the promoters of those members responsive to elicitor treatment contain one of these motifs. In addition, a sequence differing in only two nucleotides from the most upstream PcPAL-1 footprint is located directly adjacent to a 10 bp region of the potato proteinase inhibitor II promoter which has been found to bind a wound-inducible nuclear protein (Palm et al., 1990). Further analyses are required to functionally characterize these sequence motifs and the corresponding trans-acting factors, which may represent final links in the related signal transduction chains.

Plant defense gene activation is a transient process, which apparently involves receptor-mediated transmembrane signaling. Efficient regulation of the defense responses requires removal of the elicitor from its primary target site on the plasma membrane. Horn et al. (1989) followed the fate of fluorescein- and ^{125}I-labeled phytoalexin elicitors in cultured soybean cells. Although impure preparations were used in these experiments, the results were reported to indicate that the elicitors were internalized after binding to the cell surface. Several hours after application to plant cells, the elicitors were located in the major vacuole. The amount of internalized label was significantly reduced in the presence of unlabeled elicitor. Control-proteins without elicitor activity neither bound to the cell surface nor were they internalized. The ^{125}I-labeled polygalacturonic acid elicitor was reported to retain its size after internalization. Receptor-mediated endocytosis might, therefore, be a mechanism by which plant cells clear elicitor-loaded receptors from the plasma membrane to guarantee both transience of the induction process and control of the number of active receptor molecules.

VI. Conclusions

Elicitors of necrosis, callose formation and plant defense gene activation have been purified from fungal culture filtrates and crude cell wall preparations employing convenient screening systems to monitor the respective resistance response. The mode of action of the necrosis-inducing elicitors is not known. Chitosan and phytoalexin elicitors appear to stimulate fluxes of Ca^{2+}, K^+ and H^+ by different mechanisms, resulting in allosteric activation of 1,3-β-glucan synthase and plant defense gene activation, respectively. The primary event in chitosan action is probably perturbation of membrane integrity, whereas phytoalexin elicitors seem to bind to receptor-like components of the plant plasma membrane. The structural diversity of these latter elicitors precludes the definition of a consensus structure. In addition, a given plant cell appears to be capable of responding to structurally different types of phytoalexin elicitors with the same defense response. We envisage, therefore, the presence of several different elicitor target sites on the plant cell surface, which generate a primary intracellular signal upon elicitor binding. These signals may then be funneled through a unique transduction chain which involves Ca^{2+}-dependent processes, such as protein phosphorylation, and results finally in the activation of multiple defense genes.

Acknowledgements

We thank Christiane Colling (Köln), Eric Cosio (Freiburg), Klaus Hahlbrock, Wolfgang Knogge, Erich Kombrink, Wendy Sacks, Imre Somssich, and Günter Strittmatter (all Köln) for discussions and critical reading of the manuscript. The work in the authors' laboratories was supported by the Deutsche Forschungsgemeinschaft (SFB 206 and Sche 235/2-1), Fonds der Chemischen Industrie and the Max-Planck-Gesellschaft.

VII References

Alexandre J, Lassalles JP, Kado RT (1990) Opening of Ca^{2+} channels in isolated red beet root vacuole membrane by inositol 1,4,5-trisphosphate. Nature 343: 567–570

Apostol I, Heinstein PF, Low PS (1989) Rapid stimulation of an oxidative burst during elicitation of cultured plant cells. Plant Physiol 90: 109–116

Ayers AR, Ebel J, Valent B, Albersheim P (1976) Host–pathogen interactions X. Fractionation and biological activity of an elicitor isolated form the mycelial walls of *Phytophthora megasperma* var. *sojae*. Plant Physiol 57: 760–765

Barz W, Bless W, Daniel S, Gunia W, Hinderer W, Jaques U, Kessmann H, Meier D, Tiemann K, Wittkampf U (1989) Elicitation and suppression of isoflavones and pterocarpan phytoalexins in chickpea (*Cicer arietinum* L.) cell cultures. In: Kurz WGW (ed) Primary and secondary metabolism of plant cell cultures. Springer, Berlin Heidelberg New York Tokyo, pp 208–218

Becker-André M, Schulze-Lefert P, Hahlbrock K (1991) Structural comparison, modes of expression and putative *cis*-acting elements of two 4-coumarate: CoA ligase genes in potato. J Biol Chem 266: 8551–8559

Beissmann B, Reisener HJ (1990) Isolation and purity determination of glycoprotein elicitors from wheat stem rust using medium pressure liquid chromatography. J Chromatogr 521: 187–197

Berridge MJ, Irvine RF (1989) Inositol phosphates and cell signalling. Nature 341: 197–205

Berridge MJ, Downes CP, Hanley MR (1989) Neural and developmental actions of lithium: a unifying hypothesis. Cell 59: 411–419

Boller T (1989) Primary signals and second messengers in the reaction of plants to pathogens. In: Boss WF, Morré DJ (eds) Second messengers in plant growth and development. AR Liss, New York, pp 227–255

Boss WF (1989) Phosphoinositide metabolism: its relation to signal transduction in plants. In: Boss WF, Morré DJ (eds) Second messengers in plant growth and development. AR Liss, New York, pp 29–56

Bruce RJ, West CA (1989) Elicitation of lignin biosynthesis and isoperoxidase activity by pectic fragments in suspension cultures of castor bean. Plant Physiol 91: 889–897

Busa WB, Nuccitelli R (1984) Metabolic regulation via intracellular pH. Am J Physiol 246: 409–438

Chappell J, Hahlbrock K (1984) Transcription of plant defence genes in response to UV light or fungal elicitor. Nature 311: 76–78

Cheong J-J, Hahn MG (1991) A specific, high affinity binding site for the hepta-β-glucoside elicitor exists in soybean membranes. Plant Cell 3: 137–147

Cheong J-J, Birberg W, Fügedi P, Pilotti Å, Garegg PJ, Hong N, Ogawa T, Hahn MG (1991) Structure-activity relationships of oligo-β-glucoside elicitors of phytoalexin accumulation in soybean. Plant Cell 3: 127–136

Colling C, Gilles R, Cramer M, Nass N, Moka R, Jaenicke L (1988) Measurement of 3′:5′ cyclic AMP in biological samples using a specific monoclonal antibody. Second Messengers Phosphoproteins 12: 123–133

Conrath U, Domard A, Kauss H (1989) Chitosan-elicited synthesis of callose and of coumarin derivatives by parsley cell suspensions. Plant Cell Rep 8: 152–155

Cosio EG, Frey T, Ebel J (1990a) Solubilization of soybean membrane binding sites for fungal β-glucans that elicit phytoalexin accumulation. FEBS Lett 264: 235–238

Cosio EG, Frey T, Verduyn R, van Boom J, Ebel J (1990b) High-affinity binding of a synthetic heptaglucoside and fungal glucan phytoalexin elicitors to soybean membranes. FEBS Lett 271: 223–226

Cosio EG, Pöpperl H, Schmidt WE, Ebel J (1988) High-affinity binding of fungal β-glucan fragments to soybean (Glycine max L.) microsomal fractions and protoplasts. Eur J Biochem 175: 309–315

Cramer CL, Edwards K, Dron M, Liang X, Dildine SL, Bolwell GP, Dixon RA, Lamb CJ, Schuch W (1989) Phenylalanine ammonia-lyase gene organization and structure. Plant Mol Biol 12: 367–383

Cruickshank IAM, Perrin DR (1968) The isolation and partial characterization of monilicolin A, a polypeptide with phaseollin-inducing activity from Monilinia fructicola. Life Sci 7: 449–458

Dangl JL, Hauffe KD, Lipphardt S, Hahlbrock K, Scheel D (1987) Parsley protoplasts retain differential responsiveness to u.v. light and fungal elicitor. EMBO J 6: 2551–2556

Darvill AG, Albersheim P (1984) Phytoalexins and their elicitors—a defense against microbial infection in plants. Annu Rev Plant Physiol 35: 243–275

Davis KR, Hahlbrock K (1987) Induction of defense responses in cultured parsley cells by plant cell wall fragments. Plant Physiol 85: 1286–1290

Davis KR, Darvill AG, Albersheim P (1986a) Host–pathogen interactions XXXI. Several biotic and abiotic elicitors act synergistically in the induction of phytoalexin accumulation in soybean. Plant Mol Biol 6: 23–32

Davis KR, Darvill AG, Albersheim P, Dell A (1986b) Host–pathogen interactions XXX. Characterization of elicitors of phytoalexin accumulation in soybean released from soybean cell walls by endopolygalacturonic acid lyase. Z Naturforsch 41c: 39–48

Dickens CJ, Gillespie JI, Greenwell JR, Hutchinson P (1990) Relationship between intracellular pH (pH_i) and calcium (Ca_i^{2+}) in avian heart fibroblasts. Exp Cell Res 187: 39–46

Dietrich A, Mayer JE, Hahlbrock K (1990) Fungal elicitor triggers rapid, transient and specific protein phosphorylation in parsley cell suspension cultures. J Biol Chem 265: 6360–6368

Dixon RA, Harrison MJ (1990) Activation, structure, and organization of genes involved in microbial defense in plants. Adv Genet 28; 165–234

Dixon RA, Lamb CJ (1990) Molecular communication in interactions between plants and microbial pathogens. Annu Rev Plant Physiol Plant Mol Biol 41: 339–367

Doke N (1985) NADPH-dependent O_2 generation in membrane fractions isolated from wounded potato tubers inoculated with Phytophthora infestans. Physiol Plant Pathol 27: 311–322

Douglas C, Hoffmann H, Schulz W, Hahlbrock K (1987) Structure and elicitor or u.v.-light-stimulated expression of two 4-coumarate:CoA ligase genes in parsley. EMBO J 6: 1189–1195

Drøbak BK, Ferguson IB (1985) Release of Ca^{2+} from plant hypocotyl microsomes by inositol-1,4,5-trisphosphate. Biochem Biophys Res Commun 130: 1241–1246

Dron M, Clouse SD, Dixon RA, Lawton MA, Lamb CJ (1988) Glutathione and fungal elicitor regulation of a plant defense gene promoter in electroporated protoplasts. Proc Natl Acad Sci USA 85: 6738–6742

Ebel J (1986) Phytoalexin synthesis: the biochemical analysis of the induction process. Annu Rev Phytopathol 24: 235–264

Ebel J, Grisebach H (1988) Defense strategies of soybean against the fungus *Phytophthora megasperma* f. sp. *glycinea*: a molecular analysis. Trends Biochem Sci 13: 23–27

Ebel J, Cosio EG, Feger M, Grab D, Habereder H (1989) Elicitation of phytoalexin synthesis in soybean (*Glycine max*) by a fungal pathogen and a fungal β-glucan. In: Lugtenberg BJJ (ed) Signal molecules in plants and plant–microbe interactions. Springer, Berlin Heidelberg New York Tokyo, pp 203–210

Farmer EE, Helgeson JP (1987) An extracellular protein from *Phytophthora parasitica* var. *nicotianae* is associated with stress metabolite accumulation in tobacco callus. Plant Physiol 85: 733–740

Feinbaum RL, Ausubel FM (1988) Transcriptional regulation of the *Arabidopsis thaliana* chalcone synthase gene. Mol Cell Biol 8: 1985–1992

Felix G, Grosskopf DG, Regenass M, Boller T (1991) Rapid changes of protein phosphorylation are involved in transduction of the elicitor signal in plant cells. Proc Natl Acad Sci USA 88: 8831–8834

Fink J, Jeblick W, Blaschek W, Kauss H (1987) Calcium ions and polyamines activate the plasma membrane-located 1,3-β-glucan synthase. Planta 171: 130–135

Flor HH (1971) Current status of the gene-for-gene concept. Annu Rev Phytopathol 9: 275–296

Fügedi P, Birberg W, Garegg PJ, Pilotti Å (1987) Synthesis of a branched heptasaccharide having phytoalexin-elicitor activity. Carbohydr Res 164: 297–312

Grab D, Feger M, Ebel J (1989) An endogenous factor from soybean (*Glycine max* L.) cell cultures activates phosphorylation of a protein which is dephosphorylated in vivo in elicitor-challenged cells. Planta 179: 340–348

Grisebach H, Ebel J (1978) Phytoalexins, chemical defense substances of higher plants? Angew Chem Int Ed Engl 17: 635–647

Grosskopf DG, Felix G, Boller T (1990) K-252a inhibits the response of tomato cells to fungal elicitors in vivo and their microsomal protein kinase in vitro. FEBS Lett 275: 177–180

Hadwiger LA, Beckman JM (1980) Chitosan as a component of pea-*Fusarium solani* interactions. Plant Physiol 66: 205–211

Hahlbrock K, Scheel D (1987) Biochemical responses of plants to pathogens. In: Chet I (ed) Innovative approaches to plant disease control. Wiley, New York, pp 229–254

Hahlbrock K, Scheel D (1989) Physiology and molecular biology of phenylpropanoid metabolism. Annu Rev Plant Physiol Plant Mol Biol 40: 347–369

Hahn MG, Grisebach H (1983) Cyclic AMP is not involved as a second messenger in the response of soybean to infection by *Phytophthora megasperma* f. sp. *glycinea*. Z Naturforsch 38c: 578–582

Hahn MG, Cheong J-J, Birberg W, Fügedi P, Piloti Å, Garegg P, Hong N, Nakahara Y, Ogawa T (1989) Elicitation of phytoalexins by synthetic oligoglucosides, synthetic oligogalacturonides, and their derivatives. In: Lugtenberg BJJ (ed) Signal molecules in plants and plant–microbe interactions. Springer, Berlin Heidelberg New York Tokyo, pp 91–97

Hauffe KD, Hahlbrock K, Scheel D (1986) Elicitor-stimulated furanocoumarin biosynthesis in cultured parsley cells: S-adenosyl-L-methionine:bergaptol and S-adenosyl-L-methionine:xanthotoxol O-methyltransferases. Z Naturforsch 41c: 228–239

Hedrich R, Schroeder JI (1989) The physiology of ion channels and electrogenic pumps in higher plants. Annu Rev Plant Physiol Plant Mol Biol 40: 539–569

Herrmann A, Schulz W, Hahlbrock K (1988) Two alleles of the single-copy chalcone synthase gene in parsley differ by a transposon-like element. Mol Gen Genet 212: 93–98

Horn MA, Heinstein PF, Low PS (1989) Receptor-mediated endocytosis in plant cells. Plant Cell 1: 1003–1009

Hunter T (1987) A thousand and one protein kinases. Cell 50: 823–829

Jahnen W, Hahlbrock K (1988) Cellular localization of nonhost resistance reactions of parsley (*Petroselinum crispum*) to fungal infection. Planta 173: 197–204

Jin DF, West CA (1984) Characteristics of galacturonic acid oligomers as elicitors of casbene synthetase activity in castor bean seedlings. Plant Physiol 74: 989–992

Kauss H (1987) Some aspects of calcium-dependent regulation in plant metabolism. Annu Rev Plant Physiol 38: 47–72

Kauss H, Jeblick W, Domard A (1989a) The degrees of polymerization and N-acetylation of chitosan determine its ability to elicit callose formation in suspension cells and protoplasts of *Catharanthus roseus*. Planta 178: 385–392

Kauss H, Waldmann T, Jeblick W, Euler G, Ranjeva R, Domard A (1989b) Ca^{2+} is an important but not the only signal in callose synthesis induced by chitosan, saponins and polyene antibiotics. In: Lugtenberg BJJ (ed) Signal molecules in plants and plant–microbe interactions. Springer, Berlin Heidelberg New York Tokyo, pp 107–116

Kauss H, Waldmann T, Quader H (1990) Ca^{2+} as a signal in the induction of callose synthesis. In: Ranjeva R, Boudet A (eds) Signal perception and transduction in higher plants. Springer, Berlin Heidelberg New York Tokyo, pp 117–132

Keen NT, Bruegger B (1977) Phytoalexins and chemicals that elicit their production in plants. In: Hedin PA (ed) Host plant resistance to pests. American Chemical Society, Washington, DC, pp 1–26

Keen NT, Dawson WO (1992) Pathogen avirulence genes and elicitors of plant defense. In: Boller T, Meins F (eds) Genes involved in plant defense. Springer, Wien New York, pp 85–114 [Dennis ES et al (eds) Plant gene research. Basic knowledge and application]

Keen NT, Tamaki S, Kobayashi D, Stayton M, Gerhold D, Shen H, Gold S, Lorang J, Thordal-Christensen H (1989) Characterization and function of avirulence genes from *Pseudomonas syringae* pv. tomato. In: Lugtenberg BJJ (ed) Signal molecules in plants and plant–microbe interactions. Springer, Berlin Heidelberg New York Tokyo, pp 183–188

Knogge W, Hahn M, Lehnackers H, Rüpping E, Wevelsiep L (1991) Fungal signals involved in the specificity of the interaction between barley and *Rhynchosporium secalis*. In: Hennecke H, Verma DPS (eds) Advances in molecular genetics of plant–microbe interactions, vol 1. Kluwer, Dordrecht, pp 250–253

Kogel G, Beissmann B, Reisener HJ, Kogel KH (1988) A single glycoprotein from *Puccinia graminis* f. sp. *tritici* cell walls elicits the hypersensitive lignification response in wheat. Physiol Mol Plant Pathol 33: 173–185

Kogel G, Beissmann B, Reisener HJ, Kogel K-H (1991) Specific binding of a hypersensitive lignification elicitor of *Puccinia graminis* f. sp. *tritici* to the plasma membrane from wheat (*Triticum aestivum* L.). Planta 183: 164–169

Köhle H, Young DH, Kauss H (1984) Physiological changes in suspension-cultured soybean cells elicited by treatment with chitosan. Plant Sci Lett 33: 221–230

Köhle, H, Jeblick W, Poten F, Blaschek W, Kauss H (1985) Chitosan-elicited callose synthesis in soybean cells as a Ca^{2+}-dependent process. Plant Physiol 77: 544–551

Kurosaki F, Tsurusawa Y, Nishi A (1987a) The elicitation of phytoalexins by Ca^{2+} and cyclic AMP in carrot cells. Phytochemistry 26: 1919–1923

Kurosaki F, Tsurusawa Y, Nishi A (1987b) Breakdown of phosphatidylinositol during the elicitation of phytoalexin production in cultured carrot cells. Plant Physiol 85: 601–604

Lindner WA, Hoffmann C, Grisebach H (1988) Rapid elicitor-induced chemiluminescence in soybean cell suspension cultures. Phytochemistry 27: 2501–2503

Lois R, Dietrich A, Hahlbrock K, Schulz W (1989) A phenylalanine ammonia-lyase gene from parsley: structure, regulation and identification of elicitor and light responsive cis-acting elements. EMBO J 8: 1641–1648

Marmé D (1989) The role of calcium and calmodulin in signal transduction. In: Boss WF, Morré DJ (eds) Second messengers in plant growth and development. AR Liss, New York, pp 57–80

Moerschbacher B, Heck B, Kogel KH, Obst O, Reisener HJ (1986) An elicitor of the hypersensitive lignification response in wheat leaves isolated from the rust fungus *Puccinia graminis* f. sp. *tritici*. II. Induction of enzymes correlated with the biosynthesis of lignin. Z Naturforsch 41c: 839–844

Niesbach-Klösgen U (1987) Molekulare Analyse des *C2* Gens aus *Zea mays* L. und Studien zur Evolution der Chalkonsynthase in Pflanzen. Dissertation, Universität zu Köln, Cologne, Federal Republic of Germany

Nothnagel EA, McNeil M, Albersheim P, Dell A (1983) Host–pathogen interactions XXII. A galacturonic acid oligosaccharide from plant cell walls elicits phytoalexins. Plant Physiol 71: 916–926

Oguni I, Suzuki K, Uritani I (1976) Terpenoid induction in sweet potato roots by cyclic-3',5'-adenosine monophosphate. Agricult Biol Chem 40: 1251–1252

Ojalvo I, Rokem JS, Navon G, Goldberg I (1987) ^{31}P NMR study of elicitor treated *Phaseolus vulgaris* cell suspension cultures. Plant Physiol 85: 716–719

Palm CJ, Costa MA, An G, Ryan C (1990) Wound-inducible nuclear protein binds DNA fragments that regulate a proteinase inhibitor II gene from potato. Proc Natl Acad Sci USA 87: 603–607

Parker JE, Hahlbrock K, Scheel D (1988) Different cell-wall components from *Phytophthora megasperma* f. sp. *glycinea* elicit phytoalexin production in soybean and parsley. Planta 176: 75–82

Parker JE, Schulte W, Hahlbrock K, Scheel D (1991) An extracellular glycoprotein from *Phytophthora megasperma* f. sp. *glycinea* elicits phytoalexin synthesis in cultured parsley cells and protoplasts. Mol Plant Microbe Interact 4: 19–27

Preisig CL, Kuc JA (1985) Arachidonic acid-related elicitors of the hypersensitive response in potato and enhancement of their activities by glucans from *Phytophthora infestans* (Mont.) de Bary. Arch Biochem Biophys 236: 379–389

Ranjeva R, Boudet AM (1987) Phosphorylation of proteins in plants: regulatory effects and potential involvement in stimulus/response coupling. Annu Rev Plant Physiol 38: 73–93

Ricci P, Bonnet P, Huet J-C, Sallantin M, Beauvais-Cante F, Bruneteau M, Billard V, Michel G, Pernollet J-C (1989) Structure and activity of proteins from pathogenic fungi *Phytophthora* eliciting necrosis and acquired resistance in tobacco. Eur J Biochem 183: 555–563

Roby D, Gadelle A, Toppan A (1987) Chitin oligosaccharides as elicitors of chitinase activity in melon plants. Biochem Biophys Res Commun 143: 885–892

Rogers KR, Albert F, Anderson AJ (1988) Lipid peroxidation is a consequence of elicitor activity. Plant Physiol 86: 547–553

Ryan CA (1988) Oligosaccharides as recognition signals for the expression of defensive genes in plants. Biochemistry 27: 8879–8883

Scheel D, Hauffe KD, Jahnen W, Hahlbrock K (1986) Stimulation of phytoalexin formation in fungus-infected plants and elicitor-treated cell cultures of parsley. In: Lugtenberg B (ed) Recognition in microbe–plant symbiotic and pathogenic interactions. Springer, Berlin Heidelberg New York Tokyo, pp 325–331

Scheel D, Colling C, Keller H, Parker J, Schulte W, Hahlbrock K (1989) Studies on elicitor recognition and signal transduction in host and non-host plant/fungus pathogenic interactions. In: Lugtenberg BJJ (ed) Signal molecules in plants and plant–microbe interactions. Springer, Berlin Heidelberg New York Tokyo, pp 211–218

Scheel D, Colling C, Hedrich R, Kawalleck P, Parker E, Sacks WR, Somssich IE, Hahlbrock K (1991) Signals in plant defense gene activation. In: Hennecke H, Verma DPS (eds) Advances in molecular genetics of plant–microbe interactions, vol 1. Kluwer, Dordrecht, pp 373–380

Schmelzer E, Krüger-Lebus S, Hahlbrock K (1989) Temporal and spatial patterns of gene expression around sites of attempted fungal infection in parsley leaves. Plant Cell 1: 993–1001

Schmidt WE, Ebel J (1987) Specific binding of a fungal glucan phytoalexin elicitor to membrane fractions from soybean *Glycine max*. Proc Natl Acad Sci USA 84: 4117–4121

Schottens-Toma IMJ, De Wit PJGM (1988) Purification and primary structure of a necrosis-inducing peptide from apoplastic fluids of tomato infected with *Cladosporium fulvum* (syn. *Fulvia fulva*). Physiol Mol Plant Pathol 33: 59–67

Schulz W, Eiben H-G, Hahlbrock K (1989) Expression in *Escherichia coli* of catalytically active phenylalanine ammonia-lyase from parsley. FEBS Lett 258: 335–338

Sharp JK, McNeil M, Albersheim P (1984) The primary structures of one elicitor-active and seven elicitor-inactive hexa (β-D-glucopyranosyl)-D-glucitols isolated from the mycelial walls of *Phytophthora megasperma* f. sp. *glycinea*. J Biol Chem 259: 11321–11336

Sinclair JB, Shurtleff MC (eds) (1975) Compendium of soybean diseases. American Phytopathological Society, St. Paul, MN

Sommer H, Saedler H (1986) Structure of the chalcone synthase gene of *Antirrhinum majus*. Mol Gen Genet 202: 429–434

Somssich IE, Bollmann J, Hahlbrock K, Kombrink E, Schulz W (1989) Differential early activation of defense-related genes in elicitor-treated parsley cells. Plant Mol Biol 12: 227–234

Stäb MR, Ebel J (1987) Effects of Ca^{2+} on phytoalexin induction by fungal elicitor in soybean cells. Arch Biochem Biophys 257: 416–423

Strasser H, Tietjen KG, Himmelspach K, Matern U (1983) Rapid effect of an elicitor on uptake and intracellular distribution of phosphate in cultured parsley cells. Plant Cell Rep 2: 140–143

Strasser H, Hoffmann C, Grisebach H, Matern U (1986) Are polyphosphoinositides involved in signal transduction of elicitor-induced phytoalexin synthesis in cultured plant cells? Z Naturforsch 41c: 717–724

Tietjen KG, Hunkler D, Matern U (1983) Differential response of cultured parsley cells to elicitors from two non-pathogenic strains of fungi. 1. Identification of induced products as coumarin derivatives. Eur J Biochem 131: 401–407

Veluthambi K, Poovaiah BW (1986) In vitro and in vivo protein phosphorylation in *Avena sativa* L. coleoptiles. Effects of Ca^{2+}, calmodulin antagonists, and auxin. Plant Physiol 81: 836–841

Walker-Simmons M, Ryan CA (1984) Proteinase inhibitor synthesis in tomato leaves.

Induction by chitosan oligomers and chemically modified chitosan and chitin. Plant Physiol 76: 787–790

Walker-Simmons M, Hadwiger L, Ryan CA (1983) Chitosans and pectic polysaccharides both induce the accumulation of the antifungal phytoalexin pisatin in pea pods and antinutrient proteinase inhibitors in tomato leaves. Biochem Biophys Res Commun 110: 194–199

Walker-Simmons M, Jin D, West CA, Hadwiger L, Ryan CA (1984) Comparison of proteinase inhibitor-inducing activities and phytoalexin elicitor activities of a pure fungal endopolygalacturonase, pectic fragments, and chitosans. Plant Physiol 76: 833–836

West CA (1981) Fungal elicitors of the phytoalexin response in higher plants. Naturwissenschaften 68: 447–457

West CA, Bruce R, Ren Y-Y (1989) Second messengers in animals and their possible relevance for plants. In: Lugtenberg BJJ (ed) Signal molecules in plants and plant–microbe interactions. Springer, Berlin Heidelberg New York Tokyo, pp 27–40

Wingender R, Röhrig H, Höricke C, Wing D, Schell J (1989) Differential regulation of soybean chalcone synthase genes in plant defence, symbiosis and upon environmental stimuli. Mol Gen Genet 218: 315–322

Yamada T, Hashimoto H, Shiraishi T, Oku H (1989) Suppression of pisatin, phenylalanine ammonia-lyase mRNA, and chalcone synthase mRNA accumulation by a putative pathogenicity factor from the fungus *Mycosphaerella pinodes*. Mol Plant Microbe Interact 2: 256–261

Yoshikawa M, Keen NT, Wang M-C (1983) A receptor on soybean membranes for a fungal elicitor of phytoalexin accumulation. Plant Physiol 73: 497–506

Zook MN, Rush JS, Kuc JA (1987) A role for Ca^{2+} in the elicitation of rishitin and lubimin accumulation in potato tuber tissue. Plant Physiol 84: 520–525

Section IV

Plant Genes Induced in the Defense Reaction

Chapter 9

Pathogenesis-related Proteins

John R. Cutt and Daniel F. Klessig

Waksman Institute, Rutgers, The State University of New Jersey, Piscataway, NJ 08855, U.S.A.

Contents

I. Introduction

The survival of higher plants is dependent upon their ability to adapt to stress. Accordingly, plants have evolved to respond to stress by altering their normal patterns of gene expression (Sachs and Ho, 1986) and physiology. Adverse conditions that affect a plant's homeostasis are caused by environmental factors such as temperature, water, salt, mechanical damage (wounding), chemicals, UV light, and the interaction with pathogenic organisms.

Following pathogenic attack, some plants activate the expression of what are known as "defense-related" genes (Bowles, 1990). These genes are assumed to function in the inhibition of pathogen multiplication and spread. Some defense-related genes encode enzymes involved in phenyl-propanoid metabolism (Hahlbrock and Scheel, 1989), hydrolytic enzymes (chitinases and β-1,3-glucanases) (Boller, 1987), hydroxyproline-rich glyco-proteins (cell wall proteins) (Cooper et al., 1987), and proteinase inhibitors (Ryan, 1988). Others encode proteins of unknown activity and biological function. Current investigations into the molecular mechanisms underlying disease resistance are aimed at understanding the role of defense-related genes.

Over twenty years ago it was observed that tobacco cultivars resistant to infection with tobacco mosaic virus (TMV) accumulated a new set of proteins after infection (Van Loon and Van Kammen, 1970; Gianinazzi et al., 1970). Subsequent to this, it was established that the accumulation of these proteins also could be induced by other pathogens and by the exposure to certain chemical agents. The proteins were referred to as "b" proteins or "new components", until 1980 when the term "pathogenesis-related" (PR) proteins was proposed by Antoniw et al. (1980). Initial studies of the PRs were performed using native, alkaline polyacrylamide gel electro-phoresis (PAGE) systems which limited the analyses to the acidic isoforms of these proteins. After the localization of PRs in the apoplastic spaces of the leaf (Parent and Asselin, 1984), they commonly became defined as host encoded, acidic, extracellular proteins whose syntheses were coordinately induced in pathological or related situations (e.g., chemical stress). The nomenclature used to identify them was based upon their relative electro-phoretic mobility in PAGE.

The PR genes represent a subset of the defense-related genes which have been extensively characterized due to their potential involvement with disease resistance. PRs have been identified in many dicots and monocots and appear to be ubiquitous in higher plants. Enzyme activities (β-1,3-glucanase and chitinase) have been discovered for some of the PRs. It is now known that "basic" isoforms of the PRs also exist and that the localization of PRs is not restricted to the extracellular spaces. In addition, the PR genes have been shown to be developmentally regulated. As opposed to the coordinated control exhibited after pathogen infection, the individual PR genes are differentially regulated during normal development.

The PR protein nomenclature has become rather confusing due to the utilization of several gel systems by different researchers, the discovery of similar genes, related by sequence identity but not associated with patho-genesis, and the existence of multiple protein isoforms. In light of this, the classical definition and nomenclature has been reconsidered to clarify the terminology used for PRs. After the second Workshop on Pathogenesis-related Proteins in Plants (Valencia, Spain, October, 1989) a committee was organized to standardize the nomenclature of PRs. They tentatively agreed on an updated definition of PRs which permits the inclusion of newly identified proteins. Accordingly, PRs are now defined as host-encoded polypeptides which are not present in healthy plants and are synthesized only in response to pathological or related stress situations. Proteins whose synthesis is induced by chemical or elicitor treatment, natural senescence, and wounding are to be considered PRs, since these types of stress often accompany pathogenesis. Genes which are stimulated during pathogenesis but are also expressed in healthy plants are not defined as PRs. Genes which share sequence identity but not the same characteristic mode of genetic regulation are to be referred to as PR-like. The most important criterion in the definition is the induction by pathogenesis-related stress. However, even with these guidelines, the definition of a PR protein remains vague. As will be discussed, recent evidence clearly demonstrates that PR gene expression is tissue-specific and developmentally regulated in a manner analogous to other defense-related genes which do not fall into the PR subset. Thus, the classification of a PR gene based upon its "lack of expression" in a healthy plant will strongly depend on the origin and developmental state of the tissue used as a control and how rigorously this control is performed.

It is not the intention of the authors to present a comprehensive compilation of PR literature in this chapter, but instead to provide an update of the more current data on PRs at the time this chapter was written (September 1990). A more complete background of the PRs can be obtained from earlier reviews by Gianinazzi (1984), Van Loon (1985), Bol and Van Kan (1988), Bol (1988), Carr and Klessig (1989), and Lotan and Fluhr (1990). We will focus primarily on the use of tobacco (*Nicotiana tabacum*) and TMV as the model host-pathogen system and the specific tobacco PRs.

II. The Tobacco-TMV System

The host-pathogen system that has been utilized most extensively for the study of PRs is the tobacco-TMV system; therefore the PRs will be described primarily within the context of this system. In general, infection of tobacco with TMV results in one of two types of responses. Certain tobacco cultivars, such as Xanthi-nc (NN) and Samsun NN, carry the dominant resistance "N" locus originating from *Nicotiana glutinosa* (Holmes, 1938; Takahashi, 1956). These resistant cultivars display a hypersensitive response (HR) (Matthews, 1981) when infected with TMV. The HR is observed in many host–pathogen interactions and is characterized by the formation of necrotic lesions at the site of infection and restriction of pathogen multiplication and spread to the area adjacent to the lesion. Associated with the HR in tobacco is the de novo synthesis of PRs. Subsequent to the HR, leaves distal to the initial infection site acquire enhanced levels of resistance to secondary challenges with TMV or other pathogens (referred to as systemic acquired resistance or SAR) and synthesize and accumulate high levels of PRs. Underlying the systemic response must be a long distance signal trans-duction network which is capable of initiating a complex set of coordinated events resulting in a broad spectrum defense barrier. The dissection of this signaling network is currently an area of intense interest and will be discussed in later sections.

In contrast, the related cultivars Xanthi (nn) and Samsun nn, which do not carry the "N" locus, are susceptible to TMV infection and systemic spread of the virus. In these cultivars the virus spreads from the initial site of virus entry and causes severe disease symptoms. The PR genes are not induced during the systemic infection (Van Loon, 1975a; Antoniw et al., 1985)

A. Molecular and Cellular Biology of the PR Gene Families

Since the discovery of PRs in 1970 the number of these proteins reported in tobacco has increased. Van Loon et al. (1987) and Bol and Van Kan (1988) proposed that the PRs of tobacco (Xanthi and Samsun) be classified into five families or groups called PR-1 through PR-5. We will use these family designations when referring to the PRs. When referring to individual

Table 1. Pathogenesis-related proteins of tobacco

Family	Acidic isoforms		Basic isoforms		Properties
	name	M.W. (kDa)	name	M.W. (kDa)	
1	1a [1a] 1b [1b] 1c [1c]	15	–	19	unknown
2	2 [2a] N [2b] O [2c]	$(\sim 33\text{–}40)^b$	–	$(\sim 33\text{–}36)^b$	β-1,3-glucanases
3	P [3a] Q [3b]	$(\sim 27\text{–}29)^b$ $(\sim 28\text{–}30)^b$	–	33	chitinases
4	R [4] (R')[a]	13/15[c]	–	–	unknown
5	S [5] (R)[a]	23	Osmotin	26	related to Thaumatin, α-amylase/ proteinase inhibitors

Nomenclature listed is for Samsun NN PRs according to Antoniw et al. (1980), Jamet and Fritig (1986) and Van Loon et al. (1987). In brackets, nomenclature proposed by Van Loon et al. (1987)

[a] Cultivar-specific nomenclature for Xanthi-nc equivalents

[b] Molecular weight values represent a range of published values

[c] PR-R is comprised of two different polypeptides referred to as 'major' and 'minor' forms

proteins and genes we will, unfortunately, need to utilize the multiple nomenclature systems found in the literature. The different names are listed in Table 1 to assist in protein identification. Family members are serologically related to each other, possess nucleotide and amino acid sequence similarity, may exhibit similar enzymatic activities, and may have acidic or basic isoelectric points. In the following section the fundamental properties and characteristics of these five families of proteins and the genes that encode them will be described. Other proteins which conform to the definition of PRs, but do not yet have family designations, will also be discussed.

1. Family 1

The PR-1 proteins were among the first PRs described in relation to TMV infection, yet their biological function continues to be elusive. This family contains three serologically related polypeptide species, called PR-1a, -1b, and -1c, with acidic isoelectric points. They exhibit different mobilities in native, alkaline polyacrylamide gel electrophoresis (Davis, 1964) but migrate as a single species (M_r 15,000) in sodium dodecyl sulfate polyacrylamide gel electrophoresis (Antoniw et al., 1980, 1985; Matsuoka and Ohashi, 1984). Genetic analyses initially demonstrated that the three PR-1 isoforms were encoded by separate, single genes (Gianinazzi and Ahl, 1983; Ahl et al., 1982), indicating that they were part of a multigene family. Further confirmation of this came when cDNA and genomic clones became available for the PR-1s. The nucleotide sequences of cDNA clones obtained from both tobacco cultivars, Samsun NN (Cornelissen et al., 1986b; Matsuoka et al., 1987; Pfitzner and Goodman, 1987) and Xanthi-nc (Cutt et al., 1988), revealed that the genes encoding the PR-1s were > 90% identical. The open reading frames of the PR-1 genes each encode polypeptides of 168 amino acids in length with a hydrophobic domain of 30 amino acids at the N-terminus. This domain corresponds to a signal polypeptide sequence which is required for protein sorting and is cleaved from the preprotein during or after protein synthesis to yield a mature PR-1 polypeptide of 138 amino acids. No similarities have yet been found between the PR-1 proteins and any other plant proteins of known function. However, an interesting yet puzzling observation has been made that the PR-1 proteins share similarities to a white-face hornet venom allergen (Fang et al., 1988).

In addition, cDNA clones were identified by Cornelissen et al. (1987) that cross-hybridized, under low stringency conditions, to the PR-1 cDNAs encoding acidic proteins. Similar clones were isolated by differential screening of a cDNA library from cytokinin-stressed tobacco shoots (Memelink et al., 1987). The deduced open reading frame of these clones suggested that they encoded a mature polypeptide of 173 amino acids. This putative polypeptide would share 67% amino acid sequence identity to the acidic PR-1 proteins (Cornelissen et al., 1987). Two interesting observations were that the predicted polypeptides would contain a C-terminal extension and have basic amino acids substituted for many of the corresponding acidic residues in the acidic PR-1 proteins. Primer extension and Northern analyses have shown that mRNAs corresponding to the presumptive basic PR-1 protein were present in the roots of healthy tobacco plants (Memelink et al., 1987, 1990) and in TMV-infected tobacco leaves (Cornelissen et al., 1987; Dixon et al., 1991) (See Sect. II. B); however, the basic PR-1 proteins have not yet been detected. Thus, it appears that these genes encode basic isoforms of the PR-1 proteins. As will be seen, the presence of both acidic and basic isoforms appears to be a common theme in the biology of PRs. This may imply that multiple isoforms are required to perform divergent functions, or alternatively, to perform similar functions with differential affinities or specificities. Indeed, the tissue specific expression, the developmental regulation, and the subcellular localizations of the proteins suggests a requirement for a multiplicity of PR genes.

Southern blot analyses of tobacco DNA revealed that the PR-1 gene family included other members not predicted from the above data; as many as eight copies of both the acidic and the basic PR-1 genes appeared to be represented in the tobacco genome (Cornelissen et al., 1987). Acidic PR-1 genomic clones have been isolated from the tobacco cultivars Samsun NN (Cornelissen et al., 1987; Ohshima et al., 1987, 1990a, b), Xanthi-nc (Payne et al., 1988b; Cutt et al., unpubl. results), and W38 (Pfitzner et al., 1988, 1990). Characterization of these clones indicated that some members of the acidic PR-1 gene family detected by Southern analyses were pseudogenes. To date, only one genomic clone representing the basic PR-1 genes has been characterized (Payne et al., 1989). One common structural feature of the PR-1 gene family is the lack of introns in all of the genes examined so far. The characterization of cis-acting DNA sequences involved in the regulation of these genes will be discussed in Sect. II. B. 3.

The cellular location of the acidic PR-1 proteins has been extensively characterized. Parent and Asselin (1984) were the first to report their extracellular location by demonstrating their presence in the intracellular fluid (IF) extracted from TMV-infected leaves. Consistent with this, Carr et al. (1985) showed that the translation of the acidic PR-1 mRNAs occurred on membrane-bound polyribosomes indicating that the PR-1 proteins were translocated to the endoplasmic reticulum for secretion. These observations were underscored by the finding that the PR-1 proteins were secreted from chemically-treated tobacco suspension cell cultures (Ohashi and Matsuoka, 1987a) and from protoplasts prepared from TMV-infected tobacco leaves (Ohashi and Matsuoka, 1987b; Carr et al., 1987).

Immunological techniques allowed more precise localization of the PR-1 protein within electron-opaque material deposited adjacent to the outer cell wall (Benhamou et al., 1988) and within xylem elements of the vascular tissue (Carr et al., 1987). The unbound state for PR-1 proteins in the apoplast, indicated by their extraction from the IF in a soluble form, may represent a site of action for PR-1 function or a means of transport within the leaf. On the other hand, the association of PR-1 proteins with cell wall deposits implies a more structural role, possibly in the formation of a physical barrier to pathogens. It would be interesting to know the relationship between the free and bound forms.

The immunolocalization experiments of Dumas et al. (1988) and Hosokawa and Ohashi (1988) detected PR-1 proteins in the cytoplasm of mesophyll cells. Since the specific immunogold labeling was found in close association with the endoplasmic reticulum, this might reflect the nascent PR-1 polypeptides or newly translated PR-1 polypeptides being translocated to the plasma membrane for secretion. More recently, an intracellular location for acidic PR-1 proteins has been identified in the vacuoles of crystal idioblast cells of TMV-infected leaves (Dixon et al., 1991). Crystal idioblasts are specialized cells found in small numbers between the palisade and spongy mesophyll layers of tobacco leaves. These cells accumulate large quantities of calcium in the form of calcium oxalate crystals within their vacuoles (Franceschi and Horner, 1980). It has been postulated that these cells may perform an important role in the regulation of calcium ion levels in the surrounding plant tissue (Franceschi, 1989). Calcium has been shown to be an important regulatory element in signal transduction pathways in eukaryotic cells and is believed to function as a regulatory molecule in plants. It has been shown to be involved in the synthesis and secretion of α-amylase (Mitsui et al., 1984) and in the activation of callose synthetase (Fink et al., 1987) which catalyzes the formation of the β-1,3-glucan callose. Callose deposition is a well known defense response to pathogen infection.

In general, a protein destined for the endoplasmic reticulum is either targeted to the central vacuole or secreted, but not both. The extracellular and intracellular locations for the acidic PR-1 proteins raises interesting new questions concerning the nature of the signals responsible for protein sorting. Future work will hopefully determine if this differential targeting is controlled by protein modifications or by variations inherent in the specific cell types. Although the PR-1 proteins have been shown not to be glycosylated (Parent and Asselin, 1984; Pierpoint, 1986; Carr et al., 1987), the vacuolar PR-1 species represents a small percentage of the total found in the leaf and may have gone undetected.

2. Family 2

Kauffman et al. (1987) demonstrated that four serologically-related tobacco PRs have β-1,3-glucanase activity. This was one of the first enzymatic activities to be assigned to the PRs. β-1,3-glucanases or laminarinases (E.C. 3.2.1.39) hydrolyze β-1,3-glucans such as laminarin and are found in bacteria, fungi, algae, higher plants and some invertebrates (Bull and Chesters, 1966; Boller, 1987). Plant β-1,3-glucanases and chitinases represent potential antifungal hydrolases which act synergistically to inhibit fungal growth in vitro (Mauch et al., 1988a). In addition, β-1,3-glucanases release glycosidic fragments from both the pathogen and the host cell walls which could act as signals in the elicitation of host defenses (Keen and Yoshikawa, 1983; Hahn et al., 1989; Takeuchi et al., 1990). In higher plants, β-1,3-glucanase activity increases in response to pathogen infection or hormonal treatments. Several studies suggest that plant β-1,3-glucanases may be a component of a general defense system against pathogen invasion in a number of different plant species (Benhamou et al., 1989; Jondle et al., 1989; Joosten and De Wit, 1989; Kauffman et al., 1987; Kombrink and Hahlbrock, 1986; Kombrink et al., 1988; Mauch et al., 1988b; Meins and Ahl, 1989; Vögeli et al., 1988).

Three of the four proteins characterized by Kauffmann et al. (1987) were acidic PR-2a, -2b, and -2c isoforms (M_r ~ 33,000–40,000). These are also referred to as PR-2, -N, and -O, respectively, (see Table 1). The fourth was a basic isoform of the PR-2 family (M_r 33,000). Van den Bulcke et al. (1989) reported the purification and the partial amino acid sequences of three acidic glucanases isolated from IF extracts and two basic glucanases isolated from vacuoles. Each of the proteins had a unique amino acid sequence indicating that they were encoded by separate genes. The vacuolar β-1,3-glucanases were nearly identical to a hormonally regulated β-1,3-glucanase (Shinshi et al., 1988). Van den Bulcke et al. (1989) concluded that the basic, vacuolar isoforms and the acidic, extracellular isoforms represented two distinct classes of the PR β-1,3-glucanases in tobacco.

The β-1,3-glucanase cDNA clone isolated by Shinshi et al. (1988) encoded a polypeptide of 359 amino acids in length. The preproenzyme contained a 21 residue hydrophobic signal polypeptide at the N-terminus required for translocation to the ER. In addition, it contained a 22 residue C-terminal extension, with a putative N-glycosylation site, that was not found on the mature protein. Following the initial polypeptide synthesis, the preproenzyme was shown to be processed by removal of the signal polypeptide and glycosylation of the proenzyme. Subsequent deglycosylation and cleavage of the C-terminal extension generated the mature enzyme.

Another vacuolar protein, concanavalin A of jack beans (Bowles and Pappin, 1988), also contains a C-terminal extension in the preproenzyme. The basic PR-1 cDNA predicts a C-terminal extension as well, but the protein has not yet been identified. It is possible that signals for vacuolar targeting reside in these sequences. However, the localization of the acidic PR-1 proteins (Dixon et al., 1991) in the vacuoles of crystal idioblasts suggest that the C-terminal sequences are not an absolute requirement for vacuolar targeting.

More recently, cDNAs which encode the acidic PR-2a, -2b, and -2c proteins (Côté et al., 1991; Ward et al., 1991) and four novel β-1,3-glucanases (Ward et al., 1991) have been isolated. Three of these novel cDNAs could encode polypeptides with neutral or slightly basic isoelectric points which are ~ 90% similar to the acidic PR-2a, -2b, and -2c proteins. The proteins encoded by these three genes have not yet been characterized. The fourth cDNA, called PR-Q', encodes an acidic, extracellular protein which is similar to the elicitor-releasing β-1,3-glucanase from soybean (Takeuchi et al., 1990) and structurally different from the acidic PR-2a, -2b, and -2c cDNAs (~ 54–59% identity). Consequently, there is now evidence for four distinct classes of PR β-1,3-glucanases in tobacco: (*i*) the basic, vacuolar proteins, (*ii*) the acidic, extracellular PR-2a, -2b, and -2c proteins, (*iii*) the basic or neutral forms that are similar to PR-2a, -2b, and -2c, and (*iv*) the PR-Q' enzyme.

The basic, vacuolar β-1,3-glucanase genes are highly induced after TMV infection. Although these proteins are present in healthy tobacco plants, particularly in their roots (Felix and Meins, 1986; Memelink et al., 1987; Neale et al., 1990), they could be considered members of the PR-2 family. The regulation of the genes encoding the two new classes of β-1,3-glucanases (*iii* and *iv*) just described have not been examined in detail. While it is not yet known if they fit the definition of PRs, they have been discussed here because of their similarity to known PRs. For a more extensive overview of the β-1,3-glucanase genes, as well as the chitinase genes discussed next, see Meins et al. (1992).

3. Family 3

In addition to the glucanases, Fritig's group purified and characterized four more proteins which were induced by TMV infection (Legrand et al., 1987). These were all found to be serologically related and to possess chitinase activities. Chitinases (E.C. 3.2.1.14) hydrolyze the polymer chitin (poly[1 → 4-*N*-acetyl-β-D-glucosamine]), a substrate found in fungi and insects, but not plants. Despite the absence of chitin in plants, chitinases have been identified in many plant species (Boller, 1987) and have been implicated in the defense against invading pathogens (Boller, 1985). As mentioned previously, plant β-1,3-glucanases and chitinases have been shown to act synergistically to inhibit fungal growth in vitro (Mauch et al., 1988b). The association of this enzyme with a number of host–pathogen interactions (Metraux et al., 1988; Roby et al., 1988; Mauch et al., 1989; Vögeli-Lange et al., 1988) implies that it may function in vivo to inhibit pathogen attack.

Two of the chitinases described by Legrand et al. (1987) corresponded to the previously identified acidic PRs, PR-3a(P) and PR-3b(Q) (M_r 27,500 and 28,500, respectively), found in the IF of infected leaves. The other two chitinases were basic isoforms (M_r 32,000 and 34,000) of the enzyme.

Complementary DNA clones that correspond to the acidic chitinases were isolated from libraries prepared from RNAs extracted from TMV-infected leaves (Hooft van Huijsduijnen et al., 1987; Payne et al., 1990; Linthorst et al., 1990). Comparison of the amino acid sequences deduced from the cDNAs and the sequences of peptides derived from the purified proteins showed that these clones encoded PR-3a(P) and PR-3b(Q) (Payne et al., 1990; Linthorst et al., 1990). Partial characterization of genomic clones corresponding to at least one of the cDNA clones indicate that the 253 amino acid open reading frame was interrupted by two introns (Linthorst et al., 1990). Clones corresponding to the basic chitinases have also been characterized (Hooft van Huijsduijnen et al., 1987; Linthorst et al., 1990) and revealed complete identity with the sequence of a hormonally-regulated chitinase cDNA (pCHN 50) reported by Shinshi et al. (1987). Recently, Shinshi et al. (1990) described a genomic clone from tobacco which was nearly identical (99.7% similarity) to the pCHN 50 cDNA clone. This basic chitinase gene contained an open reading frame of 329 amino acids which was interrupted by two introns. A 23 amino acid signal polypeptide sequence was predicted at the N-terminus. Like the basic isoforms of the PR-1 and PR-2 families, the basic chitinases have a C-terminal extension (6 amino acids) compared to their acidic counterparts. Neuhaus and Boller (pers. comm.) recently demonstrated that this short sequence was both necessary and sufficient for intercellular retention of a normally extracellular cucumber chitinase to which it was attached. This suggests that the vacuolar targeting signal resides in this short C-terminal extension. To date, two acidic and two basic chitinase proteins have been identified. Southern blot analyses suggest that two to four genes may be present in the tobacco genome for each of these two classes (Huijsduijnen et al., 1987).

4. Family 4

The older nomenclature for families 4 and 5 is particularly confusing because it is cultivar specific. Family 4 (PR-4) corresponds to PR-R' in Xanthi-nc and PR-R in Samsun NN, while family 5 (PR-5) corresponds to PR-R in Xanthi-nc and PR-S in Samsun NN (see Table 1). We will use the numerical family designation to hopefully avoid this confusion.

The PR-4 family in Xanthi-nc is represented by one (Pierpoint, 1986) or possibly two (Parent and Asselin, 1984) members with very similar molecular weights (M_r 15,000), while in Samsun NN this family appears to have two members of slightly different sizes (M_r 13,000 and 15,000) (Van Loon et al., 1987). These proteins have low pIs and can be extracted from the IF, although their precise locations in the plant have not been defined by immunolocalization. Serological cross-reactivity has been reported between this family and the PR-2 proteins by Van Loon et al. (1987) but this was not detected by Kauffmann et al. (1987). The general consensus is that the PR-4 proteins constitute a distinct PR family, though this has yet to be confirmed at the molecular level. As with the PR-1s, no function or enzymatic activities have been assigned to the PR-4s.

5. Family 5

The PR-5 family contains at least two acidic members (M_r 23,000–25,000) that exist as major and minor isoforms in the IF of infected leaves (Pierpoint et al., 1987; Kauffmann et al., 1990). The cDNA clones for the major form from Xanthi-nc (Payne et al., 1988a) and minor form from Samsun NN (Bol, 1988) both were shown to encode polypeptides which contained signal sequences necessary for secretion to the extracellular space. Two genomic clones corresponding to the major and minor isoforms have been isolated and sequenced (van Kan et al., 1989). They exhibit 95% sequence similarity, do not have introns, and share limited sequence similarities in the 5′ upstream region with the PR-1 genes and the glycine-rich protein (GRP) genes (see below) that are coordinately induced by TMV infection.

These proteins share extensive sequence similarity ($\sim 65\%$) to the sweet-tasting protein thaumatin (Cornelissen et al., 1986a; Pierpoint et al., 1987) and are often called thaumatin-like (TL) proteins. However, they also exhibit considerable sequence similarity (57%) to the maize bifunctional trypsin/α-amylase inhibitor (Richardson et al., 1987). By analogy it has been suggested that they may function as inducible protease inhibitors directed against foraging insects (see Sect. II.C).

This family also contains basic isoforms termed osmotins (M_r 26,000) which were originally characterized through their association with adaptation to salt stress (Singh et al, 1987; LaRosa et al., 1989). Only recently has the osmotin gene been shown to be induced by TMV infection and wounding (Neale et al., 1990). Osmotins are found predominantly in vacuolar inclusion bodies; however, a low level of the protein has been detected in the cytoplasm of cells as well (Singh et al., 1987). Comparison of the amino acid sequences of osmotin, thaumatin, acidic PR-5s, a salt-induced protein from tomato (King et al., 1988), and the maize bifunctional inhibitor demonstrated that these proteins are structurally-related and each contain an unusual feature of 16 conserved cysteine residues (Pierpoint and Shewry, 1987; Singh et al., 1989). Moreover, osmotin contains a 20 amino acid C-terminal extension, not found on the acidic PR-5s. Once again, a C-terminal extension can be implicated in containing signals for protein sorting to the vacuole, but a direct involvement will have to await future experiments. The multiple isoforms of osmotin might be generated by cleavage and processing events involving the C-terminal extension. Alternatively, they might be encoded by separate genes, consistent with the two copies detected by Southern analysis (Singh et al., 1989).

6. Additional PRs

The classification of newly identified proteins/genes as PRs is often tenuous. As we gain a better understanding of the PRs, many of them will become referred to more by the characteristics of their biological activities, in accordance with other defense-related genes. Thus, it appears useful to retain the working definition of PRs to initially categorize unknown proteins associated with pathogenesis but then refer to them by their biological properties as they become known. For these reasons, we will only discuss here PRS of unknown function or proteins already labeled as PRS. No attempt will be made to incorporate well characterized defense-related genes that fit the PR gene definition into this group such as peroxidases (see review by Bowles, 1990; also Lagrimini et al., 1990).

Hogue and Asselin (1987) utilized a combination of different PAGE systems to identify a total of 23 proteins from the extracellular spaces (IF) of TMV-infected Xanthi-nc leaves. Ten to twelve of these corresponded to members of the five PR families discussed above. Two others (b_{6a} and b_{7a}) comigrated with peroxidase activities (Parent et al., 1985). The remaining proteins have not been further characterized.

Several additional PRS were also identified and purified from low pH extracts of TMV-infected Samsun NN leaves (Kauffmann et al., 1990). Two of these, designated PR-R and PR-S by the authors (M_r 24,000), corresponded to the PR-5 family. The identity of the remaining four proteins was less clear. Based on M_r values and chromatographic properties PR-r_1 (M_r 14,500) and PR-r_2 (M_r 13,000) probably correspond to previously described members of family 4. PR-s_1 and PR-s_2 may represent new PR proteins.

Several cDNA clones corresponding to TMV-induced mRNAs may also represent new PR genes. Hooft van Huijsduijnen et al. (1986b) identified six classes of cDNA clones (called clusters A–F), four of which corresponded to known PR families. The remaining two (clusters A and C) have been characterized to different extents. The mRNA corresponding to cluster A was induced by salicylic acid treatment, as well as TMV infection, and encoded a predicted 40,000 kDa protein.

The cluster C cDNA was used to isolate several genomic clones (van Kan et al., 1988). The cloned gene had an intron of 555 bp, and Southern analysis indicated it was part of a small gene family. This gene could encode a polypeptide of 109 amino acids with a 25% glycine content and a putative 26 residue N-terminal signal peptide. The predicted polypeptide product, which has not yet been detected in tobacco, was called the glycine-rich protein (GRP). Similar GRPs have been described in bean (Keller et al., 1988) and in petunia (Condit and Meagher, 1986; Varner and Cassab, 1986), where they may function as cell wall components. If the tobacco GRP turns out to be a cell wall protein it may be functionally equivalent to another class of pathogen-inducible, defense-related proteins, the hydroxyproline-rich glycoproteins (HGRPS) which are found in the plant cell walls (Lawton and Lamb, 1987; Ecker and Davis, 1987; Benhamou et al., 1990).

B. PR Gene Expression

1. In Response to Environmental Signals: Biotic and Abiotic Inducers

Pathogens.   Infection by many different pathogens induces the synthesis of PRS (see review by Carr and Klessig, 1989). PR synthesis is often, but not always, associated with disease resistance (HR) and/or systemic acquired resistance (SAR) (see Sect. II.C). For instance, with TMV PRS accumulate during the HR but not during systemic infection of susceptible plants. Other viruses which cause local, necrotizing infections in tobacco, as well as in other species such as tobacco necrosis virus (TNV) (Gianinazzi and Kassanis, 1974; Van Loon, 1975), also stimulate the accumulation of PRS. However, PR protein accumulation is not always associated with virus induced local necrosis.

Bacterial, as well as viral, infection also stimulates production of PRS in tobacco and other plant species. Infection by *Pseudomonas syringae* induced PR protein accumulation and SAR to subsequent TMV infection in tobacco (Ahl et al., 1981), while tobacco leaves infiltrated with the avirulent B1 strain of *Pseudomonas solanacearum* also synthesized PRS (Leach et al., 1983).

The induction of PRS (specifically PR-1) in response to a fungal infection was first described in the tobacco cultivar White Burley responding hypersensitively to *Thielaviopsis basicola* (Gianinazzi et al., 1980). Recently β-1,3-glucanases (PR-2), chitinases (PR-3), and several other PR proteins were detected in tobacco plants after "immunization" against blue mold by stem injections with sporangiospores of the blue mold pathogen, *Peronospora tabacina* (Tuzun et al., 1989).

Microbial Elicitors.   Treatment of plants or plant tissue with elicitors of microbial origin have been shown to induce some defense responses. Most notably is the stimulation of the pathways leading to production of phytoalexins (Albersheim and Valent, 1978; Hahlbrock and Scheel, 1989; Dixon and Lamb, 1990). Microbial elicitors have also been shown to induce PR genes. In tobacco PRS have been induced in response to cell-free extracts of the bacterium *Nocardia asteroids* (Gianinazzi and Martin, 1975) and the fungus *Stachybotrys chartarum* (Maiss and Poehling, 1983) and lipopolysaccharides extracted from *Pseudomonas solanacearum* (Leach et al., 1983). However, a glucan preparation obtained from the mycelial walls of the fungus *Phytophthora megasperma* f. sp. *glycinea*, which is known to be an effective elictor of phytoalexins in soybean, failed to stimulate PR genes in tobacco (Kopp et al., 1989). This suggests that either the elicitor is species (or race)-specific and/or different induction pathways are used for the activation of these two defense-related responses (see Sect. II.B.4).

Growth Regulators.   The expression of PR genes can be altered by changing the levels of plant hormones in tissue culture or in intact plants. In the latter case, this has been achieved by addition of exogenous hormones or by the generation of endogenous auxins and/or cytokinins via transformation with *Agrobacterium tumefaciens* T-DNA which contains genes encoding enzyme involved in hormone production (see review by Carr and Klessig, 1989). For instance, hormone imbalance resulting from expression of an

introduced cytokinin gene in tobacco caused a 10-fold or greater enhancement of expression of several PR genes including the acidic PR-2s,
acidic and basic PR-1s and the basic PR-3s (Memelink et al., 1987, 1990).
The hormone ethylene also induces some PRs and may be part of the
natural induction pathway (see Sect. II. B. 4).

Physiological Stress (Other than Infection). Several physiological stresses
stimulate PR gene expression to varying degrees. Tissue cutting and abra-
sion (Ohashi and Matsuoka, 1985; Ohshima et al., 1990) or prolonged
floatation on water (Asselin et al., 1985) stimulated low levels of PR gene
expression, as did water stress induced by mannitol (Wagih and Coutts,
1981; Pierpoint et al., 1981). In contrast, induction of PR-1 and PR-2 genes
by TMV infection or ethylene treatment was inhibited at temperatures in
excess of 28–30 °C (Van Loon, 1975; Van Loon and Antoniw, 1982; White
et al., 1983). Thus, high temperatures inhibit rather than induce PR gene
expression (Cornelissen et al., 1987).

Chemical Inducers. Several chemicals have been shown to induce PR
protein synthesis in tobacco and in some cases disease resistance as well.
Early suggestions that the PRs of tobacco might be analogous to the
interferon system of animals led Gianinazzi and Kassanis (Gianinazzi and
Kassanis, 1974) to test the effects of polyacrylic acid (PAA), an inducer of
interferon synthesis (DeClerq et al., 1970), on TMV multiplication and PR
protein accumulation in plants. Application of PAA to Xanthi-nc tobacco
induced the accumulation of PR-1 and PR-2 proteins and reduced the
number of TMV-induced lesions (Gianinazzi and Kassanis, 1974). In addition, its application increased resistance to other viruses such as TNV, potato
virus X, and potato virus Y, and the bacterial pathogen *Pseudomonas
syringae* (Kassanis and White, 1975; Antoniw and White, 1980; Ahl et al.,
1985).

A number of other chemicals have been tested for their ability to
stimulate PR protein accumulation. Compounds known to enhance resistance to viruses in animal systems, such as 2-thiouracil and dioxohydro-
triazine, were found to induce PRs in plants (White et al., 1986).
Methylbenzimidazol-2-yl-carbonate, which has cytokinin-like properties
(Fraser, 1982), and a number of amino acids and their analogues (Asselin,
et al., 1985) exhibited some degree of PR gene induction, as did several metal
ions such as silver, cadmium, manganese, and barium (Asselin et al.,1985;
White et al., 1986). Two compounds associated with phosphatidylinositol
metabolism, phytic acid (Maiss and Poehling, 1983) and arachidonic acid
(Malamy et al., 1991) also stimulate PR gene expression. Phosphatidylinositol and its breakdown products are key components in many signal
transduction pathways in animal systems (Berridge and Irvine, 1989). This
suggests the possibility that phytic acid and arachidonic acid might be
endogenous signal compounds in plants.

Another potential example in which chemical inducer may turn out to be an endogenous signal compound is salicylic acid (SA). In 1979 (White, 1979) showed that application of SA, acetylsalicylic acid (ASA; aspirin) and benzoic acid induced PR-1 and PR-2 genes and reduced TMV lesion number. 2,6-dihydroxybenzoic acid, but not other derivatives of SA, also stimulated PR protein accumulation (Van Loon, 1983). SA and many of its derivatives are also potent inducers of PR-1 protein accumulation in other *Nicotianae* species and in a broad range of both dicotyledonous and monocotyledonous plants (White et al., 1987). SA and ASA treatment have also been shown to enhance resistance to pathogens other than TMV such as alfalfa mosaic virus (AMV) (Hooft van Huijsduijnen et al., 1986a) and the fungus *Peronospora tabacina* (Ye et al., 1989).

These observations, together with the finding that SA is an endogenous regulator of heat and odor production in the inflorescences of some thermogenic plants (Raskin et al., 1987), lead Malamy and coworkers (1990) to test the possibility that SA is a component of a natural induction pathway for PR gene expression during the resistance response of tobacco to TMV infection. Endogenous SA levels in resistant, but not susceptible, cultivars were found to increase at least 20-fold in inoculated leaves. A 5–10 fold increase in SA was also seen in uninoculated leaves of TMV-infected resistant plants, consistent with a role in development of SAR. Moreover, induction of the PR-1 (and PR-2; Malamy et al., unpubl. results) genes paralleled the rise in SA levels. In keeping with a possible role for SA in development of SAR, Metraux and coworkers (1990) found that SA levels in the phloem exudates of cucumber leaves rose dramatically after infection with either TNV or with the fungal pathogen *Colletotrichum lagenarium*. This suggests that SA was moving from the inoculated tissue to distal parts of the plant and is consistent with the finding that this rise in SA levels preceded the appearance of SAR in the uninoculated leaves.

2. In Response to Developmental Cues

Fraser (1981) was the first to observe the accumulation of PR-S (acidic PR-1 and possibly acidic PR-2 families) in healthy plants; they were detected in the leaves at the onset of flowering. In tobacco floral development coincides with senescence of the lower leaves. The accumulation of these acidic PRs could be inhibited by the removal of the lower leaves or the inflorescence, suggesting the requirement for multiple developmental signals.

In contrast to the coordinated expression of the PR genes during pathogenesis, their expression appears to be differentially regulated in an organ- and tissue-specific manner during development. The basic isoforms of the PR-1, PR-2, and PR-3 genes have been shown to be expressed at low levels in the lower leaves and at high levels in the roots of healthy tobacco plants (Memelink et al., 1987, 1990; Felix and Meins, 1986; Shinshi et al., 1987; Neale et al., 1990). Memelink et al., (1990) also found moderate expression levels of the basic chitinase (PR-3) genes in stem, flower and green fruit tissues, while the basic PR-1 and the basic β-1,3-glucanase (PR-2) genes were expressed at moderate levels in the floral tissue.

Organ-specific expression of several basic isoforms of the PRs was also discovered from an entirely different avenue. During their study of floral development using the in vitro tobacco thin cell layer (TCL) explant system of Tran Thanh Van et al. (1974), Peacock and coworkers (Meeks-Wagner et al., 1989; Neale et al., 1990) found that addition of the cytokinin kinetin to TCL explants not only induced flower formation but also the expression of the genes encoding the basic isoforms of chitinase (PR-3), β-1,3-glucanase (PR-2), and osmotin (PR-5). A link between these patterns of developmental expression and cytokinin was also suggested by the induction of the basic PR-1, β-1,3-glucanase and chitinase genes by the treatment of tobacco plants with this hormone or endogenous overproduction of the hormone in plants transformed with the *Agrobacterium tumefaciens* T-*cyt* gene (Memelink et al., 1987). The expression of PR genes during flower formation in TCL explants argues for roles of these genes during normal development. Alternatively, PR gene induction could be a secondary effect of cytokinin and not required for floral differentiation. However, these genes were not induced by the cytokinin zeatin which induces vegetative shoot development in the TCL explant system. It would be interesting to examine PR gene expression during the process of flower formation initiated by other regulatory factors. The observation that pectic cell wall fragments also direct flower formation of TCL explants (Eberhard et al., 1989) indicates that oligosaccharins, which can act as elicitors of defense-related genes, can also control tissue morphogenesis. If PRs are induced in this case, it would support the concept that plants have evolved to employ stress signals as developmental cues, perhaps as means of adapting to the environment (Bowles, 1990).

The affiliation of PR gene expression with normal development is not limited to the basic isoforms of these genes. Lotan et al. (1989) found that the acidic isoforms of the PR-1 protein were present in sepals while the acidic PR-3 proteins (chitinases) accumulated in sepal, pedicel, anther and ovary tissues. In addition, they identified a novel 41 kDa glycoprotein in the style/stigma tissue that was serologically related to the acidic PR-2s. This novel protein species was immunolocalized to the vacuoles of a subset of the transmitting track cells.

Cutt and coworkers have examined the floral-specific expression of the acidic PR-1 and PR-2 family genes in more detail employing cDNA and antibody probes (Côté et al., 1991; Cutt et al., unpubl. results) and a newly developed PAGE assay for the detection of β-1,3-1glucanase activities (Côté et al., 1990). The acidic PR-1 and PR-2 family mRNAs and proteins and the corresponding β-1,3-glucanase enzyme activities were detected in the sepals. Both of these gene families were expressed at low levels in the sepal before anthesis and the expression increased dramatically after anthesis.

With the exception of the sepals, the PR-1 and PR-2 genes showed different patterns of expression in other floral tissues. A novel β-1,3-glucanase gene, related to the PR-2 genes, was expressed in the stigma/style and ovary tissues. In contrast to the results of Lotan and Fluhr (1989), PR-1 gene expression was detected in the placenta, ovary wall and seeds. In seeds, the PR-1 proteins were present at one week post anthesis, and disappeared as the seed matured.

Thus it appears that differential expression of the PR genes are controlled by developmental- and tissue-specific factors. In support of this, Lotan et al. (1989) used a homeopathic mutant of tobacco to demonstrate that PR gene expression was specific for the differentiated tissue type and not the ontogenic origin of the tissue. Differential regulation may be achieved by the utilization of alternative regulatory pathways or by the selective repression of specific genes during development. The former is supported by the ability to coordinately induce expression of several PR genes in the style tissue by injection of SA (Lotan et al., 1989).

3. Level of Gene Regulation

The inducible nature of the PR genes bestows an excellent opportunity to study gene regulation in plants at the molecular level. In addition to providing basic knowledge concerning genetic control, these studies are highlighted by the prospect of one day utilizing transgenic plant technology to induce a constitutive level of systemic resistance in agronomically important plant species.

PRs accumulate at high levels in TMV- or chemically-induced tobacco leaves. Carr et al. (1985) and Matsuoka et al. (1985) both demonstrated that the synthesis of the PR-1 proteins was regulated in part by the level of mRNAs after induction. In addition, cDNA probes corresponding to the different PR families have been used to show that high levels of PR mRNAs accumulated only after induction (Hooft van Huijsduijnen et al., 1985, 1986b; Pfitzner and Goodman, 1987; Matsuoka et al., 1988; Neale et al., 1990; Payne et al., 1990; Memelink et al., 1990; Malamy et al., 1990; Côté et al., 1991). Thus, the primary mechanism governing the expression of the PR genes is either transcriptional activation or mRNA stability. In vitro nuclear "run on" transcription analysis (J. Horowitz and R. Goldberg, pers. comm.) and promoter analyses using transgenic plants (see below) indicate that the PR-1 genes were regulated at the transcriptional level. While these types of analyses have not yet been reported for the other PR genes, it is generally believed that transcriptional activation is responsible for PR protein accumulation after induction.

The DNA sequences required for transcriptional activation of the PR-1 and the glycine-rich protein (GRP) genes have been analyzed utilizing transgenic tobacco plants (Ohshima et al., 1990c; Van de Rhee et al., 1990).

Chimeric gene constructs, containing various upstream regions of the PR gene sequences fused to the coding region of a reporter gene [β-glucuronidase (GUS) or chloramphenicol acetyltransferase (CAT)], were introduced into tobacco plants via *Agrobacterium* mediated transformation. The transgenic plants were examined for the expression of these genes in response to treatment with SA or TMV infection. Ohshima et al. (1990) found that chimeric constructs, in which the GUS coding region was fused to PR-1a gene fragments, corresponding to 300, 900, or 2400 bp 5′ of the transcriptional start site, were inducible to varying extents by either SA treatment or TMV infection. The −300 construct was only marginally induced by SA (1.5 to 2 fold) while the −900 and −2400 constructs exhibited moderate levels of inducibility. Using histochemical staining for GUS activity, they demonstrated that both the −300 and −2400 constructs were inducible by TMV infection. Although the level of induction was low, it was concluded that the cis-acting elements required for SA and TMV induction were present in the −300 to −1 region of the PR-1a gene and that additional elements enhancing the level of gene activation were present further upstream.

The results of Van de Rhee et al. (1990) conflict with those of Ohshima et al. (1990) with respect to the location of the DNA elements responsible for SA and TMV induction of the PR-1a gene. Van de Rhee et al. (1990) used a chimeric gene containing the PR-1a upstream sequences, −902 to +29 nucleotides, fused to the GUS coding region. Deletion analysis of this chimera indicated that the elements which conferred SA and TMV inducibility were present between positions −643 and −689. Furthermore, the sequences from −902 to −625 were capable of conferring SA and TMV inducibility to a chimeric gene containing the cauliflower mosaic virus (CaMV) 35S core promoter sequences (−90 to +1) fused to the GUS coding region. The identification and characterization of SA and TMV inducible trans-acting factors that interact with the PR-1a gene should assist in resolving the conflicting results reported by these two groups.

Using similar types of analyses, Van de Rhee et al. (1990) mapped the elements that were required for SA and TMV inducibility of the GRP gene of tobacco to the upstream region between positions −645 to −400. Sequence comparisons of the upstream regions of the PR-1a and GRP genes have not disclosed any significant similarities between these two genes (van Kan et al., 1989). The DNA sequence motif TGTGGAAA, which is identical to the mammalian SV40 virus core enhancer sequence, was found in the GRP gene at approximately −540; however, the PR-1a gene only contained motifs with limited similarity to this sequence.

Sequence analyses of the different TMV-inducible genes have revealed the presence of motifs known to be involved in the regulation of other well-characterized genes. The PR-1a gene contains a motif at positions −867 to −847 which is similar to the element required for viral induction of the β-interferon gene (Payne et al., 1988). This is within the upstream region mapped by Van de Rhee et al. (1990), that confers TMV inducibility on the PR-1a gene. Another DNA element also found within the PR-1a gene (−71 to -57) is similar to one necessary for the induction of heat shock genes (Pfitzner et al., 1988). The PR-1 genes, however, are not induced by heat shock. In the GRP gene van Kan et al. (1989) identified a 64 bp inverted repeat which is found in a similar position in the ribulose bisphosphate carboxylase small subunit gene and may function in light- and tissue-specific regulation. The role of these sequences has not yet been directly established but remains as a point of interest in the study of PR gene regulation.

4. Signal Transduction Pathways Involved in PR Gene Activation

Tobacco possesses the ability to either coordinately (after TMV infection) or differentially (during development or after chemical treatment) regulate the expression of the five PR families. In higher eukaryotes, cell and tissue-specific patterns of gene expression are achieved by the presence of multi-gene families (Liang et al., 1989), the interaction of various transcription factors with cis-acting DNA sequences (Maniatis et al., 1987; Mitchell and Tijan, 1989) and the utilization of multiple signal transduction pathways (Berridge and Irvine, 1984; Berridge, 1987). Here we will discuss the evidence for multiple signal transduction pathways in the control of PR gene expression.

It has been shown that the treatment of tobacco plants with an ethylene-releasing compound (ethephon) induces the accumulation of members of the acidic and basic PR-1, -2, and -3 families to varying degrees (Van Loon, 1977; Van Loon and Antoniw, 1982; Van Loon, 1983; Memelink et al., 1990) and also SAR. Furthermore, the production of ethylene has been shown to accompany the HR to TMV infection in tobacco (Ross and Pritchard, 1972; Van Loon, 1977). Hence, it was postulated that ethylene was a natural inducer in the signal transduction pathway leading to the activation of the PR genes following TMV infection. Additional support for this hypothesis was provided by the observation that the production of PRS could be partially inhibited by the action of either 1-aminoethoxyvinylgly-cine, an inhibitor of 1-aminocyclopropane-1-carboxylic acid, the natural precursor to ethylene (Van Loon, 1983), or silver thiosulfate, an inhibitor of ethylene action (Lotan and Fluhr, 1990).

Other compounds used to induce both the PR genes and acquired resistance, are the benzoic acid derivatives, SA, ASA and 2,6-dihydroxy-benzoic acid (White, 1979; Antoniw and White; 1980) (Sect. II. B. 1). SA induction of PR gene expression, however, is not complete as it only effectively induces the acidic PR-1 and PR-2 genes (Hooft van Huijsduijnen et al., 1986a).

The resistance response of tobacco to TMV infection includes the HR, SAR, and the concerted induction of the PR genes. This response is temperature-sensitive. At elevated temperatures ($> 28\,^\circ$C) the HR does not occur, PRS do not accumulate, and the plant displays a susceptible phenotype. Van Loon and Antoniw (1982) used this temperature-sensitive phenomenon to investigate the putative roles of SA and ethylene in the pathway(s) leading from TMV infection to PR gene activation. Treatment with SA or ethephon at 20 °C resulted in the production of PRs, while at 30 °C only SA treatment induced PR synthesis. In addition, treatment of tobacco plants with SA did not stimulate ethylene production (Van Loon and Antoniw, 1982).

Based upon these results, it was proposed that the pathway for induction of PRs by TMV involved the production of ethylene, which activated the synthesis of an aromatic compound (related to benzoic acid or SA) and resulted in the production of PRs (Van Loon, 1983). SA's ability to induce PRs at 30 °C implied that the temperature-sensitive step was between ethylene production and the synthesis of the SA-like compound. Given the results of Malamy et al. (1990), it now appears that SA itself is an endogenous aromatic compound involved in PR gene activation (see Sect. II.B.1). It remains to be determined whether the rise in endogenous SA levels after TMV infection is temperature sensitive and dependent upon ethylene production.

Lotan and Fluhr (1990b) have shown that xylanase, an inducer of PRs, activates only a subset of the PRs and that this activation involves a pathway which is not dependent upon the synthesis or action of ethylene. Furthermore, the pathway activated by xylanase is light-independent in contrast to the light-dependent pathway observed after thiamine treatment (Asselin et al., 1985) and TMV infection (Lotan and Fluhr, 1990). In addition, cytokinin (Memelink et al., 1990) also differentially activates PR genes (discussed in Sect. II.B.2). Moreover, the discordant, organ- and tissue-specific modes of PR gene expression that occur during normal development suggest that there are multiple means by which the plant can regulate the expression of the PR genes (Memelink et al., 1987, 1990; Lotan et al., 1989; Neale et al., 1990; Côté et al., 1991; Cutt et al., unpubl. results). Thus, the concerted activation of the tobacco PR gene families by TMV infection appears to be more the exception than the rule. The information available implies that a complex regulatory network is involved in the activation of PRs. Concerted activation of the PRs appears to require multiple pathways which may be entirely distinct from one another or may be integrated in such a way as to contain common steps.

C. Toward an Understanding of the Functions of PR Proteins

1. During Pathogenic Assault

In tobacco PR gene expression (particularly PR-1) has been associated with resistance to TMV infection. These proteins were first detected in tobacco plants during a resistance (hypersensitive) response to TMV or TNV but not in healthy plants or systemically infected, susceptible plants. Their appearance also in upper, uninoculated leaves of hypersensitively responding plants moreover correlated with SAR (Van Loon and Van Kammen, 1970; Van Loon, 1975; Kassanis et al., 1974). In addition, the HR, SAR, and PR-1 accumulation were all temperature sensitive (Gianinazzi and Kassanis, 1974), and chemical inducers such as PAA and SA stimulated both PR-1 protein accumulation and acquired resistance (Gianinazzi and Kassanis, 1974; White, 1979; White et al., 1983, 1986; Fraser and Clay, 1983). Furthermore, the interspecific hybrids of N. glutinosa × N. debneyi that constitutively produced a PR-1 protein were more resistant to TMV infection than their progenitors. These hybrids also maintained both the ability to restrict virus spread and synthesize the PR-1 protein at high temperatures (Ahl and Gianinazzi, 1982).

Nevertheless, the correlation between PR gene expression and HR or SAR is not complete. In a number of circumstances it has been demonstrated that one occurred without the other. For example, stem injection of ASA (Ye et al., 1989) or leaf application of a glucan preparation from the walls of the fungal pathogen Phytophthora megasperma (Kopp et al., 1989) induced SAR to TMV without stimulating PR protein production (for a full discussion of other, earlier experiments see the review by Carr and Klessig, 1989.)

In an attempt to determine, directly, the role of the PR-1 proteins in disease resistance, transgenic plants were constructed which constitutively expressed the PR-1b (Cutt et al., 1989) or PR-1a (Linthorst et al., 1989) genes under control of the strong 35S promoter of CaMV. These plants did not exhibit increased resistance to TMV or AMV, indicating the expression of these genes was insufficient to provide protection against viral pathogens. The transgenic plants were also as sensitive to insect infestation (e.g., by tobacco hornworm) as the nontransformed progenitor plants (Cutt and Klessig, unpubl. results; J. Bol, pers. comm.).

Thus, the role of the PR-1 proteins in disease resistance remains unresolved. Protection against viral or insect pathogens may require other host factors in addition to the PR-1 proteins. Alternatively, these proteins may function in protection against non-viral (or non-insect) pathogens even though they are induced by viral infection. Induction of a broad battery of defenses after infection with one pathogen may explain how acquired resistance to subsequent challenge by a second, unrelated pathogen is achieved. Another possibility is that the PR-1 proteins have no direct role in disease resistance but perhaps have a protective function to aid the plant in dealing with stresses associated with pathogen assault. This is consistent with their induction by some chemical agents, nutrient deprivation, and osmotic stress (Coutts and Wagih, 1983; Asselin et al., 1985; Ohashi and Matsuoka, 1985). However, not all stresses induce PR-1 genes. PR proteins were not induced by elevated temperatures (Cornelissen et al., 1987); in fact their synthesis could be prevented by incubation at temperatures of 30 °C or higher (Gianinazzi and Kassanis, 1974; Van Loon, 1975). Neither wounding nor treatment with jasmonic acid (a growth regulator associated with wounding; Mason and Mullet, 1990; C. Ryan, pers. comm.) effectively induce PR-1 gene expression (Malamy et al., 1990; Hennig and Klessig, unpubl. result).

The discovery that the tobacco PR-2 and -3 families are β-1,3-glucanases and chitinases, respectively (Kauffman et al., 1987; Legrand et al., 1987) suggest a role for these PR proteins in protection against fungal or bacterial pathogens. They may act either to destroy the cell walls of pathogens (Boller, 1987) and/or to release elicitor compounds from pathogen or host cell walls which stimulate the defense system (Boller, 1987; Keen and Yoshikawa, 1983). Neuhaus et al. (1990) recently reported that high-level expression of a basic tobacco chitinase in *Nicotiana sylvestris* did not substantially increase resistance to the fungus *Cercospora nicotianae*. Also, transgenic tobacco plants which expressed a basic, vacuolar, bean chitinase showed limited resistance to the fungal pathogen *Rhizoctonia solani* which causes damping off (R. Broglie, pers. comm.). In addition, the inhibition of endogenous glucanase genes by anti-sense RNA in transgenic tobacco did not increase the susceptibility of the plant to *Cercospora nicotianae* (F. Meins, Jr., pers. comm.). In light of the synergistic antifungal activity of chitinases and β-1,3-glucanases observed in vitro, it may be essential to overexpress both the chitinase and β-1,3-glucanase genes simultaneously to achieve enhanced resistance to fungal pathogens.

Proteins of the PR-5 family of tobacco have similarities to the bifunctional α-amylase/trypsin inhibitor of maize suggesting they might function as protease inhibitors (Richardson et al., 1987). Extracellular protease inhibitors could provide a defense against herbivorous insects or protection from proteases released by fungal and bacterial pathogens or by dying host cells during HR. However, attempts to directly demonstrate protease inhibitor activity for the PR-5 proteins have been unsuccessful (Pierpoint and Shewry, 1987; Kauffmann et al., 1990). Moreover, in contrast to the demonstration by Hilder and coworkers (1987) that transgenic plants expressing the cowpea trypsin inhibitor exhibited enhanced resistance to insects, transgenic plants constitutively producing the tobacco PR-5 protein were not protected against insects (J. Bol, pers. comm.).

The PR-5 proteins also share extensive sequence similarity to osmotins ($\sim 60\%$) from tobacco (Singh et al., 1987) and tomato (King et al., 1988). Synthesis of osmotins are stimulated during adaptation to osmotic stress. Whether the acidic PR-5 proteins are induced under similar conditions and have a function during this environmental stress is unknown.

2. During Normal Plant Development

Genes encoding the basic, but not acidic, isoforms of several PR proteins (PR-1, PR-2, PR-3, and PR-5) are constitutively expressed in roots of healthy plants (Felix and Meins, 1986; Shinshi et al., 1987; Memelink et al., 1990; Neale et al., 1990). This may represent a permanent form of defense against microbes in the rhizosphere. In contrast to the predominant extracellular localization of the acidic PRs, the basic PRs such as the β-1,3-glucanases (PR-2) (Van den Bulcke et al., 1989), osmotin (PR-5) (Singh et al., 1987), and by analogy, the basic bean chitinase (PR-3) (Boller and Vögeli, 1984; Mauch and Staehelin, 1989) are generally located in the central vacuole. This storage and sequestration of defense-related proteins may provide an important first line of defense while at the same time preventing exposure of the plant to the potential harmful effects of the activities of these proteins. The morphogenic changes occurring during plant development often necessitate disruption of existing tissues such as occurs during pollen tube growth, formation of lateral or adventitious roots and vegetative to floral meristem transition. This disruption of existing tissue may increase the plant's susceptibility to pathogens. The constitutive presence of the PRs may decrease the vulnerability of this tissue.

However, the differential expression of the various acidic PR genes in the different floral organs (see Sect. II.B.2) is harder to reconcile with a defensive role and suggests roles in the normal development or physiology. It has been suggested that the glucanases may be involved in pollen tube growth (Lotan and Fluhr, 1990). This was based on the observations that extracellular glucanases function in cell elongation (Varner and Liang-Shiou, 1989) and the association of these enzymes with the transmitting track through which the pollen tube must elongate (Lotan et al., 1989). A direct test for their involvement in developmental processes will require disruption of their normal expression through classical genetics or genetic engineering. Prevention of PR gene expression using antisense RNA (Green et al., 1986) or ribozymes (Haseloff and Gerlach, 1988) is in progress in several laboratories. Providing that these latter approaches are effective in blocking expression, these experiments should yield important clues into the roles of these proteins in disease resistance and reproductive physiology.

III. Alternative Plant Systems

The discovery of PRs in tobacco initiated a search for similar proteins in other plant species. New proteins, which accumulated after pathogen infection, often in association the HR, and shared common physical characteristics with the PRs of tobacco, were identified in a number of other host-pathogen systems (for a review see Redolfi, 1983). By 1985 PRs had been reported in 16 plant species (Van Loon, 1985). PRs have now been identified in many more plant species including both monocot and dicot families; it is likely that all higher plants possess genes related to the PRs. In this section we will briefly cover PRs from other systems (for a more detailed discussion see Carr and Klessig, 1989). Some of these PRs are similar to the tobacco PRs, while others do not have known counterparts in tobacco.

The p14 protein of tomato (Lucas et al., 1985), induced after viroid infection, was shown to have a similar amino acid sequence (~ 65%

similarity) to the PR-1 proteins of tobacco (Cornelissen et al., 1986b). The
p14 protein has been shown to exhibit both intracellular (vacuolar) and
extracellular patterns of localization (Vera et al., 1989). Other proteins,
serologically-related to the tobacco PR-1 family, have been identified in
maize, barley, potato, cowpea, and *Chenopodium amaranaticolor* (Nassuth
and Sanger, 1986; White et al., 1987; Parent et al., 1988; Nasser et al., 1988).
More recently, Klessig and coworkers (Metzler et al., 1991) cloned a PR-1-
like gene from *Arabidopsis thaliana* which has ~ 50% similarity to the
tobacco PR-1 genes. In maize, cucumber, tomato, bean, potato, soybean
and pea β-1,3-glucanases and chitinases have been identified and may be
similar to the tobacco PR-2 and PR-3 families (Nasser et al., 1988, 1990;
Kombrink et al., 1988; Mauch et al., 1988b; Metraux et al., 1988; Hedrick
et al., 1988; Awade et al., 1989; Joosten et al., 1989; Takeuchi et al., 1990).
Genes encoding both basic and acidic isoforms of chitinase have been
isolated from *A. thaliana* (Samac et al., 1990). Proteins related to the PR-5
family (osmotin) have been described in barley (Bryngelsson and Green,
1989), potato (Parent et al., 1988), and tomato (King et al., 1988), although
the tomato protein was originally characterized by its association with salt
stress.

   PRs of unknown function and which do not appear to correspond to
any of the five families of tobacco have been identified in a large number of
plants including, for example, bean (Abu-Jawdah, 1982; Redolfi and
Cantisani, 1984; DeTapia et al., 1986; Hooft van Huijsduijnen et al., 1986a),
parsley (Schmelzer et al., 1989), tomato (DeWit and Bakker, 1980; Granell
et al., 1987), cucumber (Andebrahan et al., 1980; Gessler and Kuc, 1982),
cowpea (Coutts, 1978), pea (Fristensky et al., 1985), and soybean (Roggero
and Pennazio, 1989). Defense-related proteins with known activities or
functions, which under the current definition could be classified as PRs, have
also been detected in several plants species. These include the proteinase
inhibitors (see reviews by Ryan, 1988; Bowles, 1990), proteinases
(Laskowski, 1986; Vera and Conejero, 1988), peroxidases (see review by
Bowles, 1990), thionins (Bohlmann et al., 1988; Garcia-Olmedo et al., 1992),
and some of the enzymes involved in the latter steps in the phenylpropanoid
pathway which accumulate only during pathogenesis (Hahlbrook and
Scheel, 1989).

IV. Conclusions

The alterations in gene expression that constitute the defense-related
response of higher plants to biological and environmental stresses are very
complex. The PRs represent one class of defense-related genes, which are not
necessarily mutually exclusive with other classes of defense-related genes.
Although circumstantial evidence points to a role for PRs in the plant's
defense against certain environmental and biological stresses, rigorous
demonstration of such a function has not yet been achieved. The activation

of diverse sets of PR genes in response to diverse pathogens and forms of stress makes it difficult to determine which PR gene functions are necessary in a specific stress situation. Genetic mutants in the PR genes are not currently available, and molecular genetic approaches have not yet proven fruitful in the dissection of these responses. However, it is anticipated that within the not to distant future these molecular genetic approaches will yield important information concerning the functions of these proteins. The recent rapid progress in identifying suitable pathogens such as turnip crinkle virus (A. Simon, pers. comm.), *Pseudomonas syringae* (F. Ausubel and B. Staskawicz, pers. comm.), and *Peronospora parasitica* (Koch and Slusarenko, 1990) of *Arabidopsis thaliana* (the model plant for genetic and molecular genetic studies) should greatly facilitate our understanding of PR genes.

The developmental regulation of PR genes implies that PRs also perform roles in the physiology of healthy plants. Perhaps these proteins are multifunctional, playing different roles during pathogenesis and normal plant development. Alternatively, the proteins may have similar functions in both cases. If so, studying their roles in normal development may provide clues to their function in defense and vice versa. Aside from the exciting potential roles for PR genes as critical components of the disease resistance response, PRs offer the opportunity to investigate gene structure and regulation, signal transduction mechanisms, protein targeting, and development in higher plants.

Acknowledgements

The authors thank the investigators who contributed unpublished data for this chapter. Marline Boslet is thanked for the preparation of the manuscript and Ralph Dewey, David C. Dixon, and Jocelyn Malamy are thanked for helpful comments and discussions. This work and research from this laboratory was supported by grants DMB-8703293 and DMB-9003711 from the National Science Foundation.

V. References

Abu-Jawdah Y (1982) Changes in soluble protein patterns of bean leaves upon fungal or viral infections or after chemical injury. Phytopathol Z 103: 272–279

Ahl P, Gianinazzi S (1982) *b*-Protein as a constitutive component in highly (TMV) resistant interspecific hybrids of *Nicotiana glutinosa* × *Nicotiana debneyi*. Plant Sci Lett 26: 173–181

Ahl P, Benjama A, Samson R, Gianinazzi S (1981) Induction chez le tabac par *Pseudomonas syringae* de nouvelles protéines (protéines b) associées au développement d'une résistance non spécifique à une deuxième infection. Phytopathol Z 102: 201–212

Ahl P, Cornu A, Gianinazzi S (1982) Soluble proteins as genetic markers in studies of resistance and phylogeny in *Nicotiana*. Phytopathology 72: 80–85

Ahl P, Gianinazzi S, Samson R, Benjama A (1985) Cultivar dependence of polyacrylic acid effects on *syringae* in *Nicotiana tabacum*. Plant Pathol 34: 221–227

Albersheim P, Valent BS (1978) Host–pathogen interactions in plants. Plants, when exposed to oligosaccharides of fungal origin, defend themselves by accumulating antibiotics. J Cell Biol 78: 627–643

Andebrhan T, Coutts RHA, Wagih EE, Wood RKS (1980) Induced resistance and changes in the soluble protein fraction of cucumber leaves locally infected with *Colletotrichum lagenarium* or tobacco necrosis virus. Phytopathol Z 98: 47–52

Antoniw JF, White RF (1980) The effects of aspirin and polyacrylic acid on soluble leaf proteins and resistance to virus infection in five cultivars of tobacco. Phytopathol Z 98: 331–341

Antoniw JF, Ritter CE, Pierpoint WS, van Loon LC (1980) Comparison of three pathogenesis-related proteins from plants of two cultivars of tobacco infected with TMV. J Gen Virol 47: 79–87

Antoniw JF, White RF, Barbara DJ, Jones P, Longley A (1985) The detection of PR (b) protein and TMV by ELISA in systemic and localised virus infections of tobacco. Plant Mol Biol 4: 55–60

Asselin A, Grenier J, Côté F (1985) Light-influenced extracellular accumulation of b (pathogenesis-related) proteins in *Nicotiana* green tissue induced by various chemicals or prolonged floating on water. Can J Bot 63: 1276–1283

Awade A, de Tapia M, Didierjean L, Burkard G (1989) Biological function of bean pathogenesis-related (PR3 and PR4) proteins. Plant Sci 63: 121–130

Benhamou N, Côté F, Grenier J, Asselin A (1988) Immunocytochemical localization of pathogenesis-related PR-1 proteins in TMV-infected *Nicotiana tabacum* cv. Xanthi-nc. Can J Plant Pathol 11: 185

Benhamou N, Grenier J, Asselin A, Legrand M (1989) Immunogold localization of β-1,3-glucanase in two plants infected by vascular wilt fungi. Plant Cell 1: 1209–1221

Benhamou N, Mazau D, Esquerré-Tugayé M-T (1990) Immunogold localization of hydroxyproline-rich glycoproteins in necrotic tissue of *Nicotiana tabacum* L. cv. Xanthi-nc infected by tobacco mosaic virus. Physiol Mol Plant Pathol 36: 129–145

Berridge MJ (1987) Inositol trisphosphate and diacylglycerol: two interacting second messengers. Annu Rev Biochem 56: 159–193

Berridge MJ, Irvine RR (1984) Inositol trisphosphate, a novel second messenger in cellular signal transduction. Nature 312: 315–321

Berridge MJ, Irvine RF (1989) Inositol phosphates and cell signalling. Nature 341: 197–205

Bol JF (1988) Structure and expression of plant genes encoding pathogenesis-related proteins. In: Verma DPS, Goldberg RB (eds) Temporal and spatial regulation of plant genes. Springer, Wien New York, pp 201–221 [Dennis ES et al (eds) Plant gene research. Basic knowledge and application]

Bol JF, van Kan JAL (1988) The synthesis and possible functions of virus-induced proteins in plants. Microbiol Sci 5: 47–52

Bohlmann H, Clausen S, Behnke S, Giese H, Hiller C, Reimann-Philipp U, Schrader G, Barkholt V, Apel K (1988) Leaf-specific thionins of barley—a novel class of cell wall proteins toxic to plant–pathogenic fungi and possibly involved in the defence mechanism of plants. EMBO J 7: 1559–1565

Boller T (1985) Induction of hydrolases as a defense reaction against pathogens. In: Key JL, Kosuge T (eds) Cellular and molecular biology of plant stress. AR Liss, New York, pp 247–262

Boller T (1987) Hydrolytic enzymes in plant disease resistance. In: Kosuge T, Nester EW (eds) Plant–microbe interactions, molecular and general aspects, vol 2. Macmillan, New York, pp 385–413

Boller T, Vögeli U (1984) Vacuolar localization of ethylene-induced chitinase in bean leaves. Plant Physiol 74: 442–444

Bowles DJ (1990) Defense-related proteins in higher plants. Annu Rev Biochem 59: 873–907

Bowles DJ, Pappin DJ (1988) Traffic and assembly of concanavalin A. Trends Biochem Sci 13: 60–66

Bryngelsson T, Gréen B (1989) Characterization of a pathogenesis-related, thaumatin-like protein isolated from barley challenged with an incompatible race of mildew. Physiol Mol Plant Pathol 35: 45–52

Bull AT, Chesters CGC (1966) The biochemistry of laminarin and nature of laminarinase. Adv Enzymol 28: 325–364

Carr JP, Klessig DF (1989) The pathogenesis-related proteins of plants. In: Setlow JK (ed) Genetic engineering, principles and methods, vol 11. Plenum, New York, pp 65–109

Carr JP, Dixon DC, Klessig DF (1985) Synthesis of pathogenesis-related proteins in tobacco is regulated at the level of mRNA accumulation and occurs on membrane-bound polysomes. Proc Natl Acad Sci USA 82: 7999–8003

Carr JP, Dixon DC, Nikolau BJ, Voelkerding KV, Klessig DF (1987) Synthesis and localization of pathogenesis-related proteins in tobacco. Mol Cell Biol 7: 1580–1583

Condit CM, Meagher RB (1986) A gene encoding a novel glycine-rich structural protein of petunia. Nature 323: 178–181

Conejero V, Picazo I, Segado P (1979) Citrus exocortis viroid (CEV): protein alterations in different hosts following viroid infection. Virology 97: 454–456

Cooper JB, Chen JA, van Holst G-J, Varner JE (1987) Hydroxyproline-rich glycoproteins of plant cell walls. Trends Biochem Sci 12: 24–27

Cornelissen BJC, Hooft van Huijsduijnen RAM, Bol JF (1986a) A tobacco mosaic virus-induced tobacco protein is homologous to the sweet-tasting protein thaumatin. Nature 321: 531–532

Cornelissen BJC, Hooft van Huijsduijnen RAM, Van Loon LC, Bol JF (1986b) Molecular characterization of messenger RNAs for "pathogenesis-related" proteins 1a, 1b and 1c, induced by TMV infection of tobacco. EMBO J 5: 37–40

Cornelissen BJC, Horowitz J, Van Kan JAL, Goldberg RB, Bol JF (1987) Structure of tobacco genes encoding pathogenesis-related proteins from the PR-1 group. Nucleic Acids Res 15: 6799–6811

Côté F, Ouakfaoui SE, Asselin A (1991) Detection of β-glucanases acting on various β-1,3 and β-1,4-glucans after native and denaturing polyacrylamide gel electrophoresis. Electrophoresis 12: 69

Côté F, Cutt JR, Asselin A, Klessig DF (1991) Pathogenesis-related acidic β-1,3-glucanase genes of tobacco are regulated by both stress and developmental signals. Mol Plant Microbe Interact 4: 173–181

Coutts RHS, Wagih EE (1983) Induced resistance to viral infection and soluble protein alterations in cucumber and cowpea plants. Phytopathol Z 107: 57–69

Cutt JR, Dixon DC, JP, Klessig DF (1988) Isolation and nucleotide sequence of cDNA clones for the pathogenesis-related proteins PR-1a, PR-1b and PR-1c of Nicotiana tabacum cv. Xanthi nc induced by TMV infection. Nucleic Acids Res 16: 9861

Cutt JR, Harpster MH, Dixon DC, Carr JP, Dunsmuir P, Klessig DF (1989) Disease response to tobacco mosaic virus in transgenic tobacco plants that constitutively express the pathogenesis-related PR1b gene. Virology 173: 89–97

Davis BJ (1964) Disc electrophoresis. II. Method and application to human serum proteins. Ann NY Acad Sci 121: 404–427

DeClerq E, Eckstein F, Merigan TC (1970) Structural requirements for synthetic polyanions to act as interferon inducers. Ann NY Acad Sci 173: 444–461

De Tapia M, Bergmann P, Awade A, Burkard G (1986) Analysis of acid extractable bean leaf proteins induced by mercuric chloride treatment and alfalfa mosaic virus infection. Partial purification and characterization. Plant Sci 45: 167–177

De Wit PJGM, Bakker J (1980) Differential changes in soluble tomato leaf proteins after inoculation with virulent and avirulent races of *Cladosporium fulvum* (syn. *Fulvia fulva*). Physiol Plant Pathol 17: 121–130

Dixon DC, Cutt JR, Klessig DF (1991) Differential targeting of the PR-1 pathogenesis-related proteins to the extracellular space and vacuoles of crystal idioblasts. EMBO J 6: 1317–1324

Dixon RA, Lamb CJ (1990) Molecular communication in interactions between plants and microbial pathogens. Annu Rev Plant Physiol Plant Mol Biol 41: 339–367

Dumas E, Lherminier J, Gianinazzi S, White RF, Antoniw JF (1988) Immunocytochemical location of pathogenesis-related b_1 protein induced in tobacco mosaic virus-infected or polyacrylic acid-treated tobacco plants. J Gen Virol 69: 2687–2694

Eberhard S, Doubrava N, Marfà V, Mohnen D, Southwick A, Darvill A, Albersheim P (1989) Pectic cell wall fragments regulate tobacco thin-cell-layer explant morphogenesis. Plant Cell 1: 747–755

Ecker JR, Davis RW (1987) Plant defense genes are regulated by ethylene. Proc Natl Acad Sci USA 84: 5202–5206

Fang KSY, Vitale M, Fehlner P, King TP (1988) cDNA cloning and primary structure of a white-face hornet venom allergen, antigen 5. Proc Natl Acad Sci USA 85: 895–899

Felix G, Meins F, Jr (1986) Developmental and hormonal regulation of β-1,3-glucanase in tobacco. Planta 167: 206–211

Fink J, Jeblick W, Blaschek W, Kauss H (1987) Calcium ions and polyamines activate the plasma membrane-located 1,3-β-glucan synthase. Planta 171: 130–135

Francheschi VR (1989) Calcium oxalate formation is a rapid and reversible process in *Lemna minor* L. Protoplasma 148: 130–137

Francheschi VR, Horner Jr HT (1980) Calcium oxalate crystals in plants. Bot Rev 46: 361–427

Fraser RSS (1981) Evidence for the occurrence of the "pathogenesis-related" proteins in leaves of healthy tobacco plants during flowering. Physiol Plant Pathol 19: 69–76

Fraser RSS (1982) Are "pathogenesis-related" proteins involved in acquired systemic resistance of tobacco mosaic virus? J Gen Virol 58: 305–313

Fraser RSS, Clay CM (1983) Pathogenesis-related proteins and acquired systemic resistance: causal relationship or separate effects? Neth J Plant Pathol 89: 283–292

Fristensky B, Riggleman R, Wagoner W, Hadwiger LA (1985) Gene expression in suscept-ible and disease resistant interactions of peas induced with *Fusarium solani* pathogens and chitosan. Physiol Plant Pathol 27: 15–28

Garcia-Olmedo F, Carmona MJ, Lopez-Fando JJ, Fernandez JA, Castagnaro A, Molina A, Hernandez-Lucas C, Carbonero P (1992) Characterization and analysis of thionin genes. In: Boller T, Meins F (eds) Genes involved in plant defense. Springer, Wien New York, pp 283–301 [Dennis ES et al (eds) Plant gene research. Basic knowledge and application]

Gessler C, Kuc J (1982) Appearance of host protein in cucumber plants infected with viruses, bacteria and fungi. J Exp Bot 33: 58–66

Gianinazzi S (1984) Genetic and molecular aspects of resistance induced by infections or chemicals. In: Kosuge T, Nester EW (eds) Plant–microbe interactions, molecular and genetic perspectives, vol 1. Macmillan, New York, pp 321–342

Gianinazzi S, Ahl P (1983) The genetic and molecular basis of b-proteins in the genus *Nicotiana*. Neth J Plant Pathol 89: 275–281

Gianinazzi S, Kassanis B (1974) Virus resistance induced in plants by polyacrylic acid. J Gen Virol 23: 1–9

Gianinazzi S, Martin C (1975) A naturally occurring active factor inducing resistance to virus infection in plants. Phytopathol Z 83: 23–26

Gianinazzi S, Martin C, Vallee JC (1970) Hypersensibilite aux virus, temperatures et proteines solubles chez le *Nicotiana* Xanthi-nc. Apparition de nouvelles macromolecules lors de la repression de la synthese virale. C R Acad Sci Paris D 270: 2382–2386

Gianinazzi S, Ahl P, Cornu A, Scalla R, Cassini R (1980) First report of host b-protein appearance in response to a fungal infection in tobacco. Physiol Plant Pathol 16: 337–342

Granell A, Belles JM, Conejero V (1987) Induction of pathogenesis-related proteins in tomato by citrus exocortis virioid, silver ion and ethephon. Physiol Mol Plant Pathol 31: 83–90

Green PJ, Pines O, Inouye M (1986) The role of antisense RNA in gene regulation. Annu Rev Biochem 55: 566–597

Hahlbrock K, Scheel D (1989) Physiology and molecular biology of phenylpropanoid metabolism. Annu Rev Plant Physiol Plant Mol Biol 40: 347–369

Hahn MG, Bucheli D, Cervone F, Doares SH, O'Neill RA, Darvill A, Albersheim P (1989) The roles of cells wall constituents in plant-pathogen interactions. In: Nester E, Kosuge T (eds) Plant–microbe interactions, vol 3. McGraw-Hill, New York, pp 131–181

Haseloff J, Gerlach W (1988) Simple RNA enzymes with new and highly specific endoribonuclease activities. Nature 334: 585–591

Hedrick SA, Bell JN, Boller T, Lamb CJ (1988) Chitinase cDNA cloning and mRNA induction by fungal elicitor, wounding and infection. Plant Physiol 86: 182–186

Hilder VA, Gatehouse AMR, Sheerman SE, Barker RF, Boulter D (1987) A novel mechanism of insect resistance engineered into tobacco. Nature 330: 160–163

Hogue R, Asselin A (1987) Detection of 10 additional pathogenesis-related (*b*) proteins in intercellular fluid extracts from stressed "Xanthi-nc" tobacco leaf tissue. Can J Bot 65: 476–481

Holmes FO (1938) Inheritance of resistance to tobacco mosaic disease in tobacco. Phytopathology 28: 553–561

Hooft van Huijsduijnen RAM, Cornelissen BJC, Van Loon LC, Van Boom JH, Tromp M, Bol JF (1985) Virus-induced synthesis of pathogenesis-related proteins in tobacco. EMBO J 4: 2167–2171

Hooft van Huijsduijnen RAM, Alblas SW, deRijk RH, Bol JF (1986a) Induction by salicylic acid of pathogenesis-related proteins and resistance to alfalfa mosaic virus infection in various plant species. J Gen Virol 67: 2135–2143

Hooft van Huijsduijnen RAM, van Loon LC, Bol JF (1986b) cDNA cloning of six mRNAs induced by TMV infection of tobacco and a characterization of their translation products. EMBO J 5: 2057–2061

Hooft van Huijsduijnen RAM, Kauffmann S, Brederode F T, Cornelissen BJC, Legrand M, Fritig B, Bol JF (1987) Homology between chitinases that are induced by TMV infection of tobacco. Plant Mol Biol 9: 411–420

Hosokawa D, Ohashi Y (1988) Immunochemical localization of pathogenesis-related proteins secreted into the intercellular spaces of salicylate-treated tobacco leaves. Plant Cell Physiol 29: 1035–1040

Jondle DJ, Coors JG, Duke SH (1989) Maize leaf β-1,3-glucanase activity in relation to resistance to *Exserohilum turcicum*. Can J Bot 67: 263–366

Joosten MHAJ, De Wit PJGM (1989) Identification of several pathogenesis-related proteins in tomato leaves inoculated with *Cladosporium fulvum* (syn. *Fulvia fulva*) as 1,3-β-glucanases and chitinases. Plant Physiol 89: 945–951

Kassanis B, White RF (1975) Polyacrylic acid-induced resistance to tobacco mosaic virus in tobacco cv. Xanthi. Ann Appl Biol 79: 215–220

Kassanis B, Gianinazzi S, White RF (1974) A possible explanation for the resistance of virus-infected tobacco plants to second infection. J Gen Virol 23: 11–16

Kauffmann S, Legrand M, Geoffroy P, Fritig B (1987) Biological function of pathogenesis-related proteins. Four PR proteins have β-(1-3) glucanase activity. EMBO J 6: 3209–3212

Kauffmann S, Legrand M, Fritig B (1990) Isolation and characterization of six pathogenesis-related (PR) proteins of Samsun NN tobacco. Plant Mol Biol 14: 381–390

Keen NT, Yoshikawa M (1983) β-1,3-endoglucanase from soybean releases elicitor-active carbohydrates from fungus cell walls. Plant Physiol 71: 460–465

Keller B, Sauer N, Lamb CJ (1988) Glycine-rich cell wall proteins in bean: gene structure and association of the protein with the vascular system. EMBO J 7: 3625–3633

King GJ, Turner VA, Hussey CE Jr, Wurtele ES, Lee SM (1988) Isolation and characterization of a tomato cDNA clone which codes for a salt-induced protein. Plant Mol Biol 10: 401–412

Koch E, Slusarenko A (1990) *Arabidopsis* is susceptible to infection by a downy mildew fungus. Plant Cell 2: 437–445

Kombrink E, Hahlbrock K (1986) Responses of cultured parsley cells to elicitors from phytopathogenic fungi. Plant Physiol 81: 216–221

Kombrink E, Schroder M, Hahlbrock K (1988) Several "pathogenesis-related" proteins in potato are 1,3-β-glucanases and chitinases. Proc Natl Acad Sci USA 85: 782–786

Kopp M, Rouster J, Fritig B, Darvill A, Albersheim P (1989) Host–pathogen interactions XXXII. A fungal glucan preparation protects Nicotianae against infection by viruses. Plant Physiol 90: 208–216

Lagrimini LM, Bradford S, Rothstein S (1990) Peroxidase-induced wilting in transgenic tobacco plants. Plant Cell 2: 7–18

LaRosa PC, Singh NK, Hasegawa PM, Bressan RA (1989) Stable NaCl tolerance of tobacco cells is associated with enhanced accumulation of osmotin. Plant Physiol 91: 855–861

Laskowski M Jr (1986) Protein inhibitors of serine proteinases-mechanism and classification. In: Friedman M (ed) Nutritional toxicological significance of enzyme inhibitors in foods. Plenum, New York, pp 1–17

Lawton MA, Lamb CJ (1987) Transcriptional activation of plant defense genes by fungal elicitor, wounding and infection. Mol Cell Biol 7: 335–341

Leach JE, Sherwood J, Fulton RW, Sequeira L (1983) Comparison of soluble proteins associated with disease resistance induced by bacterial lipopolysaccharide and by viral necrosis. Physiol Plant Pathol 23: 377–385

Legrand M, Kauffmann S, Geoffroy P, Fritig B (1987) Biological function of "pathogenesis-related" proteins: four tobacco pathogenesis-related proteins are chitinases. Proc Natl Acad Sci USA 84: 6750–6754

Liang X, Dron M, Cramer CL, Dixon RA, Lamb CJ (1989) Differential regulation of phenylalanine ammonia-lyase genes during plant development and by environmental cues. J Biol Chem 264: 14486–14492

Linthorst HJM, Meuwissen RLJ, Kauffmann S, Bol J (1989) Constitutive expression of pathogenesis-related proteins PR-1, GRP and PR-S in tobacco has no effect on virus infection. Plant Cell 1: 285–291

Linthorst HJM, van Loon LC, van Rossum CMA, Mayer A, Bol JF, van Roekel JSC, Meulenhoff EJS, Cornelissen BJC (1990) Analysis of acidic and basic chitinases from tobacco and petunia and their constitutive expression in transgenic tobacco. Mol Plant Microbe Interact 3: 252–258

Lotan T, Fluhr R (1990) Xylanase, a novel elicitor of pathogenesis-related proteins in tobacco, uses a non-ethylene pathway for induction. Plant Physiol 93: 811–817

Lotan T, Ori N, Fluhr R (1989) Pathogenesis-related proteins are developmentally regulated in tobacco flowers. Plant Cell 1: 881–887

Lucas J, Camacho-Henriquez A, Lottspeich F, Henschen A, Sänger HL (1985) Amino acid sequence of the "pathogenesis-related" leaf protein p14 from viroid-infected tomato reveals a new type of structurally unfamiliar proteins. EMBO J 4: 2745–2749

Maiss E, Poehling HM (1983) Resistance against plant viruses induced by culture filtrates of the fungus *Stachybotrys chartarum*. Neth J Plant Pathol 89: 323

Malamy J, Carr JP, Klessig DF, Raskin I (1990) Salicylic acid—a likely signal in the resistance response of tobacco to viral infection. Science 250: 1002–1004

Malamy J, Hennig J, Klessig (1991) Salicylic acid and disease defense response in tobacco. In: Third International Congress of the International Society for Plant Molecular Biology, Abstract No 1089

Maniatis T, Goodbourn S, Fischer JA (1987) Regulation of inducible and tissue-specific gene expression. Science 236: 1237–1245

Mason HS, Mullet JE (1990) Expression of two soybean vegetative storage protein genes during development and in response to water deficit, wounding and jasmonic acid. Plant Cell 2: 569–579

Matsuoka M, Ohashi Y (1984) Biochemical and serological studies of pathogenesis-related proteins of *Nicotiana* species. J Gen Virol 65: 2209–2215

Matsuoka M, Asou S, Ohashi Y (1985) Transcriptional step is necessary for induction of pathogenesis-related proteins. Proc Jap Acad Sci [B] 61: 486–489

Matsuoka M, Yamamato N, Kamo-Murakami Y, Tanaka Y, Ozeki Y, Hirano H, Kagawa H, Oshima M, Ohashi Y (1987) Classification and structural comparison of full-length cDNAs for pathogenesis-related proteins. Plant Physiol 85: 942–946

Matsuoka M, Asou S, Ohashi Y (1988) Regulation mechanisms of the synthesis of pathogenesis-related proteins in tobacco leaves. Plant Cell Physiol 29: 1185–1192

Matthews REF (1981) Plant virology, 2nd edn. Academic Press, New York

Mauch F, Staehelin LA (1989) Functional implications of the subcellular localization of ethylene-induced chitinase and β-1,3-glucanase in bean leaves. Plant Cell 1: 447–457

Mauch F, Hadwiger LA, Boller T (1988a) Antifungal hydrolases in pea tissue. I. Purification and characterization of two chitinases and two β-1,3-glucanases differentially regulated during development and in response to fungal infection. Plant Physiol 87: 325–333

Mauch F, Mauch-Mani B, Boller T (1988b) Antifungal hydrolases in pea tissue. II. Inhibition of fungal growth by combinations of chitinase and β-1,3-glucanase. Plant Physiol 88: 936–942

Meeks-Wagner DR, Dennis ES, Van KTT, Peacock WJ (1989) Tobacco genes expressed during in vitro floral initiation and their expression during normal plant development. Plant Cell 1: 25–35

Meins F Jr, Ahl P (1989) Induction of chitinase and β-1,3-glucanase in tobacco plants infected with *Pseudomonas tabaci* and *Phytophtora parasitica* var. *nicotianae*. Plant Sci 61: 155–161

Meins F Jr, Neuhaus J-M, Sperisen C, Ryals J (1992) The primary structure of plant pathogenesis-related glucanohydrolases and their genes. In: Boller T, Meins F (eds) Genes involved in plant defense. Springer, Wien New York, pp 245–281 [Dennis ES et al (eds) Plant gene research. Basic knowledge and application]

Memelink J, Hoge JHC, Schilperoort RA (1987) Cytokinin stress changes the developmental regulation of several defence-related genes in tobacco. EMBO J 6: 3579–3583

Memelink J, Linthorst HJM, Schilperoort RA, Hoge JHC (1990) Tobacco genes encoding acidic and basic isoforms of pathogenesis-related proteins display different expression patterns. Plant Mol Biol 14: 119–126

Metzler MC, Cutt JR, Klessig DF (1991) Isolation and characterization of a gene encoding a PR-1-like protein from *Arabidopsis thaliana*. Plant Physiol 96: 346–348

Métraux JP, Streit L, Staub T (1988) A pathogenesis-related protein in cucumber is a chitinase. Physiol Mol Plant Pathol 33: 1–9

Métraux JP, Singer H, Ryals J, Ward E, Wyss-Benz M, Gaudin J, Raschdorf K, Schmid E, Blum W, Inverardi B (1990) A transient increase in salicylic acid in the phloem of cucumber correlates with the onset of systemic induced resistance. Science 250: 1004–1005

Mitchell PJ, Tjian R (1989) Transcriptional regulation in mammalian cells by sequence-specific DNA binding proteins. Science 245: 371–378

Mitsui T, Christeller JT, Hara-Nishimura I, Akazawa T (1984) Possible roles of Ca^{2+} and calmodulin in the biosynthesis and secretion of α-amylase in rice seed scutellar epithelium. Plant Physiol 75: 21–25

Nasser W, de Tapia M, Kauffmann S, Montasser-Kouhsari S, Burkard G (1988) Identification and characterization of maize pathogenesis-related proteins. Four maize PR proteins are chitinases. Plant Mol Biol 11: 529–538

Nasser W, de Tapia M, Burkard G (1990) Maize pathogenesis-related proteins: characterization and cellular distribution of 1,3-β-glucanases and chitinases induced by brome mosaic virus infection or mercuric chloride treatment. Physiol Mol Plant Pathol 36: 1–14

Nassuth A, Sanger HL (1987) Immunological relationships between "pathogenesis-related" leaf proteins from tomato, tobacco and cowpea. Virus Res 4: 229–242

Neale AD, Wahleithner JA, Lund M, Bonnett HT, Kelly A, Meeks-Wagner DR, Peacock WJ, Dennis ES (1990) Chitinase, β-1,3-glucanase, osmotin, and extensin are expressed in tobacco explants during flower formation. Plant Cell 2: 673–684

Neuhaus J-M, Ahl-Goy P, Hinz U, Flores S, Meins Jr F (1991) High-level expression of a tobacco chitinase gene in *Nicotiana sylvestris*. Susceptibility of transgenic plants to *Cercospora nicotianae* infection. Plant Mol Biol 16: 141–151

Ohashi Y, Matsuoka M (1985) Synthesis of stress proteins in tobacco leaves. Plant Cell Physiol (Japan) 26: 473–480

Ohashi Y, Matsuoka M (1987a) Induction and secretion of pathogenesis-related proteins by salicylate or plant hormones in tobacco suspension cultures. Plant Cell Physiol (Japan) 28: 573–580

Ohashi Y, Matsuoka M (1987b) Localization of pathogenesis-related proteins in the epidermis and intercellular spaces of tobacco leaves after their induction by potassium salicylate or tobacco mosaic virus infection. Plant Cell Physiol (Japan) 28: 1227–1235

Ohshima M, Matsuoka M, Yamamoto N, Tanaka Y, Kano-Murakami Y, Ozeki Y, Kato A, Harada N, Ohashi Y (1987) Nucleotide sequence of the PR-1 gene of *Nicotiana tabacum*. FEBS Lett 225: 243–246

Ohshima M, Harada N, Matsuoka M, Ohashi Y (1990a) The nucleotide sequence of pathogenesis-related (PR) 1b protein gene of tobacco. Nucleic Acids Res 18: 181

Ohshima M, Harada N, Matsuoka M, Ohashi Y (1990b) The nucleotide sequence of pathogenesis-related (PR) 1c protein gene of tobacco. Nucleic Acids Res 18: 182

Ohshima M, Itoh H, Matsuoka M, Murakami T, Ohashi Y (1990c) Analysis of stress-induced or salicylic acid-induced expression of the pathogenesis-related 1a protein gene in transgenic tobacco. Plant Cell 2: 95–106

Parent J-G, Asselin A (1984) Detection of pathogenesis-related (PR or b) and of other proteins in the intercellular fluid of hypersensitive plants infected with tobacco mosaic virus. Can J Bot 62: 564–569

Parent J-G, Hogue R, Assellin A (1985) Glycoproteins, enzymatic activities and b proteins in intercellular fluid extracts from hypersensitive *Nicotiana* species infected with tobacco mosaic virus. Can J Bot 63: 928–931

Parent J-G, Hogue R, Asselin A (1988) Serological relationships between pathogenesis-related leaf proteins from four *Nicotiana* species, *Solanum tuberosum* and *Chenopodium amaranticolor*. Can J Bot 66: 199–202

Payne G, Middlesteadt W, Williams S, Desai N, Parks TD, Dincher S, Carnes M, Ryals J (1988a) Isolation and nucleotide sequence of a novel cDNA clone encoding the major form of pathogenesis-related protein R. Plant Mol Biol 11: 223–224

Payne G, Parks TD, Burkhart W, Dincher S, Ahl P, Metraux JP, Ryals J (1988b) Isolation of the genomic clone for pathogenesis-related protein 1a from *Nicotiana tabacum* cv. Xanthi-nc. Plant Mol Biol 11: 89–94

Payne G, Middlesteadt W, Desai N, Williams S, Dincher S, Carnes M, Ryals J (1989) Isolation and sequence of a genomic clone encoding the basic form of pathogenesis-related protein 1 from *Nicotiana tabacum*. Plant Mol Biol 12: 595–596

Payne G, Ahl P, Moyer M, Harper A, Beck J, Meins F Jr, Ryals J (1990) Isolation of complementary DNA clones encoding pathogenesis-related proteins P and Q, two acidic chitinases from tobacco. Proc Natl Acad Sci USA 87: 98–102

Pfitzner UM, Goodman HM (1987) Isolation and characterization of cDNA clones encoding pathogenesis-related proteins from tobacco mosaic virus infected tobacco plants. Nucleic Acids Res 15: 4449–4465

Pfitzner UM, Pfitzner AJP, Goodman HM (1988) DNA sequence analysis of a PR-1a gene from tobacco: molecular relationship of heat shock and pathogen responses in plants. Mol Gen Genet 211: 290–295

Pfitzner AJP, Pfitzner UM, Goodman HM (1990) Nucleotide sequences of two PR-1 pseudogenes from *Nicotiana tabacum* cv. Wisconsin 38. Nucleic Acids Res 18: 3404

Pierpoint WS (1986) The pathogenesis-related proteins of tobacco leaves. Phytochemistry 25: 1595–1601

Pierpoint WS, Shewry PR (1987) Amino acid homologies suggest function for pathogenesis-related proteins. Oxford Surv Plant Mol Cell Biol 4: 337–342

Pierpoint WS, Robinson NP, Leason MB (1981) The pathogenesis-related proteins of tobacco: their induction by viruses in intact plants and their induction by chemicals in detached leaves. Physiol Plant Pathol 19: 85–97

Pierpoint WS, Tatham AS, Pappin DJC (1987) Identification of the virus-induced protein of tobacco leaves that resembles the sweet-protein thaumatin. Physiol Mol Plant Pathol 31: 291–298

Raskin I, Ehmann A, Melander WR, Meeuse BJD (1987) Salicylic acid: a natural inducer of heat production in *Arum lilies*. Science 237: 1601–1602

Redolfi P (1983) Occurrence of pathogenesis-related (b) and similar proteins in different plant species. Neth J Plant Pathol 89: 245–254

Redolfi P, Cantisani A (1984) Preliminary characterization of new soluble proteins in *Phaseolus vulgaris* cv. *saxa* reacting hypersensitively to viral infection. Physiol Plant Pathol 25: 9–19

Richardson M, Valdes-Rodriguez S, Blanco-Labra A (1987) A possible function for thaumatin and a TMV-induced protein suggested by homology to a maize inhibitor. Nature 327: 432–434

Roby D, Toppan A, Esquerré-Tugayé M-T (1988) Systemic induction of chitinase activity and resistance in melon plants upon fungal infection or elicitor treatment. Physiol Mol Plant Pathol 33: 409–417

Roggero P, Pennazio S (1989) The extracellular acidic and basic pathogenesis-related proteins of soybean induced by viral infection. J Phytopathol 127: 274–280

Ross AF, Pritchard DW (1972) Local and systemic effects of ethylene on tobacco mosaic virus lesion in tobacco. Phytopathology 62: 786

Ryan CA (1988) Proteinase inhibitor gene families: tissue specificity and regulation. In: Verma DPS, Goldberg RB (eds) Temporal and spatial regulation of plant genes. Springer, Wien New York, pp 223–233 [Dennis ES et al (eds) Plant gene research. Basic knowledge and application]

Sachs M, Ho T-HD (1986) Alteration of gene expression during environmental stress in plants. Annu Rev Plant Physiol 37: 363–376

Samac DA, Hironaka CM, Yallaly PE, Shah DM (1990) Isolation and characterization of the genes encoding basic and acidic chitinase in *Arabidopsis thaliana*. Plant Physiol 93: 907–914

Schmelzer E, Krüger-Lebus S, Hahlbrock K (1989) Temporal and spatial patterns of gene expression around sites of attempted fungal infection in parsley leaves. Plant Cell 1: 993–1001

Shinshi H, Mohnen D, Meins F Jr (1987) Regulation of a plant pathogenesis-related enzyme. Inhibition of chitinase and chitinase mRNA accumulation in cultured tobacco tissues by auxin and cytokinin. Proc Natl Acad Sci USA 84: 89–93

Shinshi H, Wenzler H, Neuhaus J-M, Felix G, Hofsteenge J, Meins F Jr (1988) Evidence for N- and C-terminal processing of a plant defense-related enzyme: primary structure of tobacco prepro-β-1,3-glucanase. Proc Natl Acad Sci USA 85: 5541–5545

Shinshi H, Neuhaus J-M, Ryals J, Meins F Jr (1990) Structure of a tobacco endochitinase gene: evidence that different chitinase genes can arise by transposition of sequences encoding a cysteine-rich domain. Plant Mol Biol 14: 357–368

Singh NK, Bracker CA, Hasegawa PM, Handa AK, Buckel S, Hermodson MA, Pfankoch E, Regnier FE, Bressan RA (1987) Characterization of osmotin. A thaumatin-like protein associated with osmotic adaptation in plant cells. Plant Physiol 85: 529–536

Singh NK, Nelson DE, Kuhn D, Hasegawa PM, Bressan RA (1989) Molecular cloning of osmotin and regulation of its expression by ABA and adaptation to low water potential. Plant Physiol 90: 1096–1101

Takahashi WN (1956) Increasing the sensitivity of the local-lesion method of virus assay. Phytopathology 46: 654–656

Takeuchi Y, Yoshikawa M, Takeba G, Tanaka K, Shibata D, Horino O (1990) Molecular cloning and ethylene induction of mRNA encoding a phytoalexin elicitor-releasing factor, β-1,3-endoglucanase, in soybean. Plant Physiol 93: 673–682

Tran Thanh Van M, Thi Dien N, Chlyay A (1974) Regulation of organogenesis in small explants of superficial tissue of *Nicotiana tabacum* L. Planta 119: 149–159

Tuzun S, Rao MN, Vögeli U, Schardi CL, Kuć J (1989) Induced systemic resistance to blue mold: early induction and accumulation of β-1,3-glucanases, chitinases and other pathogenesis-related proteins (b-proteins) in immunized tobacco. Phytopathology 79: 979–983

Van de Rhee MD, Van Kan JAL, Gonzaléz-Jaén MT, Bol JF (1990) Analysis of regulatory elements involved in the induction of two tobacco genes by salicylate treatment and virus infection. Plant Cell 2: 357–366

Van den Bulcke M, Bauw G, Castresana C, Van Montagu M, Vandekerckhove J (1989) Characterization of vacuolar and extracellular β-(1,3)-glucanases of tobacco: evidence for a strictly compartmentalized plant defense system. Proc Natl Acad Sci USA 86: 2673–2677

Van Kan JAL, Cornelissen BJC, Bol JF (1988) A virus-inducible tobacco gene encoding a glycine-rich protein shares putative regulatory elements with the ribulose bisphosphate carboxylase small subunit gene. Mol Plant Microbe Interact 1: 107–112

Van Kan JAL, Van de Rhee MD, Zuidema D, Cornelissen BJC, Bol JF (1989) Structure of tobacco genes encoding thaumatin-like proteins. Plant Mol Biol 12: 153–155

Van Loon LC (1975a) Polyacrylamide disk electrophoresis of the soluble leaf proteins from

Nicotiana tabacum var. "Samsun" and "Samsun NN". III. Influence of temperature and virus strain on changes induced by tobacco mosaic virus. Physiol Plant Pathol 6: 289–300

Van Loon LC (1975b) Similarity of qualitative changes of specific proteins after infection with different viruses and their relationship to acquired resistance. Virology 67: 566–575

Van Loon LC (1977) Induction by 2-chloroethylphosphonic acid of viral-like lesions, associated proteins and systemic resistance in tobacco. Virology 80: 417–420

Van Loon LC (1983) The induction of pathogenesis-related proteins by pathogens and specific chemicals. Neth J Plant Pathol 89: 265–273

Van Loon LC (1985) Pathogenesis-related proteins. Plant Mol Biol 4: 111–116

Van Loon LC, Antoniw JF (1982) Comparison of the effects of salicylic acid and ethephon with virus-induced hypersensitivity and acquired resistance in tobacco. Neth J Plant Pathol 88: 237–256

Van Loon LC, Van Kammen A (1970) Polyacrylamide disc electrophoresis of the soluble leaf proteins from *Nicotiana tabacum* var. "Samsun" and "Samsun NN". II. Changes in protein constitution after infection with tobacco mosaic virus. Virology 40: 199–211

Van Loon LC, Gerritsen YAM, Ritter CE (1987) Identification, purification and characterization of pathogenesis-related proteins from virus-infected Samsun NN tobacco leaves. Plant Mol Biol 9: 593–609

Varner JE, Cassab GI (1986) A new protein in petunia. Nature 323: 110

Varner JE, Liang-Shiou (1989) Plant cell wall architecture. Cell 56: 231–239

Vera P, Conejero V (1988) Pathogenesis-related proteins of tomato. Plant Physiol 87: 58–63

Vera P, Hernández-Yago J, Conejero V (1989) "Pathogenesis-related" P1 (p14) protein. Vacuolar and apoplastic localization in leaf tissue from tomato plants infected with citrus exocortis virioid; in vitro synthesis and processing. J Gen Virol 70: 1933–1942

Vögeli-Lange R, Hansen-Gehri A, Boller T, Meins F Jr (1988) Induction of the defense-related glucanohydrolases, β-1,3-glucanase and chitinase, by tobacco mosaic virus infection of tobacco leaves. Plant Sci 54: 171–176

Vögeli U, Meins F Jr, Boller T (1988) Coordinated regulation of chitinase and β-1,3-glucanase in bean leaves. Planta 174: 364–372

Wagih EE, Coutts RHA (1981) Similarities in the soluble protein profiles of leaf tissue following either a hypersensitive reaction to virus infection or plasmolysis. Plant Sci Lett 21: 61–69

Ward ER, Payne GB, Moyer MB, Williams SC, Dincher SS, Sharkey KC, Beck JJ, Taylor HT, Ahl-Goy P, Meins F Jr, Ryals JA (1991) Differential regulation of β-1,3-glucanase messenger RNAs in response to pathogen infection. Plant Physiol 96: 390–397

White RF (1979) Acetylsalicylic acid (aspirin) induces resistance to tobacco mosaic virus in tobacco. Virology 99: 410–412

White RF, Antoniw JF, Carr JP, Woods RD (1983) The effects of aspirin and polyacrylic acid on the multiplication and spread of TMV in different cultivars of tobacco with and without the N-gene. Phytopathol Z 107: 224–232

White RF, Dumas E, Shaw P, Antoniw JF (1986) The chemical induction of PR (b) proteins and resistance to TMV infection in tobacco. Antiviral Res 6: 177–185

White RF, Rybicki EP, Von Wechmar MB, Dekker JL, Antoniw JP (1987) Detection of PR1-type proteins in *Amaranthaceae*, *Chenopodiaceae*, *Graminae* and *Solanaceae* by immunoelectroblotting. J Gen Virol 68: 2043–2048

Ye XS, Pan SQ, Ku J (1989) Pathogenesis-related proteins and systemic resistance to blue mould and tobacco mosaic virus induced by tobacco mosaic virus, *Peronospora tabacina* and aspirin. Physiol Mol Plant Pathol 35: 161–175

Chapter 10

The Primary Structure of Plant Pathogenesis-related Glucanohydrolases and Their Genes

Frederick Meins, Jr.[1], Jean-Marc Neuhaus[2], Christoph Sperisen[1], and John Ryals[3]

[1]Friedrich Miescher-Institut, CH-4002 Basel, Switzerland, [2]Botanisches Institut, Abteilung für Pflanzenphysiologie, Universität Basel, CH-4056 Basel, Switzerland, and [3]Biotechnology Research, Ciba-Geigy Corp., Research Triangle Park, NC 27709-2257, U.S.A.

With 4 Figures

Contents

Our choicest plans/have fallen through.
Our airiest castles/tumbled over,
because of lines/we neatly drew
and later neatly/stumbled over.

Piet Hein (1966)

I. Introduction

The endo-type glucanohydrolases β-1,3-glucanase (E.C. 3.2.1.39) and chitinase (E.C. 3.2.1.14) are abundant proteins widely distributed in seed-plant species (Clarke and Stone, 1962; Ballance and Manners, 1978; Powning and Irzykiewicz, 1965). The physiological functions of β-1,3-glucanase and chitinase are not known. Based on the distribution of the enzyme and its putative substrates such as callose, it has been proposed that β-1,3-glucanases may have a role in fruit ripening (Hinton and Pressey, 1980), pollen tube growth (Roggen and Stanley, 1969; Ori et al., 1990), coleoptile growth (Masuda and Wada, 1967), regulation of transport through vascular tissues (Clarke and Stone, 1962), cellulose biosynthesis (Meier et al., 1981) and cell division (Waterkeyn, 1967; Fulcher et al., 1976). Although the existence of other substrates has not been ruled out, chitin, the known substrate of chitinase, is not found in higher plants.

The current interest in β-1,3-glucanase and chitinase stems largely from the hypothesis that these enzymes have a role in plant defense. Because chitin as well as β-1,3-glucans are often major components of fungal cell walls (Wessels and Sietsma, 1981), it has been proposed that these glucano-hydrolases may play a role in defense against infection by pathogenic and potentially pathogenic fungi (Abeles et al., 1971; Pegg, 1977; Boller, 1985). There is considerable indirect evidence in support of this view. Chitinase and β-1,3-glucanase are induced when plants are infected by viral, bacterial and fungal pathogens or treated with fungal elicitors and the stress hormone ethylene (Boller, 1985). Chitinase, and particularly combinations of chitinase and β-1,3-glucanase, exhibit fungicidal activity in vitro when used at concentrations comparable to those found in extracts from induced plants (Schlumbaum et al., 1986; Mauch et al., 1988). Finally, certain plant chitinases exhibit lysozyme activity (Boller, 1988) suggesting they may offer protection against bacterial infection as well.

Understanding at the molecular level how the glucanohydrolases are regulated and their role in pathogenesis will depend on having detailed information about the sequence of the enzymes and their genes. The purpose of this chapter is to provide a framework for classifying chitinases and β-1,3-glucanases on the basis of their primary structures. We begin by reviewing the structural diversity of the enzymes, then propose classification schemes, and end with a discussion of the structure and evolution of their gene families.

II. Multiple Isoforms of β-1,3-Glucanase and Chitinase

In early studies, β-1,3-glucanases and chitinases were purified on the basis of enzyme activity. More recently, proteins or cDNA clones have been identified with interesting regulatory properties that were later shown to be β-1,3-glucanase or chitinase as judged by catalytic activity of the purified proteins or comparison of the amino acid sequences deduced from DNA clones with the sequences of the known enzymes. For example, a group of acidic proteins known as *b* proteins (Gianinazzi et al., 1970) or pathogenesis-related (PR) proteins (Antoniw et al., 1980), are induced as part of the hypersensitive reaction of tobacco leaves to tobacco mosaic virus (TMV) infection (van Loon and van Kammen, 1970; Gianinazzi et al., 1970; reviewed by Cutt and Klessig, 1992). These proteins have now been found in at least 20 plant species following infection (Carr and Klessig, 1989). Recently it was shown that several of the PR proteins induced in tobacco and potato following infection are β-1,3-glucanases and chitinases (Kauffmann et al., 1987; Legrand et al., 1987; Kombrink et al., 1988).

Boller (1988) has reviewed the chitinases and β-1,3-glucanases purified from a variety of monocotyledonous and dicotyledonous species. The enzymes are monomeric polypeptides with molecular weights in the range of ca. 20–40 kDa. They often exist as isoforms within the same species and are commonly designated acidic or basic on the basis of their measured or deduced isoelectric points. These isoforms may also be localized in different cellular compartments. Following infection with bacterial and viral pathogens or treatment of plants with salicylate or ethylene, acidic isoforms of chitinase and β-1,3-glucanase are found in the intercellular wash fluid obtained by vacuum infiltrating leaves (Parent and Asselin, 1984; van den Bulke et al., 1989). Therefore, these acidic isoforms appear to be secreted by cells into the extracellular compartment. In contrast, basic isoforms of the enzymes have been found localized predominantly in the central vacuole of the cell (Boller and Vögeli, 1984; van den Bulke et al., 1989; Mauch and Staehelin, 1989; Keefe et al., 1990; Meyer, 1990; Neuhaus et al., 1991a).

Chitinases and β-1,3-glucanases often exhibit similar patterns of regulation. For example, they are coordinately induced in potato infected by *Phytophthora infestans* (Kombrink et al., 1988), in tobacco infected with *P. parasitica* var. *nicotianae* (Meins and Ahl, 1989) or TMV (Kauffmann et al., 1987; Legrand et al., 1987; Vögeli-Lange et al., 1988), and in bean treated with ethylene (Vögeli et al., 1988). The basic isoforms of the enzymes also show similar distributions in the healthy plant. They are present at very low concentrations in upper leaves of tobacco plants and increase to high concentrations—up to 4% of the soluble-protein fraction—in lower leaves and roots (Felix and Meins, 1986; Shinshi et al., 1987; Neale et al., 1990). Both enzymes are localized predominantly in the epidermis of healthy leaves (Keefe et al., 1990). Basic isoforms of chitinase and β-1,3-glucanase are coordinately induced at the mRNA level when tobacco cells are transferred from a complete culture medium to a medium without added auxin and cytokinin. This induction is blocked by combinations of auxin and cytokinin added to the medium (Felix and Meins, 1986; Shinshi et al., 1987).

There are also examples of independent regulation of chitinases and β-1,3-glucanases. Although basic isoforms of both enzymes are induced in tobacco leaves infected with *Pseudomonas tabaci*, the kinetics of mRNA accumulation for the two enzymes differ (Meins and Ahl, 1989). Different

Table 1. Representative cloned and sequenced chitinases and β-1,3-glucanases

Species	Enzyme	Protein sequence	cDNA clones	Genomic clones	References
Alfalfa	β-1,3-glucanase		X		Maher et al., 1990
Arabidopsis	chitinase	X	X	X	Samac et al., 1990; Verburg and Huynh, 1991
	β-1,3-glucanase			X	F. Ausubel, pers. comm.
Barley	β-1,3-glucanase	X	X		Høj et al., 1989
	chitinase	X	X		Leah et al., 1987; Jacobsen et al., 1990; Kragh et al., 1990, 1991; Swegle et al., 1989
Bean	β-1,3-glucanase		X		Edington et al., 1991
	chitinase	X	X	X	Awade, 1989; Broglie et al., 1986, 1989; Lucas et al., 1985; Margis-Pinheiro et al., 1991
Beet	chitinase	X			Nielsen and Mikkelsen, 1990
Cucumber	chitinase	X	X		Métraux et al., 1989
Hevea	chitinase	X			Jekel et al., 1991; Tata et al., 1983
Maize	chitinase		X		Samac et al., 1990
N. plumbaginifolia	β-1,3-glucanase	X	X	X	Bauw et al., 1987; Castresana et al., 1990; De Loose et al., 1988; Gheysen et al., 1990
Peanut	chitinase		X		Herget et al., 1991
Pea	chitinase	X			Vad et al., 1990
Petunia	chitinase		X		Linthorst et al., 1990
Poplar	chitinase		X		Parsons et al., 1989
Potato	β-1,3-glucanase		X		Witte et al., 1990
	chitinase	X	X	X	Gaynor, 1988; Gaynor and Unkenholz, 1989; Laflamme and Roxby, 1989; Pierpoint et al., 1990

Plant				References
Pumpkin	chitinase	X		Esaka et al., 1990
Rice	chitinase		X	Huang et al., 1991; Zhu and Lamb, 1991
	β-1,3-glucanase		X	Simmons et al., 1992
Soybean	β-1,3-glucanase	X	X	Takeuchi et al., 1990
Tobacco	chitinase	X	X	Fukuda et al., 1990; Hooft van Huijsduijnen et al., 1987; Linthorst et al., 1990b; Memelink et al., 1987; Neale et al., 1990; Payne et al., 1990a; Shinshi et al., 1988, 1990; van Buuren et al., 1992
	β-1,3-glucanase	X	X	Côté et al., 1991; Linthorst et al., 1990a; Ori et al., 1990; Payne et al., 1990b; Shinshi et al., 1988; Sperisen et al., 1991; van den Bulke et al., 1989; Ward et al., 1991.
Tomato	chitinase	X	X	Durand-Tardif, 1986

kinetics of mRNA accumulation for acidic and basic isoforms of β-1,3-glucanase have been reported for tobacco after TMV infection (Ward et al., 1991). Expression of basic isoforms of β-1,3-glucanase in leaves of healthy tobacco plants decreases markedly with the onset of flowering (Felix and Meins, 1986; Neale et al., 1990); whereas the acidic isoform PR-2 is expressed in leaves with the onset of flowering (Fraser, 1981; Côté et al., 1991). A unique 41 kDa form of β-1,3-glucanase is expressed exclusively in the style of the flower; but it is not induced by substances that induce the other acidic and basic forms of the enzyme (Lotan et al., 1989). The acidic chitinases, on the other hand, are localized in the pedicel and sepals of flowers (Lotan et al., 1989). When tobacco plants are sprayed with salicylic acid, low amounts of the acidic β-1,3-glucanases PR-2 and PR-N but not the acidic forms of chitinase accumulate in leaves (Bol, 1988). Finally, basic but not acidic isoforms of chitinase and β-1,3-glucanase accumulate in cultured tobacco tissues (Antoniw et al., 1983; Shinshi and Kato, 1983a; Felix and Meins, 1985, 1986, 1987; Shinshi et al., 1987).

III. Primary Structure

A. β-1,3-Glucanases

1. Evidence for Structural Classes

Amino acid sequences and/or DNA clones have been reported for at least 26 β-1,3-glucanases from 9 plant species (Table 1). In most cases, primary structures were deduced from cDNA or genomic clones. The aligned amino acid sequences of representative β-1,3-glucanases starting at the experimentally determined or putative N-terminal end of the mature protein are shown in Fig. 1. The consensus sequence shown in the figure was calculated using as a criterion identical amino acids at a given position in a weighted average of 75% of the sequences, i.e., individual sequences were given a fractional weight so that the value for the sum of the sequences for an individual species equals 1.0. Based on this test, the sequences of the tobacco β-1,3-glucanases were conserved at 37% of the positions. Allowing structurally conservative substitutions, viz., threonine for serine and valine for isoleucine or leucine, 30.6% of the positions had identical or very similar amino acids. It is of interest that in a weighted average of 18 representative sequences from 6 species, the interspecific conservation, 42.9%, is comparable to the intraspecific conservation found in tobacco.

The most detailed sequence information available is for the more than 14 isoforms of the tobacco enzyme, of which 10 are shown in Fig. 1. The primary structure of the first basic isoform was established from overlapping partial cDNA clones and an incomplete set of sequenced tryptic peptides including the N- and C-terminal ends of the mature protein (Shinshi et al., 1988). This structure was independently confirmed by more extensive sequencing of the protein (van den Bulke et al., 1989). Comparison of deduced amino-acid sequences with partial sequences of the proteins has also been used to establish the primary structure of the acidic isoforms PR-2, PR-N, PR-O, and PR-Q' that are induced in tobacco leaves following TMV infection (Payne et al., 1990b; Côté et al., 1991; Linthorst et al., 1990a; Ward et al., 1991) and the isoforms 41a and 41b constitutively expressed in the style (Ori et al., 1990). Three additional isoforms induced by TMV infection, GL 153, GL 161 and Ci30 have been identified solely on the basis of cDNA clones (Linthorst et al., 1990a; Ward et al., 1991).

Based on sequence identity, tobacco β-1,3-glucanases can be classified into at least three structural groups (Payne et al., 1990b; Ward et al., 1991): The class I enzymes, represented by GLA in Table 2, include at least 4 basic

Table 2. Comparison of tobacco β-1,3-glucanase protein sequences

Class	Isoform	Percent of positions with identical amino acids[a]						
		GL A	PR-2	PR-O	GL 153	GL 161	Sp41 a	PR-Q'
I	GL A[b]	100	50.3	51.3	50.3	51.6	49.8	56.6
II	PR-2[c]		100	91.2	83.8	81.8	80.1	55.6
II	PR-O[c]			100	86.6	84.3	82.0	56.1
II	GL 153[c]				100	89.3	87.7	55.8
II	GL 161[c]					100	85.0	55.6
II	Sp41 a[d]						100	53.4
III	PR-Q[e]							100

[a] Sequences aligned using the Bestfit algorithm of Devereux et al., 1984
[b] Sperisen et al., 1991; [c] Payne et al., 1990b; [d] Ori et al., 1990; [e] Ward et al., 1991

Fig. 1. Consensus sequence of representative β-1,3-glucanases. Sequences starting with the putative N-terminal amino acid of the mature proteins. Source of data: ARA-1, ARA-2, ARA-3, *Arabidopsis* cDNA clones (F. Ausubel, pers. comm.); BAR-1, barley β-1,3 1,4-glucanase cDNA (Fincher et al., 1986); BAR-2, barley β-1,3-glucanase cDNA clone (Ballance and Svedsen, 1988); NPLUM, *N. plumbaginifolia* cDNA clone pBEG (De Loose et al., 1988); POT, potato cDNA clone g 46 (E. Kombrink, pers. comm.); TOB-A and TOB-B, tobacco genomic clones GL A and GL B (Sperisen et al., 1991); TOB-O, tobacco PR-O composite sequence from cDNA clone λFJ1 (Côté et al., 1991) and cDNA GL 134 (Ward et al., 1991); TOB-2, tobacco PR-2 cDNA clone GL 117 (Ward et al., 1991); TOB-N, tobacco PR-N genomic clone g 19 (Linthorst et al., 1990a); TOB-153 and TOB-161, tobacco cDNA clones (Ward et al., 1991); TOB-S41A and TOB-S41B, tobacco stylar β-1,3-glucanase cDNA clones sp41 a and sp41 b (Ori et al., 1990); TOB-Q', tobacco PR-Q' cDNA clone (Payne et al., 1990b); SOYA, soybean composite sequence from cDNA clones EG 4, EG 412, EG 488 (Takeuchi et al., 1990). . . . . No sequence; *deletion. Positions conserved in plant β-1,3-glucanases and in yeast *exo*-β-1,3-glucanase (Klebl and Tanner, 1989) are underlined in the consensus sequence; C-terminal extensions verified by protein sequencing are shaded. The class designation of tobacco β-1,3-glucanases is indicated in parentheses

```
                                                                                                    1                                                                                          110
ARA-1         **v      r n   spae   tia fkqn   qrv l s dh   dv a       vt gl   s y qsv ssqs q na   ty   mnyang rfr   s    ki   s***dsyaqf
ARA-2         ks*i  r * l qttkhpana ********   rrm l g dp   ga a        li dv  ss d erl ssqt e dk   e    qsyrdg rfr   n    k    s*****vggf
ARA-3         **i      r n   rpase   va yqqrn   rrm l a nq   et n        lv dv   p d qrl ssqa a as   rm   rnyan* tfr   s    q    s***dqaasf
BAR-1         es*i  m sa aast  vsmfkfng ksm l a nq   aa q vg tg  nvvvga d v snl aspa a as   ks i qay*pk sfr   vc   ag   g****atrm
BAR-2         s*i   v i  srsd  vq yrskg  ngm i fadg qa s       n  gli dig d q ani asts n as   n    rpyypa nik   aa   qg   g****atqs
NPLUM         s*v   m l  pasq  vq ykskn  rrm l d nq   aa q        vm gv   v dvkhi sgme h rw   r    rnfwpa kfr   a    is   vtgtssltry
POT           **l   m m  shse  iq yksrn  grl l d nq   ga n        vi gl   s dvkhi sgme h rw   k    kdfwpd kik   a    is   vtgtsltsf
TOB-A(I)      s*i   m l  nhwe  iq yksrn  grl l d nh   ga q        vm gl   s dvkhi sgme h rw   k    kdfwpd kik   a    is   vtgtsyltsf
TOB-B(I)      s*i   m l  nhwe  iq yksrn  grl l d nh   ga q        vm gl   q gslt*dps r ng   d    inhfpd kfk   a    s    tn*ngqyapf
TOB-O(II)     ....  k ia sdqd  in ynang  krm i y et   nvfn        ii dv   lq gslt*dps r ng   d    inhfpd kfk   a    s    gn*ngqyapf
TOB-2(II)     s*i   k ha sdqd  in ydang  rkm i n dt   nvfn        ii dv   lq gslt*dps r ng   d    inhfpd kfk   a    s    gn*ngqyapf
TOB-N(II)     s*i   k ha sdqd  in ynang  rkm i n dt   nvfn        ii dv   q d eal *nss i ng   d    irshfpy kfk  si   s    in*ngqysqf
TOB-153(II)   s*i   k ia seqd  in ykang  rkm i y dk   nifk  k     ii dv   q d eal *nss i ng   d    irshfpy kfk  si   k s  tn*ndqysef
TOB-161(II)   s*i   k aa sdqd  in ynang  rkl i y dk   nifk  n     ii gv   q d eal *nss i ng   d    irshfpy kfk  si   s    sn*ngqysqf
TOB-S41A(II)  sni   k ia seqd  in ykang  rkm i nsdt  nifks n     ii ev   q d eal *nss i ng   d    irshfpy kfk  si   s    tn*ngqysqf
TOB-S41B(II)  sni   e ia seqd  in ykang  rkm i y dt   nifk  n     lm gv   p d env asqa n dt   n    rmy*gm kfr   a    s    lnenskyvpv
TOB-Q'(III)   **a   r q g spad  vs cnrrn  rrm i d dq   pt e        ll di   d n rml ssqd n nk   di   knyann rfr   vs   k    e***hsfaqf
SOYA          s**   r l  tpqe   va ynqan
CONSENSUS     Q--GVCYG- -GNNLP-  V--L----  --R-Y-P--  -L-ALRGSN IE-L--PN  -L---A---  -A--WVQ-NV  ------V--   YI--VGNFV-P
```

```
              111                                                                                                                                           220
ARA-1         lv    me  dr  vla   ggr   svdmg vlgesy   k s rgdvm*v imepiir   v skns ll l   t   s ag i gq r d   tapsgi* s pprs q
ARA-2         llq   mq  en  vsg   ***e v aiatd tttdts   q r rdeyk*s flepvig   a skqs lv l       s mgdt an h d   taqst *dn pgys q
ARA-3         vl    mq  er  vssl  ****  aidtr gis*gf   s t tpefr*s fiapvis   a skqs lv n       s tg m rd r di   tapst * r gqnq r          h
BAR-1         lv    mk  vhg va    g*h  t svsqa ilgvfs   a v knay*** *mtdvarl a stga ma i       laway p samdmg   nasgt * r gayg q
BAR-2         il    mr  lna sa    g*a   sirfd evansf   q s rndvr*s fidpiig   vr rins lv i       a rd p gs s n t qpgtt rdqm nglt ts
NPLUM         ll    mr  rn  iss   qnn   ssvdmt lignsf   q s rndvr*w ftdpivg   r dtra lv i       s sg p rd s p   tapnv * q gslg r
POT           qv    lv  yk  vge   gnd   svdmt lignsy   q s rndar*w fvdpivg   r dtra lv i       s sg p gq s p s tapnv * q gsrq r
TOB-A(I)      lt    mv  yk  ige   gnn   svdmt lignsy   q s rndar*w ftdpivg   r dtra lv i       s sg p gq s p s tapnv * q gsrq r
TOB-B(I)      lt    mv  yk  ige   gnn   atysg llanty   kd si...fn*s finpiiq   a rmnl la v       ghiy t advp s   tqqea***** npag q
TOB-O(II)     vg    mq  vyn aa    qdq   atysg llantn   kd si rgefn*s finpiiq   v qhnl la v       ghif t advp s   tqqea***** npag q
TOB-2(II)     va    mq  vyn aa    qdq   atysg llanty   kd si rgefn*s finpiiq   v qhnl la v       ghif t advp s   tqqea***** npag q
TOB-N(II)     va    mq  vyn aa    qdq   atysg llanty   ka si rgefn*s finpiiq   a qmnl la v       vhis t advs s   tqrgk***** nsag q
TOB-153(II)   llh   me  vyn aas   cdk  t vtysg vlanty   er si reefk*s finpiiq   a rmnl la v       vhvs t advs s   tqqgt***** nsag q
TOB-161(II)   llq   mk  vyn aa    cdm   vtysg vlanty   kd si reefk*s finpiie   a rmnl la i       ghiy t vdvp s   nqqgt***** nstg q
TOB-S41A(II)  llh   me  vyn aa    qdk  t atysg llanty   kd si reelk*s finpiie   a rmnl la i       ghi* t vdvp s   nqqet***** nstg q
TOB-S41B(II)  llh   mk  vyn aa    qdk   aietg lttdts   kd si reelk*s finpiin   v tnra lv l       a*ia n ad k e   tssev * n ngrg r
TOB-Q'(III)   lln   mr  qt  isg   gnq   aidtg alaesf   k s ksdyrga yldgvir   v nnna mv v   s a ta p kd s d   rspsv * q gslg r
SOYA          lv    le  qr  isn   gnq v
CONSENSUS     --PA--NI-- AL--AGL--- IKVST---- -----PPS-  -G-F-----  ------FL-  ----PL--N-  -PYF-Y--N-  --I-L-YALF  ----V-V-D-  --Y-NLFD
```

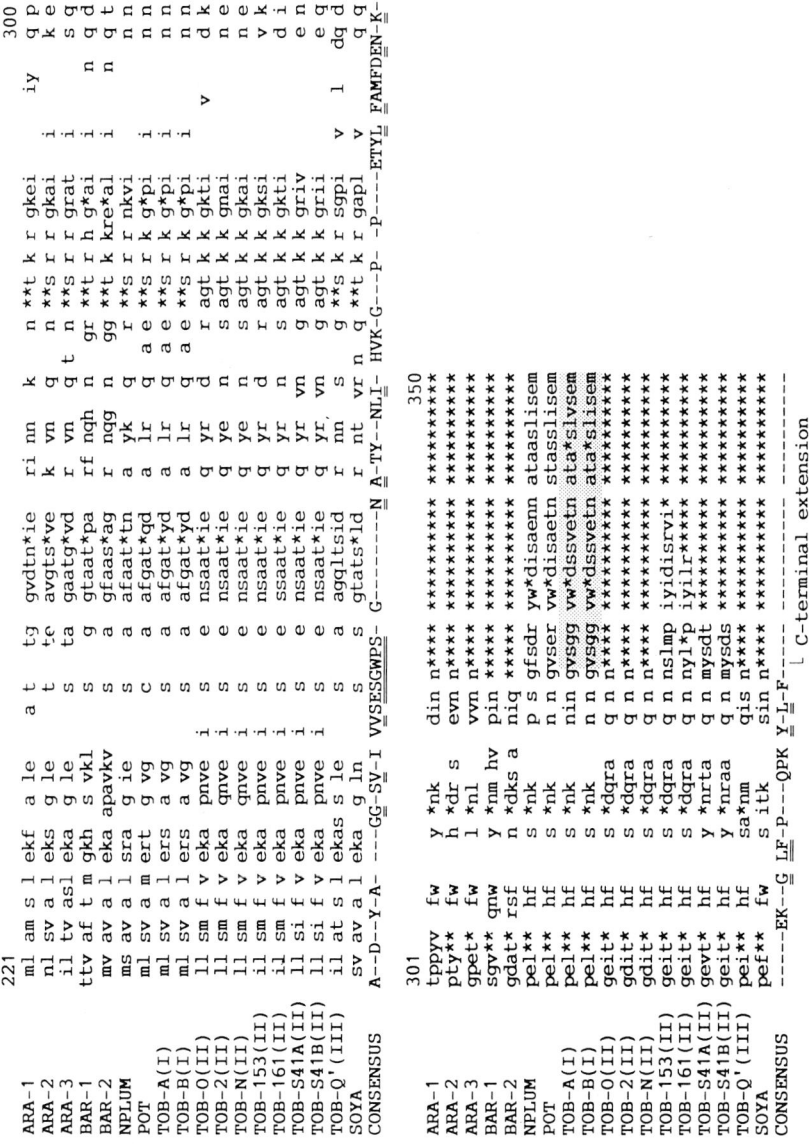

isoforms that diverge at less than 1% of the positions (Shinshi et al., 1988; Sperisen et al., 1991). The class II enzymes, for which five examples are shown in Table 2, include the acidic forms PR-2, PR-N, PR-O, GL153, GL161 and Ci30 (Ward et al., 1991; Linthorst et al., 1990a), and the two stylar isoforms, Sp41a and Sp41b (Ori et al., 1990). Within this class, the sequences differ at less than 18% of the positions. This class differs from the class I enzymes at a minimum of 48.4% of the positions. The sole representative of the class III enzymes known in tobacco is PR-Q', an acidic protein that differs by a minimum of ca. 43% in sequence from the class I and class II enzymes (Payne et al., 1990b).

The in vitro translation product obtained with mRNA for class I β-1,3-glucanases is larger than the mature enzyme (Shinshi and Kato, 1983b; Mohnen et al., 1985). Comparison of the N-terminal amino acid sequence of the mature enzyme with the sequence deduced from cDNA clones showed that these enzymes are synthesized as pre-proteins with a hydrophobic leader sequence typical of signal peptides (Shinshi et al., 1988). All of the tobacco β-1,3-glucanases for which complete DNA clones are available have hydrophobic N-terminal signal peptides 24 to 32 amino acids in length that are presumably cotranslationally processed as the nascent polypeptides enter the membrane system of the cell (Shinshi et al., 1988; Ori et al., 1990; Sperisen et al., 1991; Payne et al., 1990b; Ward et al., 1991; Côté et al., 1991; Linthorst et al., 1990a).

Sequencing of the C-terminal end of the mature protein and pulse-chase labelling experiments have established that class I isoforms of β-1,3-glucanase are synthesized as proenzymes with an N-glycosylated, C-terminal extension (Shinshi et al., 1988). During processing, the extension and N-glycan are lost to give the mature enzyme. There is evidence, however, that the mature enzyme contains 1-2 arabinoses per mole protein (Shinshi and Kato, 1983a). In contrast, the mature forms of several class II enzymes, viz., the stylar isoforms encoded by cDNAs 41a and 41b (Ori et al., 1990), GL161 and GL153 (Ward et al., 1991) have one or more putative N-glycosylation sites. Of these, only the mature stylar β-1,3-glucanases have been tested and shown to be glycosylated (Lotan et al., 1989).

Sequencing of the mature proteins has established that the class II enzymes, PR-2, PR-O, PR-N and PR-Q, end at the position corresponding to the class I processing site (van den Bulke et al., 1989; Payne et al., 1990b; Ward et al., 1991). The C-terminal sequence deduced from the cDNA clone FB7-5(2), which encodes a class I enzyme, is truncated and ends 12 residues upstream of the other class I enzymes (Neale et al., 1990). There is also considerable variation in the length of sequences downstream of the class I processing site. The deduced sequences of the class II stylar isoforms and GL161 and GL153 extend 4, 8, and 13 residues, respectively (Ward et al., 1991; Ori et al., 1990). It has not, however, been established that these sequences are C-terminal extensions removed by processing.

The basic β-1,3-glucanases are vacuolar enzymes, whereas the acidic β-1,3-glucanases PR-2, PR-O, PR-N, PR-Q' and the two stylar isoforms are secreted into the extracellular space (van den Bulke et al., 1989; Keefe et al., 1990; Meyer, 1990; Ori et al., 1990). The default pathway for proteins synthesized in the membrane system of plant cells is thought to be secretion into the extracellular compartment (Dorel et al., 1989). The vacuolar localization of the class I isoforms is correlated with the presence of a C-terminal extension. Van den Bulke et al. (1989) have suggested that positive sorting information for targeting the class I β-1,3-glucanases to the vacuole may be contained in this extension.

2. Catalytic Activity

The hydrolysis of glucans by endohydrolases involves carboxyl amino acids at the active site of the enzymes (Høj et al., 1989b). In the case of lysozyme, specific aspartyl and glutamyl residues required for catalysis have been identified (Dickerson and Geis, 1969). Inhibitor studies with β-1,3-glucanase purified from *N. glutinosa* suggest tryptophan and tyrosine residues are also required for enzyme activity (Moore and Stone, 1972). Comparison of conserved regions in different β-1,3-glucanases suggests specific amino acid residues that may be important for catalysis. The sequences of yeast exo-β-1,3-glucanase and the plant β-1,3-glucanases differ significantly in amino acid sequence (Klebl and Tanner, 1989). Only ca. 12.3% of the positions have identical amino acids when compared to the class I tobacco β-1,3-glucanase and the barley β-1,3-1,4-glucanase (Fincher et al., 1986). Nevertheless Y181, E244, W248, E296, and Y321 are identical in yeast β-1,3-glucanase and all plant β-1,3-glucanases surveyed. The two glutamates are present in the conserved sequences

---

Fig. 2. Consensus sequence of representative chitinases. A, Class I and class II chitinases starting with the putative N-terminal amino acid of the mature proteins. The consensus was calculated for 75% of the sequences shown weighted by species. Absolutely conserved positions are shaded in the consensus sequence. The C-terminal extensions verified by protein sequencing are shaded. The class designation, where known, is shown in parenthesis. B, Class III chitinases. The consensus is for a plurality of two sequences. Source of data: ARA and ARA-2, *Arabidopsis* genomic clones A2 and AC5, respectively (Samac et al., 1990); BAR-28, barley cDNA clone 28 (Leah et al., 1987); BEAN and BEAN-5b, bean genomic clones (Broglie et al., 1986, 1989); BEAN-4, bean cDNA clone (Awade, 1989; Margis-Pinheiro et al., 1991); CUCUMBER, cucumber cDNA clone (Métraux et al., 1989); HEVEA, *Hevea brasiliensis* chitinase (Jekel et al., 1991); PET, petunia cDNA clone Pach 1 (Linthorst et al., 1990a); POP-6 and POP-8, poplar cDNA clones Win 6 and Win 8 (Parsons et al., 1989); POT, potato cDNA clone RU8604 (Gaynor, 1988); POT-4, potato cDNA clone CCH 4 (Laflamme and Roxby, 1989); POT-28 and POT-35, potato cDNA clones BC 28 and BC 35 (E. Kombrink, pers. comm.); TOB-A, TOB-B, TOB-C, tobacco genomic clones λCHN 17, λCHN 200, and λCHN 14, respectively (Shinshi et al., 1990; van Buuren et al., 1992); TOB-P and TOB-Q, tobacco PR-P and PR-Q cDNA clones (Payne et al., 1990a); TOM, tomato genomic clone ch3.5 (Durand-Tardif, 1986) . . . . No sequence; *deletions

```
                                                                                                              110
        1
ARA(I)      eqcgrgqagg alcpnglccs efgwcgntep yckgpg*cgs *qctpgg*** **tpppgpt** ****gdlsgi isssq d kh daa pa   rg* n  it ks
BAR-28(II)  ********** ********** ********** ********** ********** ********** ****svssi vsraq r lh dgatqa kg* d  va aa
BEAN(I)     eqcgr*qagg alcpggnccs qfgwcgsttd ycg*pg*cgs *qc*gg***  ****ps*pa pt**dlsal isrst qa* kh dga pa  kg* rd ia kay s
BEAN-4(I)              .gycgtged ycg*tg*cqe gpc.......       .ngi i*****ngi ****dsg ag   kn*            ls lnsyts
BEAN-5b(I)  eqcgr*qagg alcpggnccs qfgwcgsttd ycg*kd*cqs *qc*kg***  ****ps*pa pt***dlsal isrst qv kh dga pa kg* d  ia kay s
PET(II)     ********** ********** ********** ********** **********  ****qnvgsi vtsdl q kn dar fa vr* d  ia ns
POP-6       .......... .......... .......... .......... .......... .......... .......... ..... ... .. ...  ... ..  ...ef* d
POP-8(I)    *qcgs*qagn atcpndlccs sggycgltva ycc*ag*cvs *qc******* ******rncf ftesm eq pn nds pg kg* d y fv tefy
POT(I)      qncgs*qggg kacasgqccs kfgwcgntnd ycg*sgncqs *qc*pgg**  *****pg*pg p*ggdlgsa isnsm q kh ens qg kn* s n in rs
POT-4(I)    snvvh*rpd* alcapglccs kfgwcgntnd ycg*pgncqs *qc*pgg**  p**sgdlggv isnsm q nh dna qg knn s n is gs
POT-28(I)   eqcgs*qagg alcasglccs kfgwcgntnd ycg*pgncqs *qc*pgg**  p***gdlggv isnsm q nh dna qg kgn s n is gs
POT-35(I)   eqcgs*qagg alcapglccs kfgwcgntnd ycg*pgncqs *qc*pgg*ps ****gdlggv isnsm q nh dna qg knm s n is gs
TOB-A(I)    eqcgs*qagg arpsglccs  kfgwcgntnd ycg*pgncqs *qc*pgg**p tptpptp*pg ***ggdlgsi isssm q kh dna qg kg* s n in rs
TOB-B(I)    eqcgs*qagg arcasglccs kfgwcgntnd ycg*pgncqs *qc*pgg**  ***ptp*pg  **ggdlgsi isssm q kh dna qg kg* s n in rs
TOB-C(I)    eqcgk*qagg arpsgmccs  nfgwcgntqd ycg*pgkcqs *qc*psgpgp tprpptptpg pst*gdisni isssm q kh dnt qg ks* n it rs r
TOB-P(II)   ********** ********** ********** ********** **********  ****qgigsi vtndl ne kn dgr pa ng* d  ia ns
TOB-Q(II)   ********** ********** ********** ********** **********  ****qgigsi vtsdl ne kn dgr pa ng* d  ia ns
TOM(I)      qncgs*qggg kvcasgqccs kfgwcgntnd hcg*sgncqs *qc*pgg**  ****pg*pg  pvtggdlgsv isnsm q kh ens qg knn s n it rs
CONSENSUS
            |_N-Cysteine-rich domain_|          |_Hinge_|                   --FD-ML --RN--C-- ----FYTY-AF --AA--FPGF
                                                                            |_Catalytic domain
```

```
                                                                                                              220
        111
ARA(I)      tt tat  k v g    wat a  ysw  fkq qmpas d*ycep***s atwpccasgkr y m ls w y     n        n a a
BAR-28(II)  rt sada ...... . ...... .     .   .   t*ycsa***t pqfpcapgqq  .     .   . ......
BEAN(I)     nt tat  r lg     wat a  yaw  fvr rnp*s t*ycsa***t pqfpcapgqq  y   is w y   n k  gv r  gv     n        s
BEAN-4(I)   rv edd**r a ahf t hf** ********* yie idgas dk....... a*ycsa***t pqfpcapgqq .h ls w f  ga s nnf g ga  et sn v v
BEAN-5b(I)  nt tat  r lg     wat a  yaw  fvr rnp*s a*ycsa***t pqfpcapgqq  y   is w y   n k  qc r  gv     n        s
PET(II)     tt dta  k g      tls *  yag  flr gnqmg n********g *****g      c lt qgs dla k eq v n atv
POP-6       nt dlm  r lg     wpd a c yaw  ylk incqp **ycdp***s snyqcvagkq y   ls w y   d  dlklp qe e  p
POP-8(I)    mt ddt  r l a q s rsii ge a ftw lvn lnpns d*ycdbktks s*ypcvad** l lr w y lc d dlklp qe ek p l
POT(I)      ts ina  r a      was a  yaw  flr rgnpg d*ycpp***s sqwpcapgrk f is h y   n pc r gv     n        p
POT-4(I)    tt ita  r la     wps a  yaw  flr qgspg d*yctp***s sqwpcapgrk f is h y   n pc r gv     n        p
POT-28(I)   tt ita  r a      was a  yaw  flr qgspg d*yctp***s nqwpcapgrk f is h y   n pc r gv     n        si
POT-35(I)   tt ita  r la     wps a  yaw  flr qgspg d*yctp***s gqwpcapgrk f is h y   n pc r gv     n        si
TOB-A(I)    ts tta  r a      wat a  yaw  wlr qgspg d*yctp***s gqwpcapgrk f is h y   n pc r gv     n        p
TOB-B(I)    ts tta  r a      wat a  yaw  wlr qgspg s*ycvq***s gqwpcapgrk y is y y   n pc r gqm    n        p
TOB-C(I)    tt ttr  r v a    wdt a  ryaw ylr qgnpp s*ycvq***s sqwpcapgqk y is Y y   n pc r gqm    n        na v
TOB-P(II)   ts dta  r k g    sls ae** ftg fvr*qndqs d*********r   lt nqn eka rq v n at
TOB-Q(II)   tt dta r k g     sls ae** ftg fvr*qndqs d*********r y lt nrm eka t gqe v n at
TOM(I)      ts ina  r a      wps a  faw  flr rgnpg d*ycsp***s sqwpcapgrk y is h y   n pc r gv     n        p
CONSENSUS   G--GD--RK -ETAAFF-QT SHETTGG--- -PDGP---GY C---E-----   Y-GRGPIQL- -N-NYG--G- AI--DLLN-P DLVATD-VIS
                                                       |_Variable region_|
```

A

```
                221                                                                                                    330
ARA(I)          a         ap *    c a   a qq q   d ra g l   y         l   r *q dgr a   f  q   nifgv np  g  d yn   rs v*ngll **eaai
BAR-28(II)      .....     ap *    . . .....   ra g v   f     .....    l   r *q dsr q   f     te         gy  n  d ys   tp g*nsll l*sdlvtsq.
BEAN(I)         s l       a s     s d   tsr t   s  va r l   y tv      al  dga*n ptt qa vny   iqlgv at   d  t  :    .  :    .   nsll l*sdlvtsq.
BEAN-4(I)       t l   y   *  ..... . .....   va r l   y         l   r *q dsr q   f  fk  dllgv gy  n  d ys   tp g*nslf l*sdlvtsq.
BEAN-5b(I)      s l       a s     s d   tsr t   s  va r l   y tv      i   k *q nar e   y     r nvsimnv ap  d  yn   rm a*ev** ********
PET(II)         t         p gm *  c d   trt a   ts n v   y         i   k *q nare   f  kk  dslgt ty  s  d yq   rp g*ygls glkdtm***
POP-6           t         k p s * c a   tnt a   le g v   y         i   q gp naane     f     tkd*g ktrqqm dy  l   dmlqv dp  d  y dn   et edngll kmvgtm***
POP-8(I)        ea l      n phstga c e   tes s   ie g k   f  ml   t n g   l   r *t dnr g   f  f     silgv tp  d  vn   rw g*nall **vdtl***
POT(I)          t l       p s *   c d   irn s   ra n l   f         l   h *s dsr q   f  f     gilgv sp  d  gm   rs g*ngll **vdtv***
POT(I)          s         p s *   c d   trq g   qa n v   f         l   h *s dsr q   f  f     gilgv sp  d  gm   rs g*ngll **vdtv***
POT-4(I)        s         p s *   c d   trq g   gt qa n v   f         l   h *s dsr q   f  f     gilgv sp  d  gm   rs g*ngll **vdtv**.
POT-28(I)       s         p s *   c d   trq g   qa n v   f         l   r *t dsr q   f  f     silgv sp  d  gm   rs g*ngll **vdtm***
POT-35(I)       s l       p s *   c d   irq g   ra n l   f         l   r *t dsr q   f  f     silgv sp  d  gm   rs g*ngll **vdtm***
TOB-A(I)        s l       a s *   c d   trt a   ra n l   y         l   h *s dar q   f  f     silgv sp  d id dm   ks n*sgll **letm***
TOB-B(I)        s         p s *   c d   dist a   qs n a   c         i   v *p naa e   y     r gmlnv ap  d  yn   rn a*qg** ********
TOB-C(I)        t         p dn *  s d   irt a   qa n v   y         i   *r nda e   y  f     gmlnv ap  e  d yn   rn g*qg** ********
TOB-P(II)       t         p dn *  s d   irt a   qa n v   f         l   r *n dnr q   f  r     gi..   :    .
TOB-Q(II)       t         p s *   c d   dirn agrsn l                                                   
TOM(I)
CONSENSUS       FK-AIWFWMT -Q-PK-PS-H -VI-G-W-PS -AD--A-R-P G-GVITNIIN GG-ECG-G-- ---V-DRIG- Y-RYC---- --G-NL-C-- -Q--F---   --- C- Extension
```

B

```
                1                                                                                      100
HEVEA           .......... iyflffiscs lskpsdasrg .........g      t tq  s  rk sy  i     nk  n  t  qi      aagg tivs ngirs qigg
ARA2            mtnmtlrkhv iyflffiscs lskpsdasrg .........g      n sa  a  gr ay  v     vk  n  t  el      aant thfg sqvkd qsrg
CUCUMBER        maahkitttl sifflssif rs**sda***a      s as  a  gm ef  i     ss  s  a  vl      dnng afls deins ksqm
CONSENSUS       ---------- ---------- GIAIYWQONG NEG-L-TC- T--Y--VN-A FL--FG-GQ- P--NLAGHCN P---C---- ----C----

                101                                                                                    200
HEVEA           i  m  l   i s tla qa   nv dyl   nf  k s      d      i d  h  t ly  d  ry   sayskqgkkv y ta       f  rylgt ln
ARA2            i  m  l   i n sig re   vi dyl   nf  k s      d      i n  l  p qh  d  rt   skfshrgrki y tg       f  rlmqs ln
CUCUMBER        v  l  i   a s sls ad   qv nfi   sy  q d      v      d s  g  qf v qe   knfqg*****v i sa       i  ahlda ik
CONSENSUS       -KV-LS-GGG -G-Y---S-- DAK-A---W N--LGG-S-S RPLG-AVLDG -DF-IE-GS- ---WD-LA--L ---------- --L-APQCP- PD-----A--

                201                                                                                    300
HEVEA           gl  y  v   qys sg in iin   r tsina g ifl      p  g  *yv  p  i ri   ei k pk      * y  dkgm  s
ARA2            kr  y  i   sys sg tq lfd   k tsiaa q ffl      p  d  *yi  p  t qi   tl k rk      * w  dkng  s
CUCUMBER        gl  s  v   mf* ad ad lls   q *afpt s lym      r  p  gfi  a  i qv   tia a sn      a *  **ng  d
CONSENSUS       T--FD-VW-Q FYNNPPC--- --N--N--S WN-WT----- --K--GLPAA -EAA-SG--- P-DVL-S--L P--K-S--YG GVMLWSK-F- D----YS-SI

                301
HEVEA           ld  v.
ARA2            la  v.
CUCUMBER        kg  ig
CONSENSUS       --S--
```

(I/V/L)(V/I)V(S/A)E(S/T)GW P(S/T) starting at position 240 and
ET(V/L/I)F(A/V)(M/L/I)(F/Y)(D/N)E(D/N) starting at position 287
of the aligned sequences (Fig. 1). Although the sequences of the acidic
isoforms PR-2, PR-N and PR-O are highly conserved, PR-O has a specific
activity 60 fold higher than PR-N; and, PR-N has a specific activity 4 fold
higher than PR-2 (Kauffmann et al., 1987). PR-O differs from PR-2 and
PR-N at 4 positions, 32, 47, 58, and 265, at which PR-O has the same
residue as the highly active basic isoforms. The least active β-1,3-glucanase,
PR-2, shows a substitution of asparagine for Y146 and K283 in PR-N and
PR-O. This suggests that amino acids at one or more of these positions also
have a role in catalysis or are important for substrate specificity.

B. Chitinases

Amino acid sequences have been reported for at least 29 chitinases from 15
plant species (Table 1). Only one chitinase, a constitutive isoform in potato
(Witte et al., 1990; E. Kombrink, pers. comm.), has been shown to be
glycosylated. Comparison of aligned sequences shows that the chitinases
fall into two major structural groups: proteins related to the basic chitinases
of bean and tobacco conserved at 43.3% of the positions in the catalytic
domain (Fig. 2A); and, a group of chitinases and chitinase/lysozymes
including enzymes from *Hevea* (Tata et al., 1983; Jekel et al., 1991), *Parthenocissus* (Bernasconi et al., 1987), *Arabidopsis* (Samac et al., 1990), tobacco
(J. Ryals, unpubl.) and cucumber (Métraux et al., 1989) that are highly
conserved with identical amino acids at 48% of the positions (Fig. 2B).
Chitinases similar to the basic isoforms from bean and tobacco can be
subdivided on the basis of sequence into additional structural groups to
give a total of three distinct classes of the enzyme (Shinshi et al., 1990).

1. Class I

The class I chitinases of tobacco include three basic isoforms: chitinase A
and B with apparent molecular weights of 34 kDa and 32 kDa, respectively, that have been purified to homogeneity (Shinshi et al., 1987), and the
33 kDa chitinase C, which has only been identified on the basis of a
genomic clone and expression of the gene-specific mRNA (van Buuren et al.,
1992). These proteins are conserved at over 83% of the positions and differ
from the acidic chitinases PR-P and PR-Q induced in TMV-infected leaves
by a minimum of ca. 40% (Table 3). The primary structures of these
polypeptides and their homologues from other plant species deduced from
DNA clones have four major domains: an N-terminal signal peptide, a
cysteine-rich domain, a main structure, and a C-terminal extension. In vitro
translation studies and comparison of the amino acid sequence deduced
from cDNA and genomic clones with the N-terminal sequence of the mature

Table 3. Comparison of tobacco class I and class II chitinase protein sequences

| Class | Isoform | Percent of positions with identical amino acids[a] | | | | |
		CHN A	CHN B	CHN C	PR-P	PR-Q
I	CHN A[b]	100	98.9	82.6	63.6	64.5
I	CHN B[c]		100	82.7	63.6	64.5
I	CHN C[d]			100	61.2	63.0
II	PR-P[e]				100	93.5
II	PR-Q[e]					100

[a] Sequences aligned using the Bestfit algorithm of Devereux et al., (1984)
[b] Shinshi et al., 1987, 1990; [c] Fukuda et al., 1990; van Buuren et al., 1992; [d] van Buuren et al., 1992; [e] Payne et al., 1990a

proteins shows that chitinase A and B are synthesized as preenzymes with a hydrophobic signal peptides 23 amino acids in length that are presumed to be removed as the nascent polypeptide enters the membrane system of the cell (Shinshi et al., 1987, 1990).

The conserved region starting at the N-terminal end of the mature enzymes is a cysteine-rich domain of ca. 40 amino acids with 8 cysteines (Shinshi et al., 1987, 1990). This domain is also found in various proteins without chitinase activity, including lectins from wheat (Wright et al., 1984; Wright and Raikhel, 1989), barley (Learner and Raikhel, 1989), rice (Wilkins and Raikhel, 1989), and stinging nettle (Broekaert et al., 1989). Wheat germ agglutinin (WGA) is a chitin-binding lectin composed of two identical subunits (Nagata and Burger, 1974). These subunits consist of 4 conserved cysteine-rich domains of 41 to 43 residues with disulfide bonds in homologous positions. The 8 cysteine residues in WGA and the tobacco class I domains are in homologous positions. The S62 and Y73 residues of the WGA cysteine-rich domain, which have been shown to interact with N-acetylated sugars (Wright et al., 1984) are also conserved in the tobacco enzyme.

This cysteine-rich module is also found in hevein, a miniprotein isolated from the latex of *Hevea brasiliensis* (Lucas et al., 1985) and in two wound-induced (WIN) proteins of potato with unknown function (Stanford et al., 1989a). Recently a cDNA clone encoding hevein has been identified (Broekaert et al., 1990). The clone encodes a larger protein with the cysteine-rich domain at its N-terminal end which is highly homologous to the potato WIN proteins. This suggests that hevein either arose from the WIN-type protein by processing or as a proteolysis artifact formed during isolation. It is of interest that cysteine-rich modules of similar length with a similar spacing of cysteines are found in the thionins (see García-Olmedo et al., 1992), which have been implicated in plant defense, as well as in various animal gene families encoding the precursor of epidermal growth factor, membrane receptors, and proteases involved in blood coagulation and fibrinolysis (Pfeffer and Ullrich, 1985).

The class I chitinases have an hypervariable region between the cysteine-rich domain and the remainder of the protein molecule defined by the need to introduce gaps in this region to align the highly conserved sequences of the protein. This region, which is rich in glycine, threonine and proline, is likely to function as a spacer or hinge joining the small, folded cysteine-rich module to the main structure of the enzymes. There is considerable inter- and intraspecific variation in the length of the spacer. It is absent in the poplar enzyme WIN-8 (Parsons et al., 1989), ranges from 8 to 18 residues in tobacco chitinases (Shinshi et al., 1990) and from 4 to 6 residues in potato chitinases (Gaynor, 1988; Gaynor and Unkenholz, 1989; Laflamme and Roxby, 1989).

The main structure of the class I chitinases, which is shared by the class II chitinases of tobacco described below, is conserved at 61% of its positions and is identical at 58.8% of the positions in the aligned sequences. This region is also conserved at 53.4% of the positions and identical at 24.7% of the positions in a compilation of 18 sequences from 8 species (Fig 2 a). The main structure is split by a second variable region of ca. 20 residues starting at position 161 of the aligned sequences. Except in the acidic chitinases of petunia and tobacco (i.e., PR-P and PR-Q), where the whole region is missing, the enzymes have two cysteines at identical positions, 164 and 175. Cysteines are generally well conserved (Thornton, 1981) suggesting that these cysteines form a small loop.

Chitinases such as the tobacco enzymes PR-P and PR-Q are enzymatically active, but have neither the cysteine-rich domain nor the hinge region. Therefore, the conserved main structure is the catalytic domain of the enzyme. Although the chitin-binding cysteine-rich domain is not required for catalytic activity, the tobacco enzymes without this domain have lower specific activities (Legrand et al., 1987) suggesting it may participate in catalysis. Several residues implicated in the catalytic activity of lysozyme (viz., glutamate and aspartate) or β-1,3-glucanase (viz., tryptophan and tyrosine) are in regions of conserved sequence and identical in all the plant chitinases surveyed. These are FYTY starting at 94, (K/R)(K/R)EIAAFF at 120, QTSHETTG at 129, Y181, NYG at 194, A(L/I)W(F/Y)W M at 224, W247, AD at 252 and ECG at 274.

The deduced C-terminal sequences of the tobacco class I chitinases extend by 6 residues beyond the end of the acidic PR-P and PR-Q proteins. Peptides from the C-terminal ends of tobacco chitinases A and B have been sequenced and shown to end at position 317 of the aligned sequences, i.e., 1 residue upstream of the end of PR-P and PR-Q (Sticher et al., in prep.). Therefore, a short C-terminal extension of 7 amino acids is removed during processing of both class I chitinases. It is interesting to note that WGA, which is a vacuolar protein with cysteine rich domains, is synthesized as a proenzyme with an N-glycosylated C-terminal extension (Raikhel and Wilkins, 1987); there is, however, no potential N-glycosylation site in the chitinase extension. Class I chitinases of bean (Mauch and Staehelin, 1989) and tobacco (Meyer, 1990; Sticher et al., in prep.) are localized in the cell vacuole; whereas the acidic chitinases PR-P and PR-Q lacking the C-terminal extension and cysteine-rich module are secreted (Legrand et al., 1987) suggesting that the extension, the cysteine-rich module or both regions contain information for sorting to the vacuolar compartment.

2. Class II

The class II chitinases are acidic proteins localized in the extracellular space. They are synthesized as preproteins with hydrophobic signal peptides that are 24 residues long in the case of tobacco enzymes PR-P and PR-Q (Payne et al., 1990a; Linthorst et al., 1990b). The catalytic domains of the class I and class II chitinases are highly conserved. The major structural distinctions between the two classes are that the class II enzymes are missing the cysteine-rich domain, the spacer, the variable region of the catalytic domain, and the C-terminal extension.

3. Class III

The last group of chitinases include lysozyme/chitinases and chitinases from papaya, *Hevea, Parthenocissus, Rubus,* cucumber, and *Arabidopsis* (Boller, 1988; Métraux et al., 1989; Samac et al., 1990; Jekel et al., 1991). These enzymes have highly conserved sequences that differ from the sequences of the class I and class II enzymes (Fig. 2b). The enzymes from *Rubus, Parthenocissus,* and cucumber appear to be localized in the extracellular compartment (Bernasconi et al., 1985, 1987; Boller and Métraux, 1988); whereas the homologue from *Hevea brasiliensis* is present in the latex as well as the lutoids, which are a vacuole-like compartment of laticifer cells (Tata et al., 1983). *Arabidopsis* (Samac et al., 1990), bean (Awade, 1989; Margis-Penheiro et al., 1991), cucumber (Métraux and Boller, 1986), tobacco (J. Ryals, unpubl.), and *Hevea* (Kush et al., 1990) have been shown to contain class I or class II chitinases as well as class III chitinases.

IV. Genes Encoding Chitinase and β-1,3-Glucanase

A. Gene Structure

1. 5'-Flanking Sequence

Genomic clones have been reported for at least 10 chitinase genes from 5 species and 7 β-1,3-glucanase genes from 3 species (Table 1). In a few cases sufficient sequence information is available to compare the structure of the different genes. Regulation of transcription often depends on *cis*-acting

elements in the 5' region flanking the coding sequence of plant genes
(Benfrey and Chua, 1989). Steady-state levels of mRNAs encoding chitinases
and β-1,3-glucanases have been measured in a number of experimental
systems (Table 4). The results show that mRNA content increases with
induction of the enzymes and decreases under conditions that down-
regulate the enzymes in response to hormone treatment and pathogen
infection. Although regulation at the level of translation and messenger
stability have not been ruled out, by analogy with other inductive systems
in plants, it is likely that chitinase and β-1,3-glucanase are regulated
primarily at the level of transcription.

Direct evidence of transcriptional regulation has been obtained by
introducing chimeric constructions of 5-flanking sequences fused to the
reporter gene β-glucuronidase (GUS) into tobacco plants by Ti-mediated
DNA transformation. A 2.0 kb 5'-sequence of the *N. plumbaginifolia* β-1,3-
glucanase gene gn 1 was sufficient for developmental regulation as well as
pathogen and chemical induction of GUS in tobacco (Castresana et al.,
1990). Similar studies have shown that all the elements needed for correct
hormonal, developmental and pathogenesis regulated expression are present in a ∼1.6 kb 5'-fragment of the tobacco β-1,3-glucanase gene GL B
(R. Vögeli-Lange et al., 1991). A 1.7 kb fragment from the 5'-region of the
bean chitinase gene 5B was sufficient for induction of GUS in response to
infection with fungal pathogens and treatment with ethylene (Broglie et al.,
1989; Roby et al., 1990). Deletion analysis of this fragment showed that the
sequence between -422 and -195 upstream of the transcriptional start
site is important both for constitutive expression and ethylene-induced
expression (Broglie et al., 1989). A similar deletion analysis of the tobacco
chitinase gene CHN 50 promoter in a transient expression system suggests
that the distal region between -788 and -345 contains sequences
important for high level expression in tobacco protoplasts whereas a
second region between -68 and -47 proximal to the TATA box functions
as a silencer (Fukuda et al., 1990). The distal region contains a 12 bp repeat
with the motif GCCGCC. This motif is also found in the 5' flanking region of
genes encoding chitinase A in tobacco (Shinshi et al., 1990), the class I
chitinases of tomato (Durand-Tardif, 1986) and bean (Broglie et al., 1989),
as well as the genes encoding the tobacco class I β-1,3-glucanases (Ohme-
Takagi and Shinshi, 1990; Sperisen et al., 1991). An important regulatory
function for the motif is suggested by the fact that it is highly conserved in
genes showing a very similar pattern of expression and is present in the
region of the bean chitinase promoter required for ethylene induction
(Broglie et al., 1989).

2. Intervening Sequences

The coding sequences of chitinase and β-1,3-glucanase genes are usually
interrupted by intervening sequences with typical intron/exon junctions

Table 4. Examples of chitinase and β-1,3-glucanase regulation at the mRNA level

Species	Regulation	Enzyme/class	References
Arabidopsis	developmental regulation; induction by ethylene treatment of leaves	GLU I, CHN I, and CHN III	F. Ausubel, pers. comm.; Samac et al., 1990
Barley	increases during seed germination	CHN	Swegle et al., 1989
Bean	induction by ethylene treatment, fungal elicitors and fungal pathogens	GLU I and CHN I	Broglie et al., 1986; Edington et al., 1991; Roby et al., 1990; Vögeli et al., 1988
Cucumber	induction by salicylate treatment and tobacco necrosis virus infection of plants	CHN III	Métraux et al., 1988
Hevea	induced in latex by wounding and ethephon	CHN I	Kush et al., 1990
Melon	*Colletotrichum lagenarium* infection	CHN	Roby et al., 1987
N. plumbaginifolia	tissue specific expression; down regulation by auxin and cytokinin in cultured cells	GLU I	Castresana et al., 1990; De Loose et al., 1988
Peanut	transfer of cultured cells to fresh medium; elicitors	CHN	Herget et al., 1991
Poplar	systemic induction by wounding	CHN	Parsons et al., 1989
Potato	induction by elicitor treatment and infection with *Phytophthora infestans*	GLU and CHN	Kombrink et al., 1990; Witte et al., 1990
Rice	induction by *P. megasperma* pv. *glycinea* elicitor	CHN I	Zhu and Lamb, 1991
Soybean	induction by ethylene treatment of seedlings	GLU	Takeuchi et al., 1990
Tobacco	induction by salicylate, thiamine and elicitor treatment; induction by infection with TMV, *Cercospora nicotianae*, *Pseudomonas tabaci*, *P. solanacearum*, and *Phytophora parasitica* var. *nicotianae*, developmental regulation and tissue specific expression; down regulation by auxin and cytokinin in cultured cells	GLU I, GLU II, GLU III, CHN I, and CHN II	Côté et al., 1991; Godiard et al., 1990; Hooft van Huijsduijnen et al., 1986, 1987; Linthorst et al., 1990a, b; Meins and Ahl, 1989; Mohnen et al., 1985; Neale et al., 1990; Ohme-Takagi and Shinshi, 1990; Ori et al., 1990; Payne et al., 1990a, b; Shinshi et al., 1987; van Buuren et al., 1992; Vögeli-Lange et al., 1988; Ward et al., 1991

Table 5. Intervening sequences in β-1,3-glucanase and chitinase genes

Species	Enzyme	Enzyme class	Gene	Intron	Codon position[a]	Size (bp)
Arabidopsis	chitinase	I	A 2[c]	I	Thr (135)	476
		III	AC 5[c]	I	NA[b]	169
				II	NA	269
Bean	chitinase	I	CH 5B[d]	none	–	–
N. plumbaginifolia	β-1,3-glucanase	I	gn 2[e]	I	Gly (-2)	975
		I	gn 1[f]	I	Gly (-2)	669
				II	Tyr (91)	199
Potato	chitinase	I	RU 8713[g]	none	–	–
Tomato	chitinase	I	ch 3.5[h]	I	Gly (136)	79
				II	His (191)	81
Tobacco	chitinase	I	CHN14[i]	I	Gly (136)	116
				II	Tyr (191)	169
			CHN48[k]	I	Gly (136)	274
				II	His (191)	269
			CHN50[i]	I	Gly (136)	272
				II	His (191)	173
		II	CHA18[l]	I	Gly (136)	195
				II	Gln (192)	184
	β-1,3-glucanase	I	GLA[m]	I	Gly (-2)	789
		I	GLB[m]	I	Gly (-2)	747
			gglb50[n]	I	Gly (-2)	787
			gglx33[n]	I	Gly (-2)	NA
				II	NA	NA
		II	gl9[n]	I	Gly(-2)	341

[a] Position of intron in aligned amino acid sequences of chitinase and β-glucanase shown in Figs. 1 and 2a. Numbering is from the putative N-terminal end of the mature proteins

[b] Data not available or not applicable

[c] Samac et al., 1990; [d] Broglie et al., 1989; [e] De Loose et al., 1988; [f] Castresana et al., 1990; [g] Gaynor and Unkenholz, 1989; [h] Durand-Tardif, 1986; [i] Fukuda et al., 1990; van Buuren et al., 1992; [k] Shinshi et al., 1990; [l] Linthorst et al., 1990b; [m] Sperisen et al., 1991; [n] Linthorst et al., 1990a

and an high A + T content required for processing of nuclear pre-mRNA introns in plants (Goodall and Filipowicz, 1989). The genes encoding the class I and class II chitinases of tobacco and tomato are interrupted by two introns at highly conserved positions (Table 5). The sequence of the introns in class I chitinase genes is also conserved. The two tobacco genes CHN 48 and CHN 50 are identical at 95.8% of the positions in IVS I, after introducing 1 gap, and identical at 51% of the positions in IVS II, after introducing 3 gaps (Shinshi et al., 1990; Fukuda et al., 1990; van Buuren et al., 1992). The basic chitinase gene of *Arabidopsis* has a single intron at the position of intron I of the tomato and tobacco genes (Samac et al., 1990). No introns were found in the genes encoding class I chitinases of potato (Gaynor and Unkenholz, 1989) and bean (Broglie et al., 1989). All known class I and class II β-1,3-glucanase genes, with the exception of the *Arabidopsis* gene 2 (F. Ausubel, pers. comm.), have an intron interrupting the coding sequence in the penultimate amino acid of the signal peptide (De Loose et al., 1988; Sperisen et al., 1991; Linthorst et al., 1990a). After introducing 14 gaps, the introns of genes GL A and GL B of tobacco were identical at 88% of the positions (Sperisen et al., 1991). *Arabidopsis* gene 1 and *N. plumbaginifolia* gene gn 1 have two introns (F. Ausubel, pers. comm.; Castresana, 1990). Two introns have also been reported for what appears to be a pseudogene with incorrect intron borders encoding a tobacco class I β-1,3-glucanase (Linthorst et al., 1990a).

3. 3'-Flanking Sequence

The 3'-untranslated regions of chitinase and β-1,3-glucanase genes have two putative polyadenylation signals (Joshi, 1987). These regions are highly conserved in homologous tobacco genes within the same class (Sperisen et al., 1991; van Buuren et al., 1992). The 3' sequences are identical at 92.7% of the positions (11 gaps) in tobacco class I β-1,3-glucanase genes GL A and GL B and at 90% of the positions (3 gaps) in tobacco class I chitinase genes CHN 48 and CHN 50. There is evidence that two polyadenylation sites are used in expression of the class I chitinase genes of *Arabidopsis* (Samac et al., 1990) and bean (Broglie et al., 1989) and the class I β-1,3-glucanase genes of tobacco (Neale et al., 1990).

B. Organization and Evolution of Gene Families

1. Sequence Conservation and Diversity

Sequence comparison and Southern blot analyses show that, with the exception of *Arabidopsis*, chitinase and β-1,3-glucanase are encoded by small gene families with at least 3 to 4 members (Table 1). By analogy to other plant and animal genes, it is likely that these families arise after the duplication of a single gene by recombination and gene conversion (Maeda and Smithies, 1986). With time, copies derived from the same gene will tend to diverge as a result of random base-pair substitutions, deletions, insertions, and rearrangements. Thus, for example, two tandem but divergent genes encoding β-1,3-glucanase have been found represented in the same genomic clone of *Arabidopsis* (F. Ausubel, pers. comm.).

Further diversity can arise in amphiploid plant species that result from the sexual hybridization of ancestors followed by chromosome duplication (Gerstel, 1966). In the case of tobacco, which is amphidiploid, the genome consists of tomentosiformis (T) and sylvestris (S) subgenomes derived from ancestral species most closely related to the present day *N. tomentosiformis* and *N. sylvestris* (Goodspeed, 1954; Gray et al., 1974; Gerstel, 1966, 1976). Southern blot analyses of genomic DNA from tobacco and the present day representatives of the two ancestral species have shown that the patterns of restriction enzyme fragments found in *N. tomentosiformis* and *N. sylvestris* are essentially additive for class I chitinases and β-1,3-glucanase. The use of restriction enzymes giving unique fragments for different members of the same gene family has provided evidence that β-1,3-glucanase gene GLA and chitinase gene CHN50 are derived from the sylvestris-type ancestor whereas β-1,3-glucanase gene GLB and chitinase gene CHN48 are derived from the tomentosiformis-type ancestor (Neale et al., 1990; Sperisen et al., 1991; van Buuren et al., 1992).

In the case of the chitinases, the origins of genes have been confirmed at the protein level by isolation and partial sequencing of the enzymes from the three species (van Buuren et al., 1992). Chitinases with the same molecular weight as the major A and B isoforms of tobacco have been purified from *N. tomentosiformis* and *N. sylvestris*, respectively. Chitinases A and B differ at one position, 37, in the N-terminal, cysteine-rich domain. Sequencing of the purified proteins showed that a proline at this position in the tomentosiformis enzyme and chitinase A is substituted for alanine in the sylvestris enzyme and chitinase B. Therefore, the class I chitinase gene family of tobacco includes at least two subfamilies representing genes in the T and S subgenomes as has been described for the gene family encoding the ribulose-1,5-bisphosphate carboxylase small subunit (Gray et al., 1974; Strøbaek et al., 1976; Jamet et al., 1987).

The sequence of members within the chitinase and β-1,3-glucanase gene families are highly conserved. The β-1,3-glucanases encoded by the tobacco genes GLA and GLB differ at only 6 positions. Three of the amino acid changes are in the signal peptide, two are in the mature protein, and one is in the C-terminal extension (Sperisen et al., 1991). The chitinases A and B encoded by genes CHN48 and CHN50, respectively, in the Havana 425 variety of tobacco differ at one position in the cysteine-rich domain and at two positions in the catalytic domain (Fig. 2). Gene CHN48 differs from the cDNA clone FB7-1(3) of Samsun nn by two frameshifts in the hypervariable spacer region, which generate 4 amino acid changes, and a single amino acid difference in the catalytic domain (Neale et al., 1990). Genomic clones representing gene CHN 50 isolated independently from the Havana 425 and BY cultivars of tobacco differ at only 5 positions in ca. 2.8 kb of aligned sequence (Fukuda et al., 1990; van Buuren et al., 1992). The two genes encode identical proteins that differ only in the signal peptide from the chitinases of Samsun nn tobacco represented by cDNA clones FB7-1(1) and FB7-1(2) (Neale et al., 1990). At the nucleotide level, the four DNA sequences encoding Havana 425 β-1,3-glucanases differ at less than 2.2% of the positions in the two exons (Sperisen et al., 1991). In contrast, these sequences diverge by 19.2% in 1065 bp from the coding sequence of the closely related species *N. plumbaginifolia* (De Loose et al., 1988) and by 47% in 351 bp from the monocot barley (Høj et al., 1989).

2. Genetic Selection and DNA Exchange

Although copies of the same gene would be expected to diverge, members of the same gene family have highly conserved sequences indicating that these families undergo concerted evolution. Two types of mechanisms, which are not mutually exclusive, have been proposed for this type of sequence conservation (Maeda and Smithies, 1986): genetic selection and the exchange of DNA between genes by recombination or gene conversion. There is evidence that both mechanisms have had a role in the concerted evolution of tobacco chitinase and β-1,3-glucanase gene families. If sequence homogeneity results from genetic selection, then there should be an high ratio of silent to non-silent nucleotide substitutions in pairwise comparisons of the sequences encoding the mature proteins (Pichersky et al., 1986). The values obtained with the β-1,3-glucanase genes GL A and GL B and the chitinase genes CHN 48 and CHN 50 are 3:1 and 7.4:1, respectively, indicating that selection is, at least in part, responsible for the high level of sequence conservation in these gene families.

The second mechanism, DNA exchange, generates alternating blocks of identical sequence between different members of the same family (Maeda and Smithies, 1986). The nucleotide differences between tobacco class I chitinase genes are distributed rather uniformly. This is in striking contrast to the patterns obtained when the cDNA clones GL 31 and GL 43 and genomic clones GL A and GL B encoding class I β-1,3-glucanases of tobacco are compared. The gene GL 31 shows large alternating blocks of sequence identical to the corresponding regions of the genes GL A and GL B (Fig. 3). The gene GL 43 exhibits a similar pattern of alternating blocks, but in an inverse order (Sperisen et al., 1991). Therefore, it appears that DNA exchange has occurred between different β-1,3-glucanase genes.

Because tobacco is an amphidiploid species, these exchanges of DNA could have occurred between genes in the same subgenome or between genes in different subgenomes. Genetic and molecular studies indicate that the T and S subgenomes and their respective ancestral genomes are very similar at both the chromosome and DNA level (Gerstel, 1960, 1963;

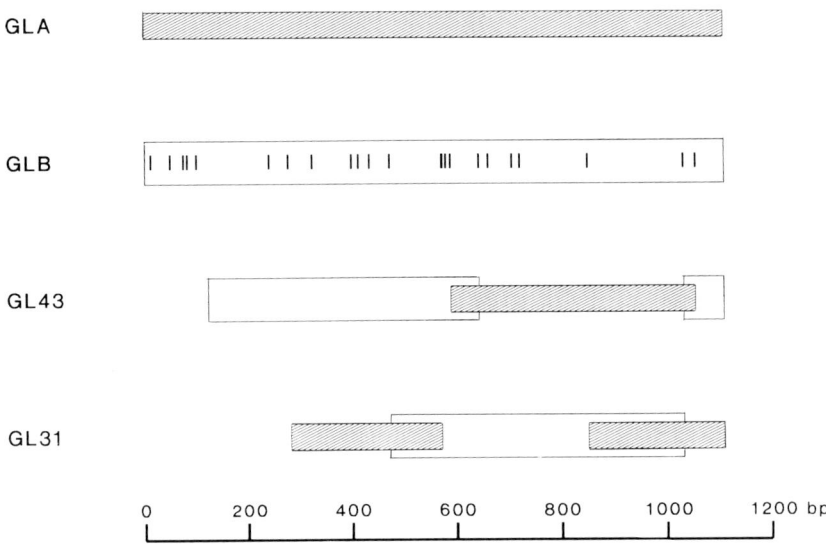

Fig. 3. Exchange of blocks of DNA in the coding sequence of tobacco class I β-1,3-glucanase genes. Data from Sperisen et al. (1991). The genes represented by genomic clones GL A and GL B are present in tobacco and have their origin in *N. sylvestris* and *N. tomentosiformis*, respectively. The genes represented by cDNA clones GL 31 and GL 43 are unique to tobacco. Solid bars, sequence identical to genomic clone GL A; open bars, sequence identical to genomic clone GL B; vertical lines, positions at which nucleotides in GL A and GL B differ

Okamuro et al., 1985; Shinshi et al., 1988, 1990; Jamet et al., 1987; Vaucheret et al., 1989). Therefore, the S and T subgenomes appear to be highly conserved and evolving independently in tobacco at roughly the same rates as the ancestral genomes. This implies that intergenomic exchange of DNA is likely to be a rare event. Nevertheless, there is genetic and cytogenetic evidence that intergenomic, i.e., homeologous, pairing of chromosomes and exchange of chromosomal segments can occur in tobacco (Goodspeed, 1954; Gerstel, 1966; Cameron, 1952).

Southern blot analyses of the clones and genomic DNA from tobacco, *N. sylvestris* and *N. tomentosiformis* using restriction enzymes that cut within the coding sequence suggest that the genes represented by GL 31 and GL 43 are unique to tobacco; whereas, the genes GL A and GL B are also present in *N. sylvestris* and *N. tomentosiformis*, respectively. Apparently GL A and GL B, or related genes, exchanged DNA during the evolution of tobacco to generate the genes represented by GL 31 and GL 43. Based on this observation, Sperisen et al. (1991) have proposed that the concerted evolution of the class I tobacco β-1,3-glucanases involved intergenomic exchange of DNA, either by homeologous recombination at the same locus or between different loci in the S and T subgenomes. The fact that the genes GL A and GL B are still present in the tobacco genome indicates that these events occurred between some copies of the genes leaving other copies in the subgenomes intact.

3. Genetic Transposition

There is indirect evidence that genetic transposition occurred in the evolution of class I and class II chitinases (Shinshi et al., 1990). Insertion of transposable genetic elements into DNA often results in the duplication of target sequences of characteristic length at the insertion site (Döring and Starlinger, 1986). The sequence encoding the cysteine-rich domain of tobacco class I chitinases is flanked by 9 bp direct repeats that are identical at 6 to 9 positions. Comparable repeats, but differing in sequence, are found immediately 5′ of the cysteine-rich domain coding region (i.e., in the region encoding the signal peptide) and at variable distances 3′ of this coding region in class I chitinase genes of tomato, potato, bean, and poplar as well as in the genes encoding the potato wound-induced proteins WIN 1 and

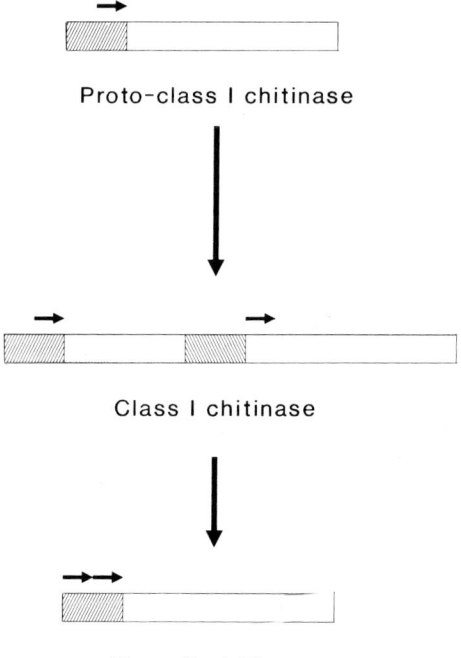

Proto-class I chitinase

Class I chitinase

Class II chitinase

Fig. 4. Proposed model for the origin of class I and class II chitinases in tobacco. Modified from Shinshi et al. (1990). Hypothetical evolution of class II chitinase via class I chitinases from an ancestral proto-class I chitinase with a single target sequence (arrow) and chitinolytic domain. Right cross-hatched bar, signal peptide; open bar, chitinolytic domain; solid bar, cysteine-rich domain; left cross-hatched bar, hinge region; arrow, imperfect 10 base pair direct repeats

WIN 2 (Broglie et al., 1986; Durand-Tardif, 1986; Gaynor, 1988; Stanford et al., 1989; Parsons et al., 1989). This suggests that the highly conserved cysteine-rich domains arose from a common ancestral gene and were introduced into different genes by a transposition event. The fact that the direct repeats differ in sequence suggests, moreover, that cysteine-rich modules were introduced as a result of transposition after the evolutionary divergence of the catalytic domains. Similar conclusions may be drawn from the observation that barley chitinases with and without the cysteine-rich module are more closely related to each other than to the class I or class II chitinases of dicots (Jacobsen et al., 1990; Leah et al., 1991).

This hypothesis provides a model for the origin of class I and class II chitinases in tobacco (Shinshi et al., 1990) (Fig. 4). Two imperfect direct repeats, very similar in sequence to the ones found flanking the region coding the cysteine-rich domain of class I chitinases, are present in genes for PR-P and PR-Q separated by 3 bp immediately 5′ to the codon for the N-terminal amino acid of a mature proteins (Payne et al., 1990a). This suggests that the class I and class II chitinases of tobacco evolved sequentially from an ancestral gene by insertion and then excision of a genetic element which includes sequences encoding a cysteine-rich domain. According to this model, a cysteine-rich module was inserted near the 5′ end of an ancestral chitinase to generate genes of the class I type with direct repeats flanking the inserted sequences. The region coding the cysteine-rich domain was subsequently excised to give class II type chitinase genes leaving as a footprint a duplication of the target sequence as has been reported for the excision of plant transposons (Döring and Starlinger, 1986).

V. Conclusion

Species of higher plants have evolved a remarkably broad range of different chitinases and β-1,3-glucanases that differ in molecular weight, isoelectric point and cellular localization. The various isoforms are commonly classified on the basis of isoelectric point. This is not a robust criterion. Comparison of deduced amino-acid sequences indicates that most chitinases and β-1,3-glucanases have relatively small net changes at neutral pH. It is likely that amino acid substitutions in non-conserved regions of the molecules could lead to acidic and basic proteins in the same structural group. Thus, for example, the tobacco β-1,3-glucanases represented by cDNA clones GL 161 and GL 153 are structurally most closely related to the acidic PR-2, N, and O enzymes but have calculated pI values that are not acidic (Ward et al., 1991). The highly acidic chitinases from poplar (Parsons et al., 1989) and bean (Awade, 1989; Margis-Pinheiro et al., 1991) are most similar in sequence to the nominally basic class I enzymes. The *Arabidopsis* chitinase encoded by gene AC5 has been designated an "acidic chitinase" because its deduced sequence is similar to the acidic cucumber enzyme

(Samac et al., 1990); nevertheless, the calculated pI of this enzyme is basic. Finally, values calculated from amino acid compositions do not necessarily provide a reliable estimate of the actual isoelectric point of a protein.

We have proposed classifying chitinases and β-1,3-glucanases on the basis of amino-acid sequence identity and structural domains. Our survey of deduced and experimentally determined primary sequences leads to the conclusion that these enzymes fall into at least three major structural classes. The chitinases show more pronounced structural differentiation than the β-1,3-glucanases. The class I isoforms are predominantly vacuolar and have a cysteine-rich domain, spacer, an hypervariable region splitting the catalytic domain, and a C-terminal extension removed by processing that are missing in the homologous, extracellular, class II isoforms. The class III isoforms appear to be a distinct structural class with no homology to the class I and class II enzymes. This classification scheme, established by comparing the tobacco enzymes, appears to apply generally to the plant chitinases studied in detail.

The classification of β-1,3-glucanases proposed is based on sequence similarity of the tobacco enzymes. Like the class I chitinases, the class I β-1,3-glucanases are localized predominantly in the vacuole and have a C-terminal extension removed by processing. The class II and class III enzymes appear to be extracellular forms and in the case of PR-2, PR-N, PR-O, and PR-Q′ lack a C-terminal extension. The remaining class II β-1,3-glucanases (GL 153, GL 161, sp41 a, and sp41 b) appear to have C-terminal extensions, but it has not been established experimentally that these extensions are removed by processing. Few isoforms of β-1,3-glucanases have been identified in species other than tobacco and further information is needed to establish the generality of the classification scheme.

Differentiation into classes is an evolutionarily conserved feature of chitinases and β-1,3-glucanases, i.e., several classes of each enzyme have been found in the same plant species. This suggests that the classes are functionally significant. Different isoforms of the enzymes differ dramatically in catalytic activity using model substrates (Kauffmann et al., 1987; Legrand et al., 1987). Thus, the isoforms may be specific for different naturally occurring substrates or act to hydrolyze different bonds in the same substrate. Another possibility is that the isoforms differ in functions unrelated to enzymatic activity. For example, recent studies show that class I β-1,3-glucanases of tobacco and *Hyoscyamus muticus* are auxin-binding proteins with high affinity ($K_D \approx 3$ μM) for the naturally occurring auxin, indole-3-acetic acid (MacDonald et al., 1991). At present it is not known whether or not different isoforms of the enzyme differ in binding affinity. The highly conserved cysteine-rich domain characteristic of class I chitinases is present in several proteins without chitinolytic activity, e.g., the WIN proteins induced in potato by wounding (Stanford et al., 1989) and lectins of wheat (Wright et al., 1984), barley (Learner and Raikhel, 1989), rice (Wilkins and Raikhel, 1989), and stinging nettle (Broekaert et al., 1989) that specifically bind polysaccharides. Stinging nettle lectin and the mini-protein hevein, which also has this cysteine-rich domain, have been reported to inhibit fungal growth in vitro (Broekaert et al., 1989; van Parijs et al., 1991). This raises the possibility that class I chitinases have domains with distinct fungicidal effects. Finally, the C-terminal extension of class I chitinases and β-1,3-glucanases might have an important role in cellular localization. Comparison of the localization of chitinase in *N. sylvestris* plants transformed with expression vectors encoding either the vacuolar, class I chitinase A of tobacco with its C-terminal extension deleted or the extracellular class III cucumber chitinase with the C-terminal extension of chitinase A added indicate that the short C-terminal extension of chitinase A is necessary and sufficient for the correct targeting of the protein to the cell vacuole (Neuhaus et al., 1991a).

It is tempting to speculate that differentiation into homologous structural classes may be a general feature of PR proteins. The "classical" PR proteins of tobacco have an acidic isoelectric point and are extracellular (see Cutt and Klessig, 1992). Recently, clones representing basic homologues of the acidic PR-1 group of tobacco PR proteins have been identified (Cornelissen et al., 1987; Payne et al., 1989). The deduced amino acid sequences of these proteins extend beyond the C-terminal end of the acidic isoforms by 36 amino acids. It will be of interest to establish whether or not this homologue is localized in the vacuole and exhibits C-terminal processing. If, as has been proposed, cysteine-rich modules can be inserted into genes by transposition (Shinshi et al., 1990), then it is likely that a family of proteins with this domain will be found within the same plant species. Indeed, potato has two types of protein with this module: class I chitinases (Gaynor, 1988) and WIN proteins (Stanford et al., 1989). By analogy with WGA, it is possible that the cysteine-rich potato lectin (Owens and North-cote, 1980), which binds N-acetylglucosamine, also has this module.

The presence of the same module in several different proteins emphasizes the danger of deducing the function of a protein from partial sequences. A conserved cysteine-rich region at the N-terminal end of a polypeptide is a poor criterion for chitinase. Another important consideration is that deduced amino acid sequence and the sequence of the protein are not necessarily collinear. In addition to the familiar removal of an N-terminal signal peptide, plant proteins undergo other forms of proteolytic processing. Well documented examples include the formation of different subunits of the same protein from one polypeptide (Crouch et al., 1983), the removal of a C-terminal extension (Chrispeels et al., 1986; Raikhel and Wilkins, 1987; Shinshi et al., 1988), and the cleavage, transposition, and ligation of a precursor to form concanavalin A (Carrington et al., 1985; Bowles et al., 1986; Chrispeels et al., 1986). The occurrence of proteolytic processing and the site of cleavage cannot be reliably established by comparing deduced amino acid sequences of homologous proteins, even when their sequences are very similar. The chitin-binding lectins from wheat (Wright et al., 1984), rice (Wilkins and Raikhel, 1989), and barley

(Lerner and Raikhel, 1989) have very similar cysteine rich domains; nevertheless, the rice polypeptide is interrupted by a single proteolytic cleavage site in the third cysteine-rich module. Finally, the cleavage sites of class I tobacco chitinases do not exactly coincide with the C-terminal ends of the homologous class II chitinases deduced from CDNA clones (Payne et al., 1990a; Sticher et al., in prep.).

Little is known about the hormonal, developmental, and pathogenesis-related regulation of chitinases and β-1,3-glucanases. The fact that their patterns of regulation are very similar but not absolutely coordinated suggests that in some cases, at least, the proteins have related but distinct signal transducing pathways. The availability of cloned genes encoding different class I and class II isoforms of both enzymes provides an experimental system particularly well suited for elucidating these pathways and for identifying specific DNA sequences important in transcriptional regulation. Finally, availability of cloned sequences make it possible to establish biological functions of the enzymes by introducing into plants the sequences under the regulation of constitutive promoters. The few experiments reported are promising. They show that chimeric genes encoding chitinases and β-1,3-glucanases can be expressed in transgenic plants to give enzymatically active proteins (Linthorst et al., 1990b; Neuhaus et al., 1991b).

Notes added in proof. Since our literature review was completed late in 1990, numerous chitinase and β-1,3-glucanase genes have been cloned. The deduced sequences of the proteins, with certain important exceptions, fall into the classes we have described. Of particular interest are two chitinases cloned from sugar beet (K. Bojsen, pers. comm.). Sequence comparisons suggest that the chitinase represented by clone CH4 and the bean PR-4 enzyme (Margis-Pinheiro et al., 1991) define a new class. Clone CH1 represents a novel chitinase with a truncated cysteine-rich lectin module linked by a long, proline-rich, 134 amino-acid long spacer to a catalytic domain related to the class I and II enzymes. Another intriguing structural variation is the precursor of stinging-nettle lectin (D. Lerner and N. Raikhel, pers. comm.). Based on sequence deduced from a cDNA clone, this protein consists of two N-terminal cysteine-rich lectin modules in tandem linked to a chitinase catalytic domain.

Acknowledgements

We thank the many colleagues cited in this chapter who kindly provided us with their unpublished data; Sue Thomas and Susanne Krueger-Lebus for help in preparing the manuscript; and our colleagues at the Friedrich Miescher-Institut for their comments.

VI. References

Abeles FB, Bosshart RP, Forrence LE, Habig WH (1971) Preparation and purification of glucanase and chitinase from bean leaves. Plant Physiol 47: 129–134

Antoniw JF, Ritter CE, Pierpoint WS, Van Loon LC (1980) Comparison of three pathogenesis-related proteins from plants of two cultivars of tobacco infected with TMV. J Gen Virol 47: 79–87

Antoniw JF, Ooms G, White RF, Wullems GJ, Vloten-Doting L (1983) Pathogenesis-related proteins in plants and tissues of *Nicotiana tabacum* transformed by *Agrobacterium tumefaciens*. Plant Mol Biol 2: 317–320

Awade A (1989) Les protéines PR (pathogenesis-related) de haricot: induction par infection virale on traitement chimmique. Purification, propriétés sérologiques, activités biologiques et structure primaire. Doctoral Dissertation, Université Louis Pasteur, Strasbourg, France

Ballance GM, Manners DJ (1978) Partial purification and properties of an endo-1,3-β-D-glucanase from germinated rye. Phytochemistry 17: 1539–1543

Ballance GM, Svedsen I (1988) Purification and amino acid sequence determination of an endo-1,3-β-glucanase from barley. Carlsberg Res Comm 53: 411–419

Bauw G, DeLoose M, Inzé D, Van Montagu M, Vandekerckhove J (1987) Alterations in the phenotype of plant cells studied by NH_2-terminal amino acid-sequence analysis of proteins electroblotted from two-dimensional gel-separated total extracts. Proc Natl Acad Sci USA 84: 4806–4810

Benfrey PN, Chua N-H (1989) Regulated genes in transgenic plants. Science 244: 174–181

Bernasconi P, Pilet PE, Jolles P (1985) A one-step purification of a plant lysozyme from in vitro cultures of *Rubus hispidus*. FEBS Lett 186: 263–266

Bernasconi P, Locher R, Pilet PE, Jolles J, Jolles P (1987) Purification and N-terminal amino-acid sequence of a basic lysozyme from *Parthenocissus quinquifolia* cultured in vitro. Biochim Biophys Acta 915: 254–260

Bol JF (1988) Structure and expression of plant genes encoding pathogenesis-related proteins. In: Verma DPS, Goldberg RB (eds) Temporal and spatial regulation of plant genes. Springer, Wien New York pp 201–221

Boller T (1985) Induction of hydrolases as a defense reaction against pathogens. In: Key JL, Kosuge T (eds) Cellular and molecular biology of plant stress. AR Liss, New York, pp 247–262 (UCLA Symp Mol Cell Biol NS, vol 22)

Boller T (1987) Hydrolytic enzymes in plant disease resistance. In: Kosuge T, Nester EW (eds) Plant–microbe interactions, vol 2. Macmillan, New York, pp 385–413

Boller T (1988) Ethylene and the regulation of antifungal hydrolases in plants. Oxford Surv Plant Mol Cell Biol 5: 145–174

Boller T, Métraux J-P (1988) Extracellular localization of chitinase in cucumber. Physiol Mol Plant Pathol 33: 11–16

Boller T, Vögeli U (1984) Vacuolar localization of ethylene-induced chitinase in bean leaves. Plant Physiol 74: 442–444

Boller T, Gehri A, Mauch F, Vögeli U (1983) Chitinase in bean leaves: induction by ethylene, purification, properties, and possible function. Planta 157: 22–31

Bowles DJ, Marcus SE, Pappin DJC, Findlay JBC, Eliopoulos E, Maycox PR, Burgess J (1986) Posttranslational processing of concanavalin A precursors in jackbean cotyledons. J Cell Biol 102: 1284–1297

Broekaert W, van Paris J, Leyns F, Joos H, Peumans WJ (1989) A chitin-binding lectin from stinging nettle rhizomes with antifungal properties. Science 245: 1100–1102

Broekaert W, Lee H-I, Kush A, Chua N-H, Raikhel N (1990) Wound-induced accumulation of mRNA containing a hevein sequence in laticifers of rubber free (*Hevea brasiliensis*). Proc Natl Acad Sci USA 87: 7633–7637

Broglie KE, Gaynor JJ, Broglie RM (1986) Ethylene-regulated gene expression: molecular cloning of the genes encoding an endochitinase from *Phaseolus vulgaris*. Proc Natl Acad Sci USA 83, 6820–6824

Broglie KE, Biddle P, Cressman R, Broglie R (1989) Functional analysis of DNA sequences responsible for ethylene regulation of a bean chitinase gene in transgenic tobacco. Plant Cell 1: 599–607

Cameron DR (1952) Inheritance in *Nicotiana tabacum*. XXIV. Intraspecific differences in chromosome structure. Genetics 37: 288–296

Carr JP, Klessig DF (1989) The pathogenesis-related proteins of plants. Genet Engineer 11: 65–109

Carrington DM, Auffret A, Hanke DE (1985) Polypeptide ligation occurs during post-translational modification of concanavalin A. Nature 313: 64–67

Castresana C, de Carvalho F, Gheysen G, Habets M, Inzé D, van Montagu M (1990) Tissue-specific and pathogen-induced regulation of a *Nicotiana plumbaginifolia* β-1,3-glucanase gene. Plant Cell 2: 1131–1143

Chrispeels MJ, Hartl PM, Sturm A, Faye L (1986) Characterization of the endoplasmic reticulum-associated precursor of concanavalin A. J Biol Chem 261: 10021–10024

Clarke AE, Stone BA (1962) β-1,3-Glucan hydrolases from the grape vine (*Vitis vinifera*) and other plants. Phytochemistry 1: 175–188

Cornelissen BJC, Hooft van Huijsduijnen RAM, van Loon LC, van Boom JH, Tromp M, Bol JF (1985) Virus-induced synthesis of messenger RNAs for precursor of pathogenesis-related proteins in tobacco. EMBO J 4: 2167–2171

Cornelissen BJC, Horowitz J, van Kan JAL, Goldberg RB, Bol JF (1987) Structure of tobacco genes encoding pathogenesis-related proteins from the PR-1 group. Nucleic Acids Res 15: 6799–6811

Côté F, Cutt JR, Asselin A, Klessig DF (1991) Pathogenesis-related acidic β-1,3-glucanase genes of tobacco are regulated by both stress and developmental signals. Mol Plant Microbe Interact 4: 173–181

Crouch ML, Tenbarge KM, Simon AE, Ferl R (1983) cDNA clones for *Brassica napus* seed storage proteins: evidence from nucleotide sequence analysis that both subunits of napin are cleaved from a precursor polypeptide. J Molec Appl Genet 2: 273–283

Cutt JR, Klessig DF (1992) Pathogenesis-related proteins. In: Boller T, Meins F (eds) Genes involved in plant defense. Springer, Wien New York, pp 209–243 [Dennis ES et al (eds) Plant gene research. Basic knowledge and application]

De Loose M, Alliotte T, Gheysen G, Genetello C, Gielen J, Soetaert P, Van Montagu M, Inzé D (1988) Primary structure of a hormonally regulated β-glucanase of *Nicotiana plumbaginifolia*. Gene 70: 12–23

Dickerson RE, Geis I (1969) The structure and function of proteins. Harper and Row, New York

Dorel C, Voelker TA, Herman EM, Chrispeels MJ (1989) Transport of proteins to the plant vacuole is not by bulk flow through the secretory system and requires positive sorting information. J Cell Biol 198: 327–337

Döring H-P, Starlinger P (1986) Molecular genetics of transposable elements in plants. Annu Rev Genet 20: 175–200

Durand-Tardif M (1986) Etude de l'induction, par l'ethephon, de l'expression du gène codant pour la chitinase chez la tomate et analyse de la structure de ce gène. Doctoral Dissertation, Université de Paris Sud, Paris, France

Edington BV, Lamb CJ, Dixon RA (1991) cDNA cloning and characterization of a putative 1,3-β-D-glucanase transcript by fungal elicitor in bean cell suspension cultures. Plant Mol Biol 16: 18–94

Esaka M, Enoki K, Kouchi B, Sasaki T (1990) Purification and characterization of abundant secreted protein in suspension-cultured pumpkin cells. Plant Physiol 93: 1037–1041

Felix G, Meins F Jr (1985) Purification, immunoassay and characterization of an abundant, cytokinin-regulated polypeptide in cultured tobacco tissues. Evidence the protein is a β-1,3-glucanase. Planta 164: 423–428

Felix G, Meins F Jr (1986) Developmental and hormonal regulation of β-1,3-glucanase in tobacco. Planta 167: 206–211

Felix G, Meins F Jr (1987) Ethylene regulation of β-1,3-glucanase in tobacco. Planta 172: 386–392

Fincher GB, Lock PA, Morgan MM, Lingelbach K, Wettenhall REH, Mercer JFB, Brandt A, Thomsen KK (1986) Primary structure of the (1 → 3,1 → 4)-β-D-glucan 4-glucohydrolase from barley aleurone. Proc Natl Acad Sci USA 83: 2081–2085

Fraser RSS (1981) Evidence for the occurrence of the 'pathogenesis-related' proteins in leaves of healthy tobacco plants during flowering. Physiol Plant Pathol 19: 69–76

Fukuda Y, Ohme M, Shinshi H (1990) Gene structure and expression of a tobacco endochitinase gene in suspension cultured tobacco cells. Plant Mol Biol 16: 1–10

Fulcher RG, McCully ME, Setterfield G, Sutherland J (1976) β-1,3-Glucans may be associated with cell plate formation during cytokinesis. Can J Bot 54: 459–542

Garcia-Olmedo F, Carmona MJ, Lopez-Fando JJ, Fernandez JA, Castagnaro A, Molina C, Hernandez-Lucas C, Carbonero P (1992) Characterization and analysis of thionin genes. In: Boller T, Meins F (eds) Genes involved in plant defense. Springer, Wien New York, pp 283–301 [Dennis ES et al (eds) Plant gene research. Basic knowledge and application]

Gaynor JJ (1988) Primary structure of an endochitinase mRNA from *Solanum tuberosum*. Nucleic Acids Res 16: 5210

Gaynor JJ, Unkenholz KM (1989) Sequence analysis of a genomic clone encoding an endochitinase from *Solanum tuberosum*. Nucleic Acids Res 17: 5855–5856

Gerstel DU (1960) Segregation in new allopolyploids of *Nicotiana*. I. Comparison of 6 × (*N. tabacum × tomentosiformis*) and 6 × (*N. tabacum × otophora*). Genetics 45: 1723–1734

Gerstel DU (1963) Segregation in new allopolyploids of *Nicotiana*. II. Discordant ratios from individual loci in 6 × (*N. tabacum × N. sylvestris*). Genetics 48: 677–689

Gerstel DU (1966) Evolutionary problems in some polyploid crop plants. Hereditas [Suppl] 2: 481–504

Gerstel DU (1976) Tobacco. In: Simmonds NW (ed) Evolution of crop plants. Longman, London, pp 273–277

Gheysen G, Inzé D, Soetaert P, van Montagu M, Castresana C (1990) Sequence of a *Nicotiana plumbaginifolia* β(1,3)-glucanase gene encoding a vacuolar isoform. Nucleic Acids Res 18: 6685

Gianinazzi S, Martin C, Vallee JC (1970) Hypersensibilité aux virus, températures et proteines solubles chez le *Nicotiana* Xanthi-nc. Apparition de nouvelles macromolécules lors de la répression de synthèse virale. C R Acad Sci Paris D 270: 2383–2386

Godiard L, Ragueh F, Froissard D, Lequay J-J, Grosset J, Chartier Y, Meyer Y, Marco Y (1990) Analysis of the synthesis of several pathogenesis-related proteins in tobacco leaves infiltrated with water with compatible and incompatible isolates of *Pseudomonas salanacearum*. Mol Plant Microbe Interact 3: 207–213

Goodall GJ, Filipowicz W (1989) The AU-rich sequences present in introns of plant nuclear pre-mRNAs are required for splicing. Cell 58: 473–483

Goodspeed TH (1954) The genus *Nicotiana*. Chronica Botanica, Waltham, MA

Gray JC, Kung SD, Wildman SG, Sheen SJ (1974) Origin of *Nicotiana tabacum* L. detected by polypeptide composition of fraction I protein. Nature 252: 226–227

Hein P (1966) Grooks. The MIT Press, Cambridge

Herget T, Schell J, Schreier PH (1990) Elicitor-specific induction of one member of the chitinase gene family in *Arachis hypogaea*. Mol Gen Genet 224: 469–476

Hinton DM, Pressey R (1980) Glucanase in fruits and vegetables. J Amer Soc Hort Sci 105: 499–502

Høj PB, Hartman DJ, Morrice NA, Doan DNP, Fincher GB (1989a) Purification of (1 → 3)-β-glucan endohydrolase isozyme II from germinated barley and determination of its primary structure from a cDNA clone. Plant Mol Biol 13: 31–42

Høj PB, Rodriguez EB, Stick RV, Stone BA (1989b) Differences in active site structure in a family of β-glucan endohydrolases deduced from the kinetics of inactivation by epoxyalkyl β-oligoglucosides. J Biol Chem 264: 4939–4947

Hooft van Huijsduijnen RAM, van Loon LC, Bol JF (1986) cDNA cloning of six mRNAs induced by TMV infection of tobacco and a characterization of their translation products. EMBO J 5: 2057–2061

Hooft van Huijsduijnen RAM, Kauffmann S, Brederode FT, Cornelissen BJC, Legrand M, Fritig B, Bol JF (1987) Homology between chitinases that are induced by TMV infection of tobacco. Plant Mol Biol 9: 411–420

Huang J-K, Wen L, Swegle M, Tran H-C, Thin TH, Naylor HM, Muthukrishnan S, Reeck GR (1991) Nucleotide sequence of a rice genomic clone that encodes a class I endochitinase. Plant Mol Biol 16: 479–480

Jacobsen S, Mikkelsen JD, Hejgaard J (1990) Characterization of two antifungal endochitinases from barley grain. Physiol Plant 79: 554–562

Jamet E, Durr A, Fleck J (1987) Absence of some truncated genes in amphidiploid *Nicotiana tabacum*. Gene 59: 213–221

Jekel PA, Hartman JBH, Beintema JJ (1991) The primary structure of hevamine, an enzyme with lysozyme/chitinase activity from *Hevea brasiliensis* latex. Eur J Biochem 200: 123–130

Joshi CP (1987) Putative polyadenylation signals in nuclear genes of higher plants: compilation and analysis. Nucleic Acids Res 15: 9627–9640

Kauffmann S, Legrand M, Geoffroy P, Fritig B (1987) Biological function of 'pathogenesis-related' proteins: four PR proteins of tobacco have 1,3-β-glucanase activity. EMBO J 6: 3209–3212

Keefe D, Hinz U, Meins F Jr (1990) The effect of ethylene on the cell-type-specific and intracellular localization of β-1,3-glucanase and chitinase in tobacco leaves. Planta 182: 43–51

Klebl F, Tanner W (1989) Molecular cloning of a cell wall exo-β-1,3-glucanase from *Saccharomyces cerevisiae*. J Bacteriol 171: 6259–6264

Kombrink E, Schröder M, Hahlbrock K (1988) Several "pathogenesis-related" proteins in potato are 1,3-β-glucanases and chitinases. Proc Natl Acad Sci USA 85: 782–786

Kombrink E, Beerhues L, Schröder M, Witte B, Hahlbrock K (1990) Local and systemic gene activation in potato leaves infected with *Phytophthora infestans*. In: Göttfert M, Hennecke H, Paul H (eds) Abstracts of the 5th International Symposium on the Molecular Genetics of Plant–Microbe Interactions. Eidgenössische Technische Hochschule, Zürich, p 203

Kragh KM, Jacobsen S, Mikkelsen JD (1990) Induction, purification and characterization of barley leaf chitinase. Plant Sci 71: 55–68

Kragh KM, Jacobsen S, Mikkelsen JD, Nielsen KA (1991) Purification and characterization of three chitinases and one β-1,3-glucanase accumulating in the medium of cell suspension cultures of barley (*Hordeum vulgare* L.). Plant Sci 76: 65–77

Kush A, Goyvaerts E, Chye M-L, Chua N-H (1990) Laticifer-specific gene expression in *Hevea brasiliensis* (rubber tree). Proc Natl Acad Sci USA 87: 1787–1790

Laflamme D, Roxby R (1989) Isolation and nucleotide sequence of cDNA clones encoding potato chitinase genes. Plant Mol Biol 13: 249–250

Leah R, Mikkelsen J, Mundy J, Svendsen IB (1987) Identification of a 28 000 Dalton endochitinase in barley endosperm. Carlsberg Res Comm 52: 31–37

Leah R, Tommerup H, Svendsen IB, Mundy J (1991) Biochemical and molecular characterization of three barley seed proteins with anti-fungal properties. J Biol Chem 266: 1564–1573

Legrand M, Kauffmann S, Geoffroy P, Fritig B (1987) Biological function of pathogenesis-related proteins: four tobacco pathogenesis-related proteins are chitinases. Proc Natl Acad Sci USA 84: 6750–6754

Lerner DR, Raikhel NV (1989) Cloning and characterization of root specific barley lectin. Plant Physiol 91: 124–129

Linthorst HJM, Melchers LS, Mayer A, van Roekel JSC, Cornelissen BJC, Bol JF (1990a) Analysis of gene families encoding acidic and basic β-1,3-glucanases of tobacco. Proc Natl Acad Sci USA 87: 8756–8760

Linthorst HJM, van Loon LC, van Rossum CMA, Mayer A, Bol JF, van Roekel JSC, Meulenhoff JS, Cornelissen BJC (1990b) Analysis of acidic and basic chitinases from tobacco and petunia and their constitutive expression in transgenic tobacco. Mol Plant Microbe Interact 3: 252–258

Lotan T, Ori N, Fluhr R (1989) Pathogenesis-related proteins are developmentally regulated in tobacco flowers. Plant Cell 1: 881–887

Lucas J, Henschen A, Lottspeich F, Vögeli U, Boller T (1985) Amino-terminal sequence of ethylene-induced bean leaf chitinase reveals similarities to sugar-binding domains of wheat germ agglutinin. FEBS Lett 193: 208–210

MacDonald H, Jones AM, King PJ (1991) Photoaffinity labelling of soluble auxin-binding proteins. J Biol Chem 266: 7393–7399

Maeda N, Smithies O (1986) The evolution of multigene families: human haptoglobin genes. Annu Rev Genetics 20: 81–108

Maher EA, Lamb CJ, Dixon RA (1990) Molecular analysis of defense-related hydrolases from alfalfa. In: Göttfert M, Hennecke H, Palu H (eds) Abstracts of the 5th International Symposium Molecular Genetics of Plant–Microbe Interactions. Eidgenössische Technische Hochschule, Zürich, p 215

Margis-Pinheiro M, Metz-Boutigue MH, Awade A, de Tapia M, le Ret M, Burkard G (1991) Isolation of a complentary DNA encoding the bean PR4 chitinase: an acidic enzyme with an amino-terminus cysteine-rich domain. Plant Mol Biol 17: 243–253

Masuda Y, Wada S (1967) Effect of β-1,3-glucanase on the elongation growth of oat coleoptile. Bot Mag 80: 100–102

Mauch F, Staehelin LA (1989) Functional implications of the subcellular localization of ethylene-induced chitinase and β-1,3-glucanase in bean leaves. Plant Cell 1: 447–457

Mauch F, Mauch-Mani B, Boller T (1988) Antifungal hydrolases in pea tissue II. Inhibition of fungal growth by combinations of chitinase and β-1,3-glucanase. Plant Physiol 88: 936–942

Meier H, Buchs L, Buchala AJ, Homewood T (1981) (1 → 3)-β-D-Glucan (callose) is a probable intermediate in biosynthesis of cellulose fibres. Nature 289: 821–822

Meins F Jr, Ahl P (1989) Induction of chitinase and β-1,3-glucanase in tobacco leaves infected with *Pseudomonas tabaci* and *Phytophthora parasitica* var. *nicotianae*. Plant Sci 61: 155–161

Memelink J, Hoge JHC, Schilperoort RA (1987) Cytokinin stress changes the developmental regulation of several defence-related genes in tobacco. EMBO J 6: 3579–3583

Métraux J-P, Boller T (1986) Local and systemic induction of chitinase in cucumber plants in response to viral, bacterial and fungal infections. Physiol Mol Plant Pathol 28: 161–169

Métraux J-P, Burkhart W, Moyer M, Dincher S, Middlesteadt W, Williams S, Payne G, Carnes M, Ryals J (1989) Isolation of a complementary DNA encoding a chitinase with structural homology to a bifunctional lysozyme/chitinase. Proc Natl Acad Sci USA 86: 896–900

Meyer AD (1990) Vacuolar localization of β-1,3-glucanase and chitinase in tobacco. Diplomarbeit Universität Basel, Basle, Switzerland

Mohnen D, Shinshi H, Felix G, Meins F Jr (1987) Hormonal regulation of β-1,3-glucanase messenger RNA levels in cultured tobacco tissues. EMBO J 4: 1631–1635

Moore AE, Stone BA (1972) A β-1,3-glucan hydrolase from Nicotiana glutinosa. II. Specificity, action pattern and inhibitor studies. Biochim Biophys Acta 258: 248–264

Nagata Y, Burger MM (1974) Wheat germ agglutinin. Molecular characteristics and specificity for sugar binding. J Biol Chem 249: 3116–3112

Neale AD, Wahleithner JA, Lund M, Bonnett HT, Kelly A, Meeks-Wagner DR, Peacock WJ, Dennis ES (1990) Chitinase, β-1,3-glucanase, osmotin, and extensin are expressed in tobacco explants during flower formation. Plant Cell 2: 673–684

Neuhaus J-M, Sticher L, Meins F Jr, Boller T (1991a) A short C-terminal sequence is necessary and sufficient for the targeting of chitinases to the plant vacuole. Proc Natl Acad Sci USA 88: 10362–10366

Neuhaus J-M, Ahl-Goy P, Hinz U, Flores S, Meins F Jr (1991b) High-level expression of a tobacco chitinase gene in Nicotiana sylvestris. Susceptibility of transgenic plants to Cercospora nicotianae infection. Plant Mol Biol 16: 141–151

Nielsen K K, Mikkelsen JD (1990) Purification and characterization of chitinases and β-1,3-glucanases from Beta vulgaris leaves infected with Cercospora beticola. In: Göttfert M, Hennecke H, Paul H (eds) Abstracts of the 5th International Symposium on the Molecular Genetics of Plant–Microbe Interactions. Eidgenössische Technische Hochschule: Zürich p 220

Ohme-Takagi M, Shinshi H (1990) Structure and expression of a tobacco β-1,3-glucanase gene. Plant Mol Biol 15: 941–946

Okamuro JK, Goldberg RB (1985) Tobacco single-copy DNA is highly homologous to sequences present in the genomes of its diploid progenitors. Mol Gen Genet 198: 290–298

Ori N, Sessa G, Lotan T, Himmelhoch S, Fluhr R (1990) A major stylar matrix polypeptide (sp41) is a member of the pathogenesis-related proteins superclass. EMBO J 9: 3249–3436

Owens RJ, Northcote DH (1980) The purification of potato (Solanum tuberosum) cultivar King-Edwards lectin by affinity chromatography on a fetuin sepharose matrix. Phytochemistry 19: 1861–1862

Parent JG, Asselin A (1984) Detection of pathogenesis-related proteins (PR or b) and of other proteins in the intercellular fluid of hypersensitive plants infected with tobacco mosaic virus. Can J Bot 62: 564–659

Parsons TJ, Bradshaw HD Jr, Gordon MP (1989) Systemic accumulation of specific mRNAs in response to wounding in poplar trees. Proc Natl Acad Sci USA 86: 7895–7899

Payne G, Middlesteadt W, Desai N, Williams S, Dincher S, Carnes M, Ryals J (1989) Isolation and sequence of a genomic clone encoding the basic form of pathogenesis-related protein 1 from Nicotiana tabacum. Plant Mol Biol 12: 595–596

Payne G, Ahl P, Moyer M, Harper A, Beck J, Meins F Jr, Ryals J (1990a) Isolation of complementary DNA clones encoding pathogenesis-related proteins P and Q, two acidic chitinases from tobacco. Proc Natl Acad Sci USA 87: 98–102

Payne G, Ward E, Gaffney T, Ahl-Goy P, Moyer M, Harper A, Meins F Jr, Ryals J (1990b) Evidence for three structural classes of β-1,3-glucanase in tobacco. Plant Mol Biol 15: 797–808

Pegg GF (1977) Glucanohydrolases of higher plants: a possible defence mechanism against parasitic fungi. In: Solheim B, Raa J (eds) Cell wall biochemistry related to specificity in host–pathogen relationships. Universitetsforlaget, Tromso, pp 305–345

Pfeffer S, Ullrich A (1985) Is the precursor a receptor? Nature 313: 184

Pichersky E, Bernatzky R, Tanksley SD, Cashmore AR (1986) Evidence for selection as a mechanism in the concerted evolution of *Lycopersicon esculentum* (tomato) genes encoding the small subunit of ribulose-1,5-bisphosphate carboxylase/oxygenase. Proc Natl Acad Sci USA 83: 3880–3884

Pierpoint WS, Jackson PJ, Evans RM (1990) The presence of a thaumatin-like protein, a chitinase and a glucanase among the pathogenesis-related proteins of potato (*Solanum tuberosum*). Physiol Mol Plant Pathol 36: 325–338

Powning RF, Irzykiewicz H (1965) Studies on the chitinase system in bean and other seeds. Comp Biochem Physiol 14: 127–133

Raikhel NV, Wilkins TA (1987) Isolation and characterization of a cDNA clone encoding wheat germ agglutinin. Proc Natl Acad Sci USA 84: 6745–6749

Roby D, Esquerre-Tugaye M-T (1987) Induction of chitinases and translatable mRNA for these enzymes in melon plants infected with *Colletotrichum lagenarium*. Plant Sci 52: 175–185

Roby D, Broglie K, Cressman R, Biddle P, Chet I, Broglie R (1990) Activation of a bean chitinase promoter in transgenic tobacco plants by phytopathogenic fungi. Plant Cell 2: 999–1007

Roggen HP, Stanley RG (1969) Cell-wall hydrolysing enzymes in wall formation as measured by pollen-tube extension. Planta 84: 295–303

Samac DA, Hironaka CM, Yallaly PE, Shah DM (1990) Isolation and characterization of the genes encoding basic and acidic chitinase in *Arabidopsis thaliana*. Plant Physiol 93: 907–914

Schlumbaum A, Mauch F, Vögeli U, Boller T (1986) Plant chitinases are potent inhibitors of fungal growth. Nature 324: 365–367

Shinshi H, Kato K (1983a) Physical and chemical properties of β-1,3-glucanase from cultured tobacco cells. Agricult Biol Chem 47: 1455–1460

Shinshi H, Kato K (1983b) In vitro synthesis of a larger precursor of tobacco β-1,3-glucanase. Agricult Biol Chem 47: 1275–1280

Shinshi H, Mohnen D, Meins F Jr (1987) Regulation of a plant pathogenesis-related enzyme: inhibition of chitinase and chitinase mRNA accumulation in cultured tobacco tissues by auxin and cytokinin. Proc Natl Acad Sci USA 64: 89–93

Shinshi H, Wenzler H, Neuhaus J-M, Felix G, Hofsteenge J, Meins F Jr (1988) Evidence for N- and C-terminal processing of a plant defense-related enzyme: primary structure of tobacco prepro-β-1,3-glucanase. Proc Natl Acad Sci USA 85: 5541–5545

Shinshi H, Neuhaus J-M, Ryals J, Meins F Jr (1990) Structure of a tobacco endochitinase gene: evidence that different chitinases genes can arise by transposition of sequences encoding a cysteine-rich domain. Plant Mol Biol 14: 357–368

Simmons CR, Litts JC, Huang N, Rodriguez RL (1992) Structure of a rice β-1,3-glucanase gene regulated by ethylene, cytokinin, wounding salicylic acid and fungal elicitors. Plant Mol Biol 18: 33–45

Sperisen C, Ryals J, Meins F Jr (1991) Comparison of cloned genes provides evidence for intergenomic exchange of DNA in the evolution of a tobacco β-1,3-glucanase gene family. Proc Natl Acad Sci USA 88: 1820–1824

Stanford A, Bevan M, Northcote D (1989) Differential expression within a family of novel wound induced genes in potato. Mol Gen Genet 215: 200–208

Strøbaek S, Gibbons GC, Haslett B, Boulter D, Wildman SG (1976) On the nature of the polymorphism of the small subunit of ribulose-1,5-diphosphate carboxylase (EC 4.1.1.39) in the amphidiploid *Nicotiana tabacum*. Carlsberg Res Comm 41: 335–343

Swegle M, Huang J-K, Lee G, Muthukrishnan S (1989) Identification of an endochitinase cDNA clone from barley aleurone cells. Plant Mol Biol 12: 403–412

Takeuchi Y, Yoshikawa M, Takeba G, Tanaka K, Shibata D, Horino O (1990) Molecular cloning and ethylene induction of mRNA encoding a phytoalexin elicitor-releasing factor, β-1,3-endoglucanase, in soybean. Plant Physiol 93: 673–682

Tata SJ, Beintema JJ, Balabaskaran S (1983) The lysozyme of *Hevea brasiliensis* latex: isolation, purification, enzyme kinetics and a partial amino-acid sequence. J Rubber Res Inst Malaysia 31: 35–48

Thornton JM (1981) Disulfide bridges in globular proteins. J Mol Biol 151: 261–288

Vad K, Mikkelsen JD, Collinge DB (1991) Induction, purification and characterization of chitinase isolated from pea leaves inoculated with *Ascochyta pisi*. Planta 184: 24–29

van Buuren M, Neuhaus J-M, Shinshi H, Ryals J, Meins F Jr (1992) The structure and regulation of homeologous tobacco endochitinase genes of *Nicotiana sylvestris* and *N. tomentosiformis* origin. Mol Gen Genet 232: 460–469

van den Bulke M, Bauw G, Castresana C, van Montagu M, Vandekerckhove J (1989) Characterization of vacuolar and extracellular β(1,3)-glucanases of tobacco: Evidence for a strictly compartmentalized plant defense system. Proc Natl Acad Sci USA 86: 2673–2677

van Loon LC, van Kammen A (1970) Polyacrylamide disc electrophoresis of the soluble leaf proteins from *Nicotiana tabacum* var. "Samsun" and "Samsun NN". II. Changes in protein constitution after infection with tobacco mosaic virus. Virology 40: 199–211

van Parijs J, Broekaert WF, Goldstein IJ, Peumans WJ (1991) Hevein: an antifungal protein from rubber-tree (*Hevea brasiliensis*) latex. Planta 183: 258–264

Vaucheret H, Kronenberger J, Rouzé P, Caboche M (1989) Complete nucleotide sequence of the two homeologous tobacco nitrate reductase genes. Plant Mol Biol 12: 597–600

Verburg JG, Huynh QK (1991) Purification and characterization of an antifungal chitinase from *Arabidopsis thaliana*. Plant Physiol 95: 450–455

Vögeli-Lange R, Hansen-Gehri A, Boller T, Meins F Jr (1988) Induction of the defense-related glucanohydrolases, β-1,3-glucanase and chitinase, by tobacco mosaic virus infection of tobacco leaves. Plant Sci 54: 171–176

Vögeli-Lange R, Hart C, Nagy F, Meins F Jr (1991) Regulation of the β-1,3-glucanase B promoter in transgenic tobacco. In: Hallick RB (ed) Program and abstracts of the Third International Congress of Plant Molecular Biology, University of Arizona, Tucson, Abstract 288

Vögeli U, Meins F Jr, Boller T (1988) Co-ordinated regulation of chitinase and β-1,3-glucanase in bean leaves. Planta 174: 364–372

Ward ER, Payne GB, Moyer MB, Williams SC, Dincher SS, Sharkey K, Beck J, Taylor HT, Ahl-Goy P, Meins F Jr, Ryals J (1991) Differential regulation of β-1,3-glucanase mRNAs in response to pathogen infection. Plant Physiol 96: 390–397

Waterkeyn L (1967) Sur l'existence d'un "stade callosique," presente par la paroi cellulaire, au cours de la cytokinese. C R Acad Sci Paris 265: 1792–1794

Wessels JGH, Sietsma JH (1981) Fungal cell walls: a survey. In: Tanner W, Loewus FA (eds) Plant carbohydrates II. Springer, Berlin Heidelberg New York Tokyo, pp 352–394 [Pirson A, Zimmermann MH (eds) Encyclopedia of plant physiology, NS, vol 13B]

Wilkins T, Raikhel N (1989) Expression of rice lectin is governed by two temporally and statically regulated mRNA in developing embryos. Plant Cell 1: 541–549

Witte B, Beerhues L, Hahlbrock K, Kombrink E (1990) Differential expression and localization of chitinases and 1,3-β-glucanase in potato. In: Göttfert M, Hennecke H, Paul

H (eds) Abstracts of the 5th International Symposium on the Molecular Genetics of Plant–Microbe Interactions. Eidgenössische Technische Hochschule, Zürich, p 218

Wright CS, Gavilines F, Peterson DL (1984) Primary structure of wheat germ agglutinin isolectin 2. Peptide order deduced from X-ray structure. Biochemistry 23: 280–287

Wright CS, Raikhel NV (1989) Sequence variability in three wheat germ agglutinin isolectins: products of multiple genes in polyploid wheat. J Mol Evol 28: 327–336

Zhu Q, Lamb CJ (1991) Isolation and characterization of a rice gene encoding a basic chitinase. Mol Gen Genet 226: 289–296

Chapter 11

Characterization and Analysis of Thionin Genes

F. Garcia-Olmedo, M.J. Carmona, J.J. Lopez-Fando, J.A. Fernandez, A. Castagnaro, A. Molina, C. Hernandez-Lucas, and P. Carbonero

Cátedra de Bioquimica y Biologia Molecular, E.T.S. Ingenieros Agrónomos, Universidad Politécnica de Madrid, E-28040 Madrid, Spain

With 12 Figures

Contents

I. Introduction

The general designation of thionins has been proposed for a family of homologous proteins that have been isolated from different tissues in a wide range of plant taxa and have been variously named purothionins, viscotoxins, crambins, etc. (see Garcia-Olmedo et al., 1989). The possible involvement of thionins in plant defense was first suggested, on the basis of their in vitro toxicity to plant pathogens, by Fernandez de Caleya et al., (1972). Those observations had been prompted by earlier reports concerning the antimicrobial properties of these polypeptides (Stuart and Harris, 1942; Balls and Harris, 1944). Work on the thionins, which has been actively pursued over the past half-century, has been recently reviewed in detail (Garcia-Olmedo et al., 1989). For this reason, earlier work will only be partially summarized in the present chapter, which will focus on recent developments concerning thionin genes and their potential role in plant defense mechanisms.

II. Thionin Types

The construction of an unrooted phylogenetic tree, following the criteria of Feng and Doolittle (1987), allowed the classification of the available amino acid sequences from the thionins (either directly determined or deduced from cDNAs) into at least four types (Garcia-Olmedo et al., 1989). The main structural features of these four types, together with those of a fifth type which has been recently discovered in our laboratory (Castagnaro et al., 1992), are summarized in Table 1.

The first type, which corresponds to that of the original purothionins isolated from wheat endosperm (Balls et al., 1942a, b), has four disulphide bridges and is highly basic, with no negatively charged residues. At least seven sequences are known of this type, and all of them have 45 amino-acid residues, 8 of which are in the central disulphide loop.

The second type is represented by thionins identified first in the leaves of the parasitic plant *Pyrularia pubera* (Vernon et al., 1985) and then in those of barley (Gausing, 1987; Bohlmann and Apel, 1987). This type has four disulphide bridges at the same positions as those of type I, but the molecules are less basic, with some negatively-charged residues, and their central disulphide loop has one or two more amino acid residues. As will be discussed later, the claim that barley leaf thionins can be divided into two structural groups (Reimann-Philipp et al., 1989b) is not adequately supported by the available evidence.

The third type includes the viscotoxins and phoratoxins, mainly characterized by Samuelsson and coworkers in leaves and stems of the mistletoes (Loranthaceae), and has the following distinctive features: three disulphide bridges which are conserved with respect to the previous types; fewer basic amino acid residues; sequence with 46 residues, 9 of which are in the central disulphide loop.

A fourth type corresponds to the crambins isolated from the Abyssinian cabbage (Cruciferae) and has the same sequence length and disulphide-bridge arrangement as type II, but the molecules are neutral, with a low proportion of charged amino acid residues.

The fifth type is divergent from the other four types, both in its amino acid sequence and with respect to the disulphide structure: the second and eighth cysteines of type I are missing, through point mutation and deletion, respectively, thus disrupting the first and second disulphide bridges and potentially allowing the formation of a new bridge between the unmatched cysteines. This new type, which is also neutral, has been identified in a cDNA

Table 1. Thionin types

Thionin type	Charges (+/−)	Residues (total/loop)	Source		References
			tissue	species	
I	10/0	45/8	endosperm	*Triticum aestivum* *Hordeum vulgare* *Avena sativa*	Othani et al., 1975, 1977 Mak and Jones, 1976; Jones and Mak 1977 Jones et al., 1972 Ozaki et al., 1980; Ponz et al., 1986 Hernandez Lucas et al., 1986 Bekes and Latztity, 1981
II	9/3 7/1	47/10 46/9	leaves	*Pyrularia pubera* *Hordeum vulgare*	Vernon et al., 1985 Gausing, 1987 Bohlman and Apel, 1987
III	6/0-2 5/1-2	46/9	leaves and stems	*Viscum album* *Phoradendron* spp. *Dendrophora clavata*	Samuelsson, 1969, 1966, 1974 Samuelsson et al., 1968 Samuelsson and Pettersson, 1970 Samuelsson and Jayarvardene, 1974 Mellstrand, 1974 Mellstrand and Samuelsson, 1973, 1974 Thurnberg and Samuelsson, 1982 Samuelsson and Pettersson, 1977
IV	2/2	46/9	cotyledons	*Crambe abyssinica*	Teeter et al., 1981 Vermeulen et al., 1987
V	2/2	38/9	endosperm	*Triticum aestivum*	Castagnaro et al., 1992

Only charges that are at invariant positions in the different variants of a given type are represented
The central loop is that between the 4th and 5th cysteines in type I and II, or between the 3rd and 4th in the other three types

library derived from developing wheat kernels and seems to be also present in barley.

It is to be noted that types I and II are closer to each other than to the other types, not only in their gross architecture but also at the level of the amino acid sequences, and that the same is true for types III and IV. Although the different types have been mostly investigated in particular tissues and organs of certain species or groups of related species, the observed divergence between the types can not be correlated with the evolutionary relationships among the taxa because the types defined here do not represent equivalent subsets of the protein family. Thus, types I, II, and III coexist in a single species, and the thionin isolated from *Pyrulaira pubera*, a parasitic plant closely related to the mistletoes, is closer to the barley-leaf thionins than to the viscotoxins and phoratoxins.

The thionins are among the best characterized proteins with respect to their three-dimensional structure, both in crystals and in solution. Because of their peculiar features they have become model molecules in the development and refinement of novel methods for the elucidation of three-dimensional structures. Thus, an X-ray diffraction method based on the anomalous scattering of sulphur was specifically developed to solve the structure of crambin (Hendrickson and Teeter, 1981). This molecule was also used to test the utility of molecular dynamics with interproton distance restraints for structure determination (Brünger et al., 1986; Clore et al., 1986a, b), and to show that structures obtained for a protein in solution with NMR data can be used to solve the crystal structure of the same protein by molecular replacement (Brünger et al., 1987). This wealth of information, which has been recently reviewed (Garcia-Olmedo et al., 1989), indicates that types I, III, and IV, in spite of their extreme divergence, have essentially the same three-dimensional shape (Fig. 1), which resembles the Greek capital letter gamma (Γ). The molecules are quite rigid and present very similar three-dimensional structures in solution and in crystal form. It will be of interest to elucidate the structure of type V thionins.

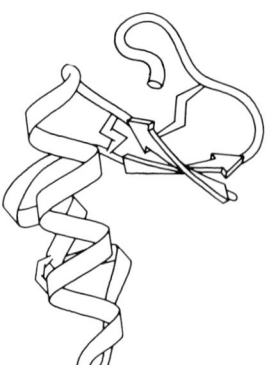

Fig. 1. Schematic drawing of the backbone of crambin. (From Whitlow and Teeter, 1985, with permission)

III. Biosynthesis and Subcellular Location

Biosynthesis of thionins has been studied in some detail only in the case of type I thionins. More specifically, it has been established that barley endosperm thionins are synthesized by membrane-bound polysomes as much larger precursors that undergo at least two processing steps (Ponz et al., 1983). Using monospecific antibodies raised against the mature protein, two types of precursors were identified: one was detected as an in vitro translation product that could not be detected in vivo, and the other was detected by in vivo labelling (Fig. 2). Pulse-labelling experiments showed conversion of the second precursor into the mature protein (Fig. 3). On the basis of these experiments, Ponz et al. (1983) postulated the cotranslational excision of a signal peptide that would convert the precursor observed by in vitro translation into that observed by in vivo labelling, and at least a second, postranslational processing leading to the mature protein. As predicted from the study of their biosynthesis, the

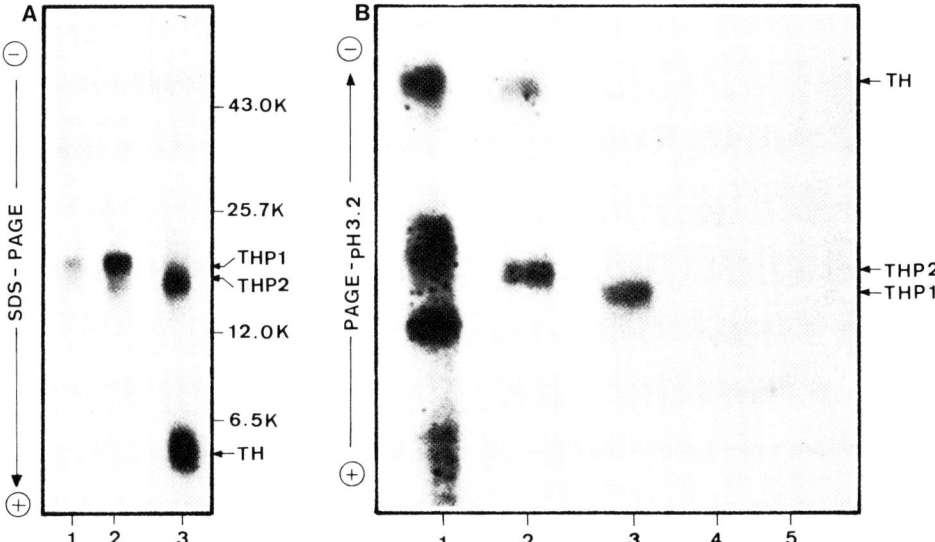

Fig. 2. Comparison of in vivo and in vitro products selected with monospecific antibodies and displacement of these products from the antigen-antibody complex by purified thionin. A, Sodium dodecyl sulphate-polyacrylamide gel electrophoresis (SDS-PAGE) of alkylated products: 1, in vitro precursor THP 1 labelled with [^{35}S]methionine; 2, as in 1, but labelled with [^{35}S]cysteine; 3, in vivo products, THP 2 and TH. B, Polyacrylamide gel electrophoresis (PAGE) at pH 3.2 of reduced, non-alkylated products: 1, total in vivo extracts; 2, in vivo products, THP 2 and TH; 3, in vitro precursor THP 1; 4, as in 2, plus 5 µg of unlabelled thionin; 5, as in 3, plus 5 µg of unlabelled thionin. Displacement by non-radioactive thionin of precursors THP 1 and THP 2 (4 and 5) indicates that the antibodies recognize the same antigens in the three molecules. (From Ponz et al., 1983, with permission)

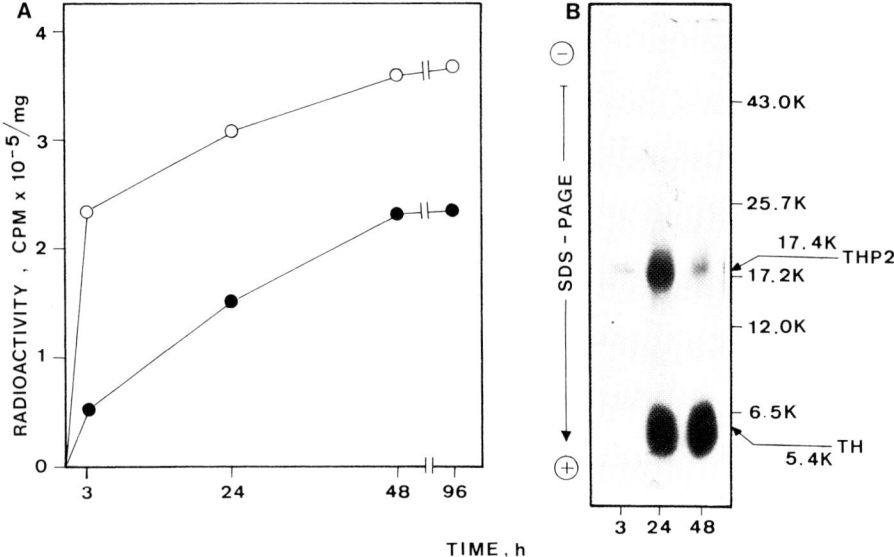

Fig. 3. Pulse-chase experiment. Time-course of [^{35}S]SO$_4^{2-}$ incorporation into proteins of 20 day barley endosperm. Samples were collected at 3, 24, 48, and 96 h after label was added. A, Total (○—○) and trichloroacetic acid insoluble (●—●) ^{35}S incorporated into endosperm at different times. B, Proteins immunoprecipitated with monospecific antibodies at the successive stages were alkylated and subjected to SDS-PAGE and fluorography. Purified thionin (TH) was also alkylated and run in parallel. The apparent M$_r$ of the thionin precursor (THP2) in its alkylated form is 17,400 (K = × 1000). (From Ponz et al., 1983, with permission)

nucleotide sequences of the cDNAs corresponding to α and β thionins from barley endosperm were found to encode precursors that were much larger than the mature proteins (Ponz et al., 1986; Hernández-Lucas et al., 1986). The deduced structures of these precursors consisted of an N-terminal signal peptide, followed by the mature protein and a C-terminal acidic protein, as shown in Fig. 4. The same precursor structure was later found for type-II thionins (Gausing, 1987; Bohlmann and Apel, 1987) and, more recently, for type V thionins (Castagnaro et al., 1992), which strongly suggests that at least these two types of thionins have the same biosynthetic pathway as type I thionins.

Preliminary cellular fractionation studies carried out with developing barley endosperm by Ponz et al. (1983), using a variety of homogenization buffers, led to the conclusion that type I thionins were intracellular and in a labile association with the particulate fraction, which could be disrupted by increasing the salt concentration or by treatment with low concentrations of non-ionic detergents. More recent localization studies, using immuno-gold detection by electron microscopy, have shown that type I thionins are in the periphery of the protein bodies (Carmona et al., unpubl.).

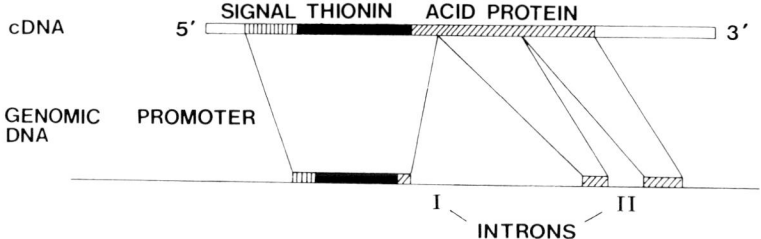

Fig. 4. Structure of the thionin precursor (Ponz et al., 1986) and of the α-thionin gene (*Hth-1*) from barley endosperm. (From Rodriguez-Palenzuela et al., 1988, with permission)

The final cellular localization of type II thionins remains to be established with certainty. It was first claimed that leaf thionins are exclusively located in the cell wall, based on electron micrographs of immunogold-labelled thin sections, using an antibody raised against a fusion protein expressed in *E. coli* (Bohlmann et al., 1988). These authors also salt-extracted thionins from cell walls after extensive washing. Subsequently, they found that about 98% of leaf thionins were intracellular and claimed that there were two distinct groups of leaf thionins: those present in the cell walls and those that are intracellular (Reiman-Philipp et al., 1989b). This claim is not sufficiently supported by the available evidence for the following reasons: (*i*) The partial N-terminal amino acid sequences reported by Reiman-Philipp et al. (1989b) for the putative intracellular and the cell-wall thionins do not differ significantly. (*ii*) Antibodies raised against the intracellular thionins cross-reacted with the putative cell-wall ones (Reiman-Philipp et al. (1989b). (*iii*) In our experience, thionins are soluble in a wide variety of buffers, aqueous solutions of alcohols, and organic solvents. This can lead to relocation of the thionins during fractionation or histological preparation procedures. Moreover, the extracted thionins would be expected to bind to negatively charged groups on the cell wall from which they could be later extracted with solutions containing high concentrations of salt.

IV. Structure and Chromosomal Location of Thionin Genes

The availability of different wheat aneuploids, and of chromosomal addition lines of wheat-rye and wheat-barley, has been extremely useful for mapping genes encoding different proteins in wheat and related species (Garcia-Olmedo et al., 1982, 1984). Three genes (*Pur-A1*, *Pur-B1* and *Pur-D1*), which corresponded to the β, α1, and α2 thionin variants from wheat endosperm (type I), respectively, were identified in the long arms of chromosomes 1A, 1B and 1D, through the electrophoretic analysis of the appropriate aneuploids and the characterization of the isolated proteins (Fernandez de Caleya et al., 1976). In a similar manner, a gene for an endosperm thionin was located in the long arm of chromosome 1R of rye (Sanchez-Monge et al., 1979). Southern-blot analyses of genomic DNAs of wheat and barley, using type I, cDNA probes, were consistent with the presence of 1–2 gene copies of this type per haploid genome (Rodriguez-Palenzuela et al., 1988; unpubl. data). Type II genes have been located in chromosome 6H of barley by Southern-blot analysis of DNAs from wheat-barley addition lines (Bohlmann et al., 1988); but there are discrepancies as to the number of copies of this type present: while Bohlmann et al. (1988) estimated about 100 genes/haploid genome, Gausing (1988) gave a lower estimate of 9–11 genes. More recently, type V genes have been located within a few kb (kilobase) of the type I genes in wheat, and their copy number per haploid genome seem to be also 1–2 (Castagnaro et al., 1992). Although the amino acid sequence of the mature type V thionin is quite different from the other types, the C-terminal acidic peptide of the corresponding precursor is less divergent than the mature protein and closer to the type I than to the type II peptide.

The gene for α-hordothionin, a type I thionin from barley endosperm, has been cloned and its complete sequence has been published (Rodriguez-Palenzuela et al., 1988). This gene has two introns, 420 and 91 nucleotides long, that interrupt the sequence encoding the C-terminal, acidic peptide of the precursor (Fig. 4). Although genomic clones of type II thionins have not been described in detail, it seems that they also have two introns (Bohlmann et al., 1988).

To date genes encoding the type III and IV thionins have not been mapped nor have cDNA and genomic clones been identified.

V. Gene Expression

Thionin accumulation in developing barley endosperm, as judged from the intensity of stained electrophoretic bands (Fig. 5) and from pulse-labelling with $[^{35}S]SO_4^{2-}$, appeared to start and to level off at earlier stages than the major storage proteins, the hordeins. This was confirmed by dot-blot hybridization analysis of the corresponding mRNA (Fig. 6), which showed a maximum steady state concentration of the messenger between 13 and 16 days after anthesis. Thus, synthesis of these proteins seems to take place during the cell-proliferation phase of endosperm development and to cease at the beginning of the cell-enlargement phase. Type I genes not only seem to be specifically expressed in barley endosperm (Fig. 6); fusions of the corresponding promoter with the β-glucuronidase (GUS) reporter gene are also specifically expressed in tobacco endosperm (Fig. 7) (Fernández et al., unpubl.). When fusions of the 35S promoter with the α-hordothionin gene (coding regions and introns) are expressed in transgenic tobacco, the size of the mRNA generated seems to be the same as that of the mRNA resulting from

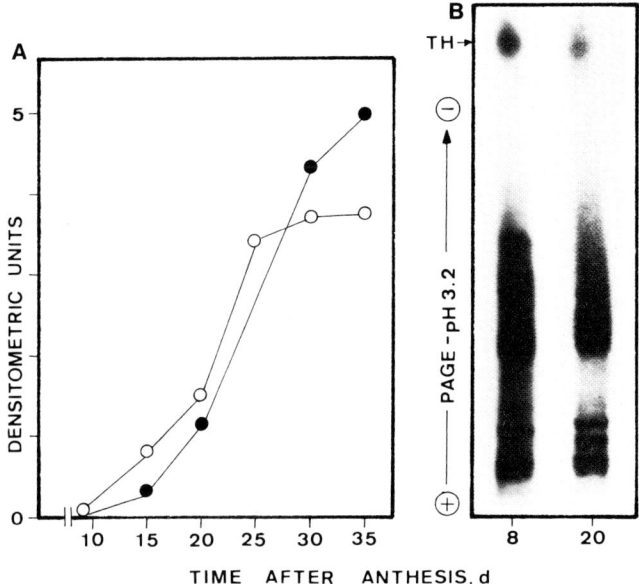

Fig. 5. Thionin synthesis during endosperm development. A, Relative amounts of thionin (○—○) were quantitated by densitometry of the stained band after PAGE pH 3.2 in endosperm samples collected at different times after anthesis. Hordeins (●—●) were similarly quantitated after SDS-PAGE in the same samples. Different arbitrary scales have been used in each case. B, PAGE pH 3.2 and fluorography of endosperm extracts. Ears were collected at 8 and 20 days after anthesis, labelled for 48 h with $[^{35}S]SO_4^{2-}$ and freeze-dried. Endosperms were separated by hand-dissection

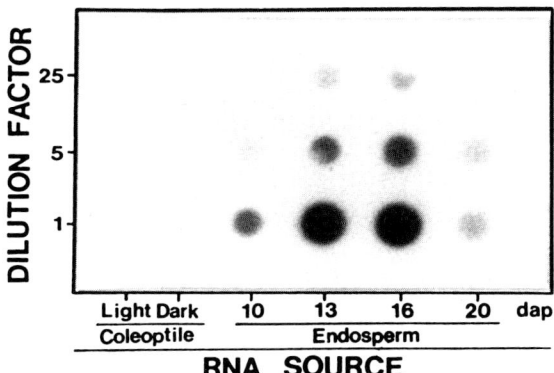

Fig. 6. Endosperm-specific expression of the *Hth-1* gene. Dot-blot hybridization of RNAs from the indicated sources with the nick-translated insert of clone pTHG 1 (Rodriguez-Palenzuela et al., 1988). Equal amounts (2 µg) of each RNA were spotted

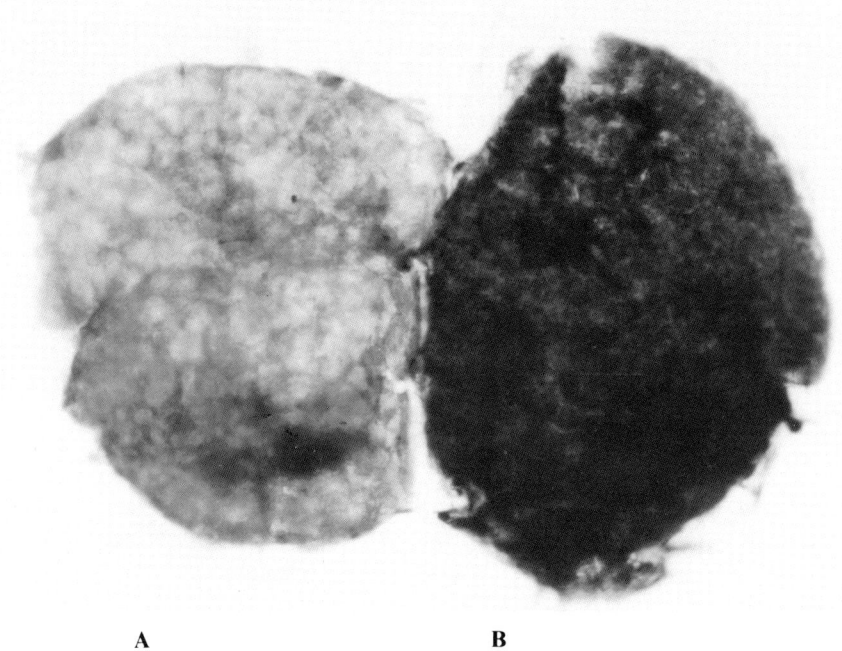

A   **B**

Fig. 7. Histochemical staining for GUS activity of hand-dissected tobacco endosperms from:
A, Non-transformed control. B, Plants transformed with a fusion of the α-hordothionin
promoter (2 kb) with the structural part of the GUS gene. Embryos, leaves, stems, and roots
did not have any GUS activity in the transformed plant (not shown)

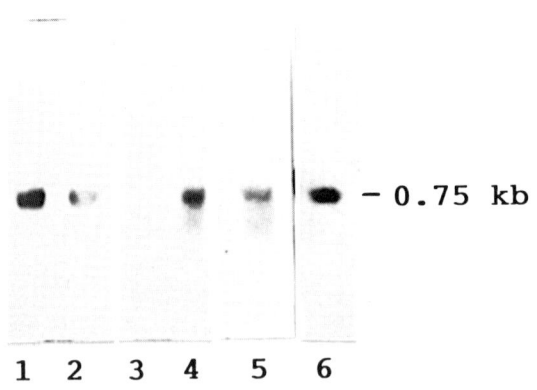

1   2   3   4   5   6

Fig. 8. Northern-blot analysis of RNA from leaves of tobacco plants transformed with
fusions of the 35S promoter and α-hordothionin gene sequences. 1 and 2, plants transformed
with the coding region of an α-hordothionin cDNA; 3, non-transformed control; 4 and 5,
plants transformed with the coding region of α-hordothionin genomic DNA; 6, barley
endosperm. The hybridization probe used was α-hordothionin cDNA.

fusions from the cDNA, which suggests that the introns from this monocot gene are properly spliced in a dicot (Fig. 8) (Carmona et al., unpubl.).

The expression of type II thionin genes has been investigated in barley leaves and a number of interesting responses of these genes to external stimuli have been described. Large amounts of messenger for type II thionins were detected in dark-grown barley seedlings (Gausing, 1987; Bohlmann and Apel, 1987). Steady state messenger levels seemed to be higher in the lower 1/3 of the leaf (younger cells) than in the upper 2/3 (older cells) and to decline sharply upon illumination (Gausing, 1987). The effect of light has been further investigated by Reimann-Philipp et al. (1989a), who have postulated the mediation of two photoreceptors, phytochrome and a blue-light-absorbing photoreceptor. Synthesis of thionins concomitantly ceased upon illumination, but the previously accumulated thionin was rather stable (Reiman-Philipp et al., 1989a). The inhibitory effect of light can be overcome by stress- and pathogen-induced signals, as it has been shown that fungal infection induces a transient expression of the thionin genes in the leaves (Bohlmann et al., 1989; Ebrahim-Nesbat et al., 1989) and that the chlorides of divalent cations (Mg^{2+}, Mn^{2+}, Cd^{2+}, Zn^{2+}) elicit a more permanent response (Fisher et al., 1989).

VI. Antimicrobial Properties and Other In Vitro Activities of Thionins

The toxicity of thionins to different kinds of organisms and to cells in culture has been investigated for several decades. Gram-positive bacteria and, to a lesser extent, Gram-negative bacteria, bakers yeast, and some human pathogenic fungi were found to be sensitive to a crystalline mixture of type I thionins from wheat endosperm; whereas the mycelial fungi tested were found to be insensitive (Stuart and Harris, 1942). After these initial findings, the toxicity to bacteria (Fernández de Caleya et al., 1972), to yeast (Balls and Harris, 1944; Nose and Ichikawa, 1968; Okada et al., 1970; Okada and Yoshizumi, 1970, 1973; Hernández-Lucas et al., 1974), and to fungi (Bohlmann et al., 1988; Reiman-Philipp et al., 1989b) has been further demonstrated for thionins of types I and II. Type I thionins were also found to be toxic to mice, guinea-pigs and rabbits when injected intravenously or intra-peritoneally, but not upon oral administration (Coulson et al., 1942). Type III thionins, isolated from the leaves of the mistletoes and related species, were also found to be toxic on parenteral administration to mice and cats (see Samuelsson, 1974). At sublethal doses they produced hypotension, bradycardia and a negative inotropic effect on the heart muscle. Intraarterial administration, in higher doses, produced vasoconstriction in arteries of skin and skeletal muscle (see Samuelsson, 1974). Cytotoxic effects on cultured mammalian cells have been reported for thionins of type I (Nakanishi et al., 1979), types I and III (Carrasco et al., 1981), type II (Vernon et al., 1985) and type III (Konopa et al., 1980). It has also been observed that type I thionins can reversibly block myogenic differentiation of chick embryonic muscle cells in culture (Kwak et al., 1989).

Several in vitro effects of thionins, which might account for their toxic properties, have been reported; (*i*) alteration of membrane permeability; (*ii*) inhibition of macromolecular biosynthesis; and (*iii*) participation in red-ox reactions in connection with thioredoxins.

A. Alteration of Membrane Permeability

Leakage of intracellular material upon exposure to thionin from wheat endosperm was demonstrated in bacteria (Fernandez de Caleya, 1973). A similar effect was described in yeast by Okada and Yoshizumi (1973), while investigating the mode of action of a toxic principle from wheat and barley that was later shown to be a mixture of thionins (Ohtani et al., 1975, 1977). They further showed that this factor not only induced leakage of phosphate ions, nucleotides, amino acids, and potassium ions, but also inhibited the incorporation of sugars. The toxic effect could be reverted by certain divalent cations, such as Ca^{2+}, Zn^{2+}, or Fe^{2+} (Okada and Yoshizumi, 1973).

A study of the effects of endosperm thionin variants and viscotoxins on cultured mammalian cells indicated that at the minimum cytotoxic concentrations leakage of Rb^{1+} and of uridine occurred (Figs. 9 and 10). Concentrations of thionins that had no detectable effects on the cultured cells lead to inhibition of translation by antibiotics such as hygromycin B that do not normally cross the plasma membrane (Carrasco et al., 1981). As in the case of yeast, Ca^{2+} and Mg^{2+} could revert the action of thionin.

The observed effects on the contraction of smooth muscle from the uterus of the guinea pig (Coulson et al., 1942) and of the flight muscle from insects (Kramer et al., 1979), or the sensitivity to thionins of A31 cells infected with the Moloney strain of murine leukemia virus (Tahara et al., 1979) and the blocking of myogenic differentiation in chick embryonic cells (Kwak et al., 1989), are all probably related to interactions of thionins with the cell membrane.

B. Inhibitory Properties

Apart from a partial inhibition of the milk-clotting power of papain, possibly due to interference with essential SH-groups (Balls et al., 1942), and the inhibition of α-amylase through competition for Ca^{2+} (Jones and Meredith, 1982), no strong inhibition of enzyme activity has been reported for the thionins. However, they are able to inhibit macromolecular synthesis. Nakanishi et al. (1979) reported that thionins could specifically kill cells during DNA synthesis (S phase), but had little effect during the G_0 phase; and Ishii and Imamoto (cited by Ozaki et al., 1980) demonstrated

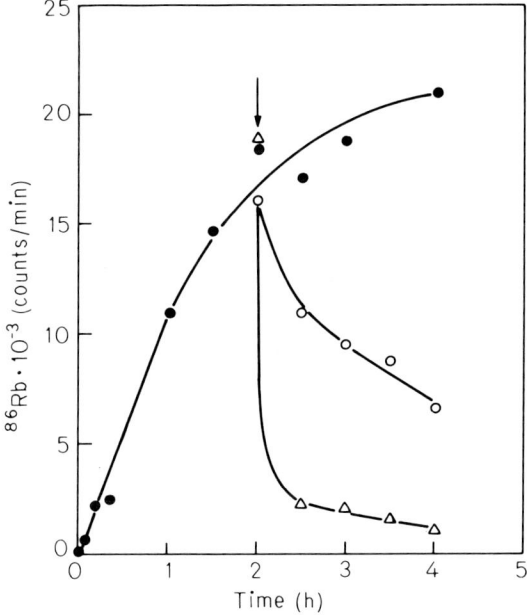

Fig. 9. Effect of thionins on the ^{86}Rb$^+$ content of BHK cells. ^{86}Rb$^+$ was estimated in the cells as indicated in Carrasco et al. (1981). The arrow indicates the time when the thionins were added. (●—●), control; (○—○), 1 μg/ml α1α2β purothionin; (△—△), 1 μg/ml α1α2β purothionin

that thionins also inhibit the transcription of phage in *Escherichia coli*. The effects of endosperm thionins and viscotoxins on the synthesis of DNA, RNA, and proteins in cultured mammalian cells have been investigated by Carrasco et al. (1981). Protein synthesis was more sensitive in these cells than RNA synthesis, which in turn was more sensitive than DNA synthesis (Fig. 11). Inhibition of protein synthesis was correlated with leakage of Rb^{1+} from the different cell variants tested suggesting that inhibition is a direct consequence of the induced leakyness. Eucaryotic cell-free translation systems derived from wheat germ and from rabbit reticulocytes were inhibited by thionins, but at higher concentrations than in intact mammalian cells (Garcia-Olmedo et al., 1983). The inhibitory concentration varied linearly with the amount of exogenous mRNA added, which suggested a direct interaction of the toxin with the RNA (Garcia-Olmedo et al., 1983). This would be in line with the reported interaction between DNA and viscotoxins (Woynarowski and Konopa, 1980).

C. Possible Participation of Thionins in Thioredoxin-related Reactions

Thioredoxin, a hydrogen carrier protein that functions in DNA synthesis and in the transformation of sulphur metabolites, has been also found to serve

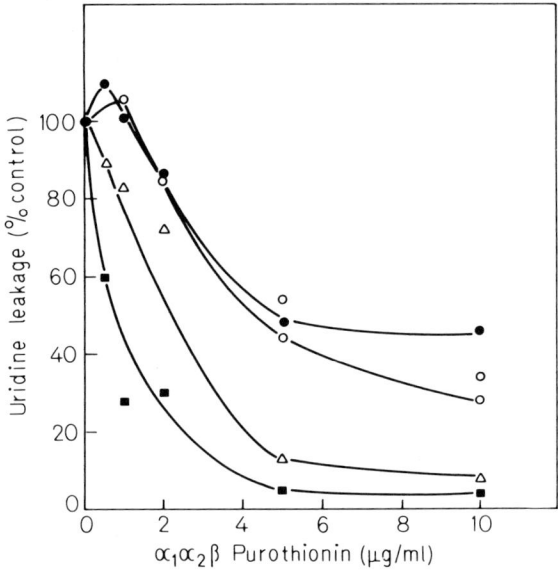

Fig. 10. Effect of thionins on the uridine pool. BHK cells grown in Linbro dishes in medium without Ca^{2+} and Mg^{2+}. 1 μCi [5,6 ^{3}H]uridine (48 Ci/mmol; 1 mCi/ml) was added per well and incubated 3 h at 37 °C in the presence of 10 μg/ml actinomycin D. The indicated concentration of α1α2β purothionin was added, the incubation continued for 1 h and then the ^{3}H content as TCA soluble fraction of the cells measured (●—●) (100%: 11942 counts/min), 2 h (○—○) (100%: 8641 counts/min), or 4 h (△—△) (100%: 7923 counts/min); (■—■) 4 h incubation in the presence of purified β-thionin (100%: 8437 counts/min). (From Carrasco et al., 1981, with permission)

as a regulatory protein in linking light to the activation of enzymes during photosynthesis (Buchanan et al., 1979). Thionin from wheat endosperm can substitute for thioredoxin f from spinach chloroplasts in the dithiothreitol-linked activation of chloroplast fructose-1,6-bisphosphatase (Wada and Buchanan, 1981). Under the standard assay conditions, thionin was only 2% as active as authentic thioredoxin f. Nevertheless, activity could be improved by increasing the time of preincubation and the concentration of reductant suggesting that the thionin could be effectively reduced by thioredoxin f (Wada and Buchanan, 1981). This led to experiments which implicate thionins in plant redox metabolism. Johnson et al. (1987) have reported a thioredoxin system, consisting of a homogeneous preparation of thioredoxin h and partially purified thioredoxin reductase (NADPH), which effectively reduced thionin with NADPH as the hydrogen donor. The reduced thionin, in turn, was capable of activating fructose-1,6-bisphosphatase. These results suggest a possible role of thionins as secondary thiol messengers in the redox regulation of enzymes. In the opinion of these authors, the redox properties of thionins could also explain their toxicity.

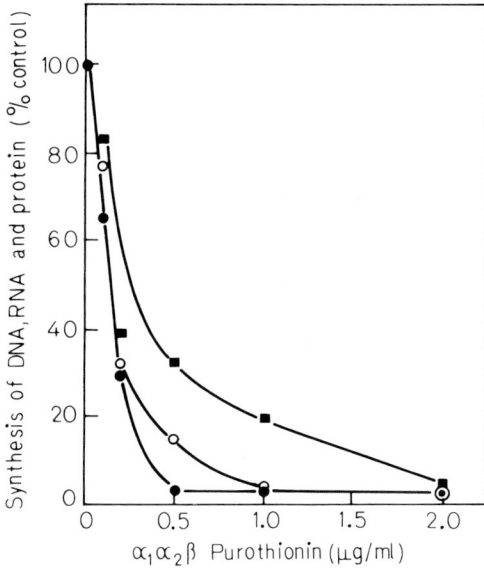

Fig. 11. Effect of thionins on macromolecular synthesis in BHK cells. Thionins were added to BHK cells in (type) medium without Ca^{2+} and Mg^{2+}. After 4 h of incubation. DNA synthesis (■—■) (100%: 67586 counts/min), RNA synthesis (○—○) (100%: 14287 counts/min) and protein synthesis (●—●) (100%: 39934 counts/min) were estimated. (From Carrasco et al., 1981, with permission)

VII. Possible Implication of Thionins in Plant Defense

The hypothesis that thionins might play a role in the protection of plants against pathogens was proposed by Fernandez de Caleya et al. (1972), who investigated the susceptibility to wheat endosperm thionins of phyto-pathogenic bacteria in the genera *Pseudomonas*, *Xanthomonas*, *Agrobacterium*, *Erwinia*, and *Corynebacterium*. Minimal inhibitory concentrations (MIC) ranged from 2×10^{-7} M to 10^{-5} M and the minimal bactericidal concentrations were usually twice the MIC. Purified genetic variants of these thionins differed in activity and showed some degree of specificity. More recently, Bohlmann et al. (1988) have shown that both endosperm (type I) and leaf (type II) thionins from barley inhibit the fungi *Thielaviopsis paradoxa*, a pathogen of sugar cane, and *Drechslera teres*, a pathogen of barley, but at concentrations of 5×10^{-4} M, i.e., several orders of magnitude higher than the concentration required for the most sensitive bacteria. Recently, we surveyed the sensitivity of fungal pathogens to purified genetic variants of type I thionins and found MIC values in the 10^{-6}–10^{-5} M range, i.e., concentrations similar to those found in certain plant tissues (Fig. 12) (Molina and Fraile, unpubl.).

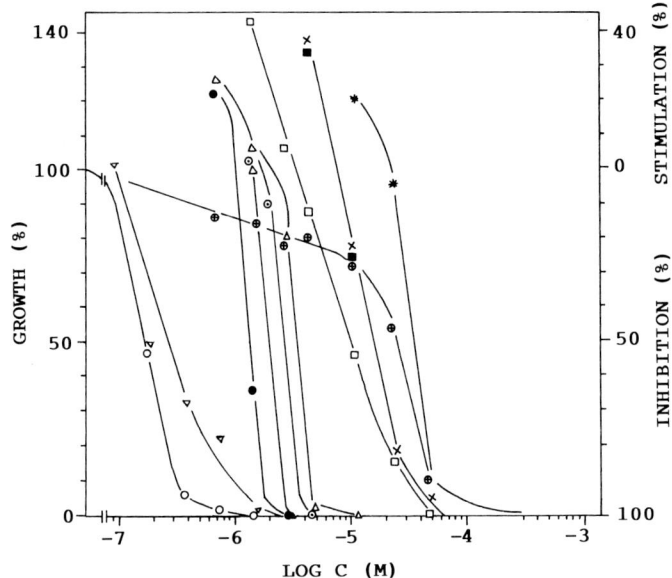

Fig. 12. Effect of thionins on bacterial and fungal pathogens. (○) *Corynebacterium sepedonicum.* (▽) *Pseudomonas solanacearum,* (●) *Rosellinia necatrix,* (▲) *Trichoderma viride,* (⊙) *Aspergillus nidulans,* (△) *Fusarium solani,* (□) *Fusarium* sp. 78, (■) *Fusarium* sp. 72, (×) *Botrytis* sp.,. (⊕) *Botrytis* sp. B100, (∗) *Rhizoctonia solani* (Molina and Fraile, unpubl.). A mixture of α- and β-thionins from wheat endosperm was used. Growth was determined by measuring absorbance at 492 nm and expressed as % of untreated controls. Stimulation at low thionin concentrations was observed

Direct evidence of a defense role for the thionins is lacking at present. Although thionin mRNA is transiently induced in barley upon infection with *Erysiphe graminis* (Bohlmann et al., 1989) and a slightly different localization of thionins seems to occur in the cell walls of susceptible and resistant barley cultivars (Ebrahim-Nesbat et al., 1989), the resistance gene and the thionin genes are located in different chromosomes and the thionin mRNA is induced to similar levels in both susceptible and resistant cultivars. Furthermore, the pre-induction levels of thionins seem to be quite high, due to their low turn-over (Reiman-Philipp et al., 1989a).

VIII. Conclusion and Perspectives

Different types of thionins are quite abundant in different tissues of a given plant and the available distribution data suggest that this family of proteins might be ubiquitous in the plant kingdom. The fact that some thionin types are toxic to both bacterial and fungal pathogens and that thionin genes can be induced under stress conditions, including microbial infection, suggests that this protein family might have a role in plant defense. However, more direct evidence for this hypothesis will have to come either from the demonstration that some resistance trait and thionin genes cosegregate or from the study of transgenic plants expressing foreign thionin genes. If this protein family is ubiquitous, then the potential of thionin genes as targets for manipulation in breeding for disease resistance would largely depend on the specificity of natural and artificial protein variants, as well as in our ability to express them under different developmental programs and environmental situations.

The possibility of a non-defense biological function for thionins remains open and their suggested role as secondary thiol messengers merits further attention, especially after the experiments of Johnson et al. (1987), in which these proteins were assayed under more physiological conditions.

Acknowledgements

We thank D. Lamoneda and J. Garcia for technical assistance, and the Fundación Ramón Areces for support of our current work on the subject of this review.

IX. References

Balls AK, Harris TH (1944) The inhibitory effect of a protamine from wheat flour on the fermentation of wheat mashes. Cereal Chem 21: 74–79

Balls AK, Hale WS, Harris TH (1942a) A crystalline protein obtained from a lipoprotein of wheat flour. Cereal Chem 19: 279–288

Balls AK, Hale WS, Harris TH (1942b) Further observations on a crystalline wheat protein. Cereal Chem 19: 840–844

Bohlmann H, Apel K (1987) Isolation and characterization of cDNAs coding for leaf-specific thionins closely related to the endosperm-specific hordothionin of barley (*Hordeum vulgare* L.). Mol Gen Genet 207: 446–454

Bohlmann H, Clausen S, Behnke S, Giese H, Hiller C, Reimann-Philipp U, Schrader G, Barkholt V, Apel K (1988) Leaf-specific thionins of barley—a novel class of cell wall proteins toxic to plant–pathogenic fungi and possibly involved in the defense mechanism of plants. EMBO J 7: 1559–1565

Brunger AT, Clore GM, Gronenborn AM, Karplus M (1986) Three-dimensional structure of proteins determined by molecular dynamics with interproton distance restraints: application to crambin. Proc Natl Acad Sci USA 83: 3801–3805

Brunger AT, Campbell RL, Clore GM, Gronenborn AG, Karplus M, Petsko GA, Teeter NM (1987) Solution of a protein crystal structure with a model obtained from NMR interproton distance restraints. Science 235: 1049–1053

Buchanan BB, Wolosiuk RA, Schurmann P (1979) Thioredoxin and enzyme regulation. Trends Biol Sci 4: 93–96

Carrasco L, Vazquez D, Hernandez-Lucas C, Carbonero P, Garcia-Olmedo F (1981) Thionins: plant peptides that modify membrane permeability in cultured mammalian cells. Eur J Biochem 116: 185–189

Castagnaro A, Maranua C, Carbonero P, Garcia-Olmedo F (1992) Extreme divergence of a novel wheat thionin generated by a mutational burst specifically affecting the mature protein domain of the precursor. J Mol Biol (in press)

Clore GM, Brunger AT, Karplus M, Gronenborn AM (1986a) Application of molecular dynamics with interproton distance restraints to three-dimensional protein structure determination. A model study of crambin. J. Mol. Biol. 191: 523–551

Clore GM, Nilges M, Sukumaran DK, Brunger AT, Karplus M, Gronenborn AM (1986b) The three-dimensional structure of α1-purothionin in solution: combined use of nuclear magnetic resonance, distance geometry and restrained molecular dynamics. EMBO J 5: 2729–2735

Coulson EJ, Harris TH, Axelrod B (1942) Effect on small laboratory animals of the injection of the crystalline hydrochloride of a sulfur protein from wheat flour. Cereal Chem 19: 301–307

Ebrahim-Nesbat F, Behnke S, Kleinhofs A, Apel K (1989) Cultivar-related differences in the distribution of cell-wall-bound thionins in compatible and incompatible interactions between barley and powdery mildew. Planta 179: 203–210

Feng D-F, Doolittle RF (1987) Progressive sequence alignment as a prerequisite to correct phylogenetic trees. J Mol Evol 25: 351–360

Fernandez de Caleya R (1973) Caracterización química y propiedades antimicrobianas de purotioninas. PhD Thesis, Universidad Politécnia de Madrid, Madrid, Spain

Fernandez de Caleya R, Gonzalez-Pascual B, Garcia-Olmedo F, Carbonero P (1972) Susceptibility of phytopathogenic bacteria to wheat purothionins in vitro. Appl Microbiol 23: 998–1000

Fernandez de Caleya R, Hernandez-Lucas C, Carbonero P, Garcia-Olmedo F (1976) Gene expression in alloploids: genetic control of lipopurothionins in wheat. Genetics 83: 687–699

Fisher R, Behnke S, Apel K (1989) The effect of chemical stress on the polypeptide composition of the intercellular fluid of barley leaves. Planta 178: 61–68

Garcia-Olmedo F, Carbonero P, Jones BL (1982) Chromosomal locations of genes that control wheat endosperm proteins. Adv Cereal Sci Technol 5: 1–47

Garcia-Olmedo F, Carbonero P, Hernandez-Lucas C, Paz-Ares J, Ponz F, Vicente O, Sierra JM (1983) Inhibition of eukaryotic cell-free protein synthesis by thionins from wheat endosperm. Biochim Biophys Acta 740: 52–56

Garcia-Olmedo F, Carbonero P, Salcedo G, Aragoncillo C, Hernandez-Lucas C, Paz-Ares J, Ponz F (1984) Chromosomal location and expression of genes encoding low molecular weight proteins in wheat and related species. Kulturpflanze 32: 21–32

Garcia-Olmedo F, Rodriguez-Palenzuela P, Hernandez-Lucas C, Ponz F, Maraña C, Carmona MJ, Lopez-Fando J, Fernandez JA, Carbonero P (1989) The thionins: a protein family that includes purothionins, viscotoxins and crambins. Oxford Surv Plant Mol Cell Biol 6: 31–60

Gausing K (1987) Thionin genes specifically expressed in barley leaves. Planta 171: 241–246

Hendrickson WA, Teeter MM (1981) Structure of the hydrophobic protein crambin determined directly from the anomalous scattering of sulphur. Nature 290: 107–113

Hernandez-Lucas C, Fernandez de Caleya R, Carbonero P (1974) Inhibition of brewer's yeasts by wheat purothionins. Appl Microbiol 28: 165–168

Hernandez-Lucas C, Royo J, Paz-Ares J, Ponz F, Garcia-Olmedo F, Carbonero P (1986) Polyadenylation site heterogeneity in mRNA encoding the precursor of the barley toxin α-hordothionin. FEBS Lett 200: 103–105

Johnson TC, Wada K, Buchanan BB, Holmgren A (1987) Reduction of purothionin by the wheat seed thioredoxin system. Plant Physiol 85: 446–451

Jones BL, Meredith P (1982) Inactivation of alpha-amylase activity by purothionins. Cereal Chem 59: 321

Konopa J, Woynarowsky JM, Lewandowska-Gumieniak M (1980) Isolation of viscotoxins. Cytotoxic basic polypeptides from *Viscum album* L. Hoppe Seylers Z Physiol Chem 361: 1525–1533

Kramer KJ, Jones BL, Speirs RD, Klassen LW and Kammer AE (1979) Toxicity of purothionin and its homologues to the tobacco hornworm, *Manduca sexta* (L.) (Lepidoptera: Sphingidae). Toxicol Appl Pharmacol 48: 179–183

Kwak KB, Lee YS, Suh SW, Chung CS, Ha DB, Chung CH (1989) Purothionin from wheat endosperm reversibly blocks myogenic differentiation of chick embryonic muscle cells in culture. Exp Cell Res 183: 501–507

Nakanishi T, Yoshizumi H, Tahara S, Hakura A, Toyoshima K (1979) Cytotoxicity of purothionin-A on various animal cells. Gann 70: 323–326

Nose Y, Ichikawa M (1968) Studies on the effects of flour extract on baker's yeast. J Ferment Technol 46: 915–925

Ohtani S, Okada T, Kagamiyama H, Yoshizumi H (1975) The amino acid sequence of purothionin A, a lethal toxic protein for brewer's yeasts from wheat. Agricult Biol Chem 39: 2269–2270

Ohtani S, Okada T, Yoshizumi H, Kagamiyama H (1977) Complete primary structures of two subunits of purothionin A, a lethal protein for brewer's yeast from wheat flour. J Biochem 82: 753–767

Okada T, Yoshizumi H (1970) A lethal toxic substance for brewing yeast in wheat and barley. II. Isolation and some properties of toxic principle. Agricult Biol Chem 34: 1089–1094

Okada T, Yoshizumi H (1973) The mode of action of toxic protein in wheat and barley on brewing yeast. Agricult Biol Chem 37: 2289–2294

Okada T, Yoshizumi H, Terashima Y (1970) A lethal toxic substance for brewing yeast in wheat and barley. I. Assay of toxicity on various grains, and sensitivity of various yeast strains. Agricult Biol Chem 34: 1084–1088

Ozaki Y, Wada K, Hase T, Matsubara H, Nakanishi T, Yoshizumi H (1980) Amino acid sequence of a purothionin homolog from barley flour. J Biochem 87: 549–555

Ponz F, Paz-Ares J, Hernandez-Lucas C, Carbonero P, Garcia-Olmedo F (1983) Synthesis and processing of thionin precursors in developing endosperm from barley (*Hordeum vulgare* L.). EMBO J 2: 1035–1040

Ponz F, Paz-Ares J, Hernandez-Lucas C, Garcia-Olmedo F, Carbonero P (1986) Cloning and nucleotide sequence of a cDNA encoding the precursor of the barley toxin α-hordothionin. Eur J Biochem. 156: 131–135

Reimann-Philipp U, Behnke S, Batschauer A, Schafer E, Apel K (1989a) The effect of light on the biosynthesis of leaf-specific thionins in barley, *Hordeum vulgare*. Eur J Biochem 182: 283–289

Reimann-Philipp U, Schrader G, Martinoia E, Barkholt V, Apel K (1989b) Intracellular thionins of barley. A second group of leaf thionins closely related to but distinct from cell wall-bound. J Biol Chem 264: 8978–8984

Rodriguez-Palenzuela P, Pintor-Toro JA, Carbonero P, Garcia-Olmedo F (1988) Nucleotide sequence and endosperm-specific expression of the structural gene for the toxin α-hordothionin in barley (*Hordeum vulgare* L.). Gene 70: 271–281

Samuelsson G (1974) Mistletoe toxins. Syst Zool 22: 566–569

Sanchez-Monge R, Delibes A, Hernandez-Lucas C, Carbonero P, Garcia-Olmedo F (1979) Homoeologous chromosomal location of the genes encoding thionins in wheat and rye. Theor Appl Genet 54: 61–63

Stuart LS, Harris TH (1942) Bactericidal and fungicidal properties of a crystalline protein isolated from unbleached wheat flour. Cereal Chem 19: 288–300

Tahara S, Hakura A, Toyoshima K, Nakanishi T, Yoshizumi H (1979) A new method for titration of murine leukemia virus using purothionin A. Virology 94: 470–473

Vernon LP, Evett GE, Zeikus RD, Gray WR (1985) A toxic thionin from *Pyrularia pubera*: purification, properties, and amino acid sequence. Arch Biochem Biophys 238: 18–29

Wada K, Buchanan BB (1981) Purothionin: a seed protein with thioredoxin activity. FEBS Lett 124: 237–240

Whitlow M, Teeter MM (1985) Energy minimization for tertiary structure prediction of homologous proteins: α-purothionin and viscotoxin A3 models from crambin. J Biochem Struct Dynam 2: 831–848

Woynarowski JM, Konopa J (1980) Interaction between DNA and Viscotoxins. Cytotoxic basic polypeptides from *Viscum album* L. Hoppe Seylers Z Physiol Chem 361: 1535–1545

Chapter 12

Regulatory Elements Controlling Developmental and Stress-induced Expression of Phenylpropanoid Genes

Jeffery L. Dangl

Max-Delbrück-Laboratorium in der Max Planck Gesellschaft, D-W-5000 Köln 30, Federal Republic of Germany

Contents

I. Introduction

Differentiation of several specialized plant cell types, and the response of plant cells to pathogen or abiotic stress, often requires the utilization of numerous compounds elaborated from L-phenylalanine. The enzymatic machinery of phenylpropanoid metabolism has long held the interest of plant biochemists. The identification of an astounding array of functional phenylpropanoid end products attests to both the ubiquity of this pathway, and to the fortitude of those involved in early and current chemical structural and functional analyses.

The general phenylpropanoid pathway consists of three enzymatic steps, whereby L-phenylalanine is converted into various coenzyme A esters of cinnamate derivatives. The key steps are the first, catalyzed by

L-phenylalanine ammonia-lyase (PAL), and the last, catalyzed by 4-coumarate:CoA ligase (4CL). PAL activity controls entry of L-phenyl-alanine pools into the pathway, while 4CL activity dictates the removal of CoA esters into end-product specific metabolic branches. Through the activity of each branch pathway, an immense structural diversity of phenyl-propanoid end-products is realized. Their wide ranging functions include: various flavonoids acting as flower and fruit pigments. Uv protectants, and activators or inhibitors of plant–microbe interactions; isoflavonoids and furanocoumarins serving as anti-microbial phytoalexins; and other classes with activities as potential cytokinins and insect attractants or repellants. As well, critical complex structural constituents of differentiated cell types, such as lignin and suberin, are products of phenylpropanoid branches.

Inherent in this spectrum of functional diversity is the requirement for exquisite temporal and spatial control of genes encoding the relevant enzymes. This is reflected in cell-type specific accumulation of end products in response to both developmental cues and the environmental signals superimposed upon them. A detailed understanding of the structure and regulatory elements of several phenylpropanoid genes is emerging. The extent of our knowledge regarding the *cis*-linked promoter elements con-trolling gene-specific responses to both developmental and environmental cues, and an evaluation of the molecular genetic tools used to divine these data, are the scope of this review.

I hope to stress several aspects of phenylpropanoid pathway gene regulation which not only highlight current research, but which may also be seen as paradigms driving the hypotheses of the next few years. One of these is the continuing need for parallel development of many systems. At a time when plant molecular biology seems to be concentrating on a select few species, it is somewhat ironic that the strength of current understanding of phenylpropanoid gene regulation lies in the complementary nature of data garnered from several species. The particular functional requirements for phenylpropanoid products among different species is reflected by funda-mentally different organization and control of the relevant genes. In this vein, another goal of this review is to illuminate the way that genome plasticity in plants may allow for the diversification of both plant gene promoter activity and enzymatic activity. This is especially apparent in specific expression of members of the PAL multi-gene family from French bean, as discussed below. Also of interest is the relationship between *cis*-elements which control "normal" tissue specific expression and those determining stress responses, which often override the "normal" state. Finally, we may begin to address whether genes encoding the various phenylpropanoid enzymes in fact share common regulatory sequences, allowing the rapid and coordinate transcriptional responses that are hall-marks of this pathway. I will concentrate on the genes and/or gene families encoding PAL, 4CL and the key enzymes of the flavonoid specific branch pathway, chalcone synthase (CHS) and chalcone isomerase (CHI). Detailed aspects of the biochemistry of phenylpropanoid metabolism are covered in other chapters of this volume, and, as this area has become quite topical, several other recent reviews are also recommended (Bailey 1987; Hahlbrock and Scheel, 1989; Lamb et al., 1989; Dangl et al., 1989; Dixon and Harrison, 1990). As well, lists of cloned phenylpropanoid genes can be found in Mol et al. (1988) and Dixon and Harrison (1990).

II. L-Phenylalanine Ammonia-Lyase (PAL)

A. Regulatory Specialization and Enzymatic Polymorphism

The first level of complexity in PAL regulation is dictated by the enzyme's tetrameric structure, and the recent recognition that multiple PAL isoforms are encoded by multigene families. In the best analyzed cases, French bean and parsley, analysis of cDNA and genomic clones points to the existence of three and four PAL genes, respectively (Cramer et al., 1989; Lois et al., 1989). Yet, up to 11 isoforms of bean PAL were observed after in vivo labelling and immunoprecipitation, and several native tetramers could be resolved by chromatofocussing (Bolwell et al., 1985; Liang et al., 1989a). These data suggested that enzyme polymorphism, combined with post-translational modification, may play a major role in bringing PAL enzymes of differing catalytic ability to bear in specific developmental situations in french bean. Analysis of transcriptional activation of PAL genes after environmental and pathogen stress (Cramer et al., 1985a, b; Edwards et al., 1985; Lawton and Lamb, 1987; Lois et al., 1989) proved, however, that de novo mRNA synthesis is the major regulatory mechanism in both cases. The conclusion that gene specific transcriptional regulation and post-translational control combine to generate functionally diverse bean PAL enzymes is strengthened by three additional observations. First, isoforms of low K_m are specifically induced by wounding of bean hypocotyls and by fungal elicitor stimulation of cultured bean cells (Bolwell et al., 1985; Liang et al., 1989a). This is interpreted to imply a rapid shuttling of phenylalanine derivatives through the pathway under stress conditions. Secondly, cDNA sequences from bean show large coding divergences (59% similarity in exon I and 74% in exon II between two bean cDNAs) (Cramer et al., 1989). Finally, promoter sequences between two of the bean PAL genomic clones are highly divergent (cited in Bevan et al., 1989).

The four parsley PAL cDNA sequences are remarkably similar. The promoter sequences available (for PAL-1 and PAL-4) are, however, very divergent (Lois et al., 1989; K. Hahlbrock, pers. comm.). Thus, in parsley, the functional need for enzyme polymorphism is not expected to be a major regulatory determinant. Nonetheless, the contrasts between the two species suggest that genome plasticity allows regulatory diversification when such is adaptatively advantageous. A potentially interesting experiment to pursue this theme would be the introduction of a divergent PAL enzyme, and its regulatory region, from bean into parsley. Assuming that it is correctly expressed, would its expression effect or disrupt the normal functions of the parsley PAL tetramer? Preliminary analysis of overexpression of a bean PAL coding region from the CAMV 35S promoter in transgenic tobacco suggests that such manipulations have profound effects of normal PAL expression (cited in Dixon and Harrison, 1990). Observations of multi-gene families for PAL may all be generalizable, as multiple isoforms exist in alfalfa (see Dixon and Harrison, 1990) and potato (Fritzemeier et al., 1987). In this latter case, an astonishing number of PAL genes, around 40 copies per haploid genome, exist (K. Hahlbrock et al., pers. comm.).

B. Gene Specific Expression: Temporal, Spatial, and
Stress Regulation

The above discussion implies that, in certain species, specialized PAL enzyme tetramers may be derived from gene specific regulation in particular cell types or under certain stress conditions. This hypothesis has been addressed through determination of PAL mRNA distribution during development, by analyses of single members of both bean and parsley PAL multigene families using gene specific probes, and by observation of cell-type specific expression of bean PAL promoters in transgenic plants.

Immuno-histochemical analysis of young parsley leaf buds localized PAL to oil duct epithelial cell (a site of constitutive furanocoumarin synthesis), developing xylem (a site of lignin deposition) and epidermal cells (a site of flavonoid biosynthesis) (Jahnen and Hahlbrock, 1988; Schmelzer et al., 1988). In situ mRNA hybridization studies confirm this cell type specificity (Schmelzer et al., 1989). Both analyses also showed large, highly localized, tissue unspecific increases in PAL protein and mRNA amounts surrounding the site of penetration of *Phytophthora megasperma*, a soybean pathogen against which parsley generates a non-host resistance reaction (Schmelzer et al., 1989). Tissue non-specific increases in PAL mRNA amount were also observed in bean hypocotyls infected with *Colletotrichum lindemuthianum* (Bell et al., 1986). In these experiments, tissue at various sites surrounding attempted fungal penetration was dissected, and PAL mRNA levels were measured. In an incompatible interaction, as in the non-host case cited above, PAL mRNA was located in sharply defined tissue directly adjacent to the infection site. In the compatible interaction, PAL mRNA accumulated in a larger diffuse area. Importantly, in all three examples, pathogen induced PAL gene activation overrides the normal cell-type specificity associated with these tissues. This allows for rapid PAL enzyme accumulation in all cell types surrounding an infection site.

Gene specific probes have been utilized in both the parsley and bean systems to assess the contribution of each multi-gene family member to tissue- and/or stress-specific PAL expression. In parsley (Lois et al., 1989), at least three of the four PAL genes are each activated by UV light or fungal elicitor in cultured cells, and each is induced in wounded roots. Interestingly, only the parsley PAL-3 gene is more active than the others in wounded leaves. These data are based on analysis of total mRNA amount in a given tissue, and cannot address which cell types are responding. It is also difficult, in the absence of genes-specific in situ hybridization methods, to answer which gene or set of genes is induced in a cell-type independent manner in response to fungal penetration (Schmelzer et al., 1989). It could be that PAL-3, due to its would responsiveness in leaf, is the only gene activated by fungus. In this scenario, the primary regulatory determinant is the unique capability of PAL-3 to be expressed in leaf tissue. Alternatively, all three PAL genes would respond to fungal penetration, as in the elicitor treated cultured cells. In this case, recognition of fungal elicitor would stimulate an inductive pathway leading to common activation of all four PAL genes, overriding normal tissue specificity.

RNase protection with gene specific probes was used to determine differential response of the three bean PAL genes to development and environmental signals (Liang et al., 1989a). The bean PAL-1 gene is expressed in nearly all tissues: in roots, shoots, stems, and leaves. PAL-2 is expressed in roots, shoots and, at very high levels, in flower petals. PAL-3 is only expressed in roots. This distribution changes significantly in hypocotyls exposed to various stress treatments: mRNA from all three genes accumulates after wounding; high intensity illumination of etiolated hypocotyls induces PAL-1 and PAL-2 activity above already detectable levels; and fungal infection (either compatible or incompatible) stimulates PAL-1 and PAL-3 transcription. Interestingly, although PAL-2 mRNA accumulation is triggered in cell cultures by fungal elicitor treatment, it is not induced by fungal infection. Liang et al. (1989) speculate that PAL-2 is activated in infected hypocotyls, but very locally, precluding its detection in RNA from whole hypocotyls. Alternatively, lack of PAL-2 expression may reflect cultivar specific differences (Liang et al., 1989a; Tepper et al., 1989).

The limitations of mRNA analyses from whole tissue, and the technical infeasibility of gene specific in situ measurements have been overcome by cellular detection of the β-glucuronidase (GUS) reporter gene expressed from specific PAL promoters in transgenic plants (Jefferson et al., 1987; Beven et al., 1989; Liang et al., 1989b; M. Bevan et al., pers. comm.). These elegant, yet laborious, studies show that observations made on the mRNA level in various tissues are largely an accurate portrayal of PAL tissue specificity and stress inducibility. What they add, however, is a much refined definition of the specialized cell types expressing bean PAL promoters. As well, these limited analyses suggest strongly that differential gene expression of bean PAL genes is a function of exclusive promoter utilization.

In transgenic tobacco, histochemical detection of GUS activity driven by the bean gPAL-2 promoter reveals high expression in pigmented petal tissue and leaf vascular areas. Significant GUS activity is also reported for anthers, stigmas, trichomes, epidermal cells and root hairs (Bevan et al., 1989; Liang et al., 1989b). In both cases, very high gPAL-2 activity was observed in meristematic tissues of root and shoot spices. This could belie a role in rapidly dividing tissue for phenylpropanoid derivatives with hormone like activity (see also Binns et al., 1987).

A striking finding in both analyses was the relationship of gPAL-2 promoter activity to xylem differentiation and lignification of mature tracheary elements. In immature, primary xylem, gPAL-2-GUS expression is localized in discrete cell groups. As this tissue matures and xylem cell files elongate, radial expression from the gPAL-2 promoter extends from younger, cambial cells inward to mature xylem. These cells are in the process of lignification, and the most mature tracheal cells appear to no longer express the gPAL-2 promoter (Bevan et al., 1989; Liang et al., 1989b). This cell-type specificity is also observed when the parsley 4CL-1 promoter is used to drive GUS expression in transgenic tobacco (see below).

The gPAL-2 promoter is also sufficient for response to environmental stimuli in transgenic tobacco (in contrast to the parsley 4CL-1 promoter, see below). Dark adaption for 3 days and subsequent illumination with continous light leads to significant expression of GUS in leaf (whole tissue) and stem epidermal cells (Liang et al., 1989b). As well, wounding of transgenic tobacco leaves, stems and roots result in clear induction over variable background levels (Liang et al., 1989b). This tissue unspecific nature of this response in leaves has also been demonstrated (M. Bevan, pers. comm.)

Bevan et al. (1989) show convincingly that the gPAL-2 promoter is also wound responsive in tubers of transgenic potatoes. In this case, the spatial response to wounding changes with time. Cells directly at the affected surface express gPAL-2 and synthesize protective suberin 40 h after wounding. At 100 h, however, these cells are fully suberized and appear to no longer express the gPAL-2 promoter. Rather, cells several layers deeper in the tuber tissue are seen to express the gPAL-2-GUS gene and to synthesize a secondary suberin layer.

That exclusive promoter utilization determines specificity of expression among various bean PAL promoters is vividly demonstrated by comparing the expression of gPAL-3-GUS fusions in transgenic tobacco and potato with that described above for gPAL-2 (M. Bevan, pers. comm.). The cell-type specificity of the gPAL-3 promoter is completely different, being moderately high in pith cells and a very narrow layer (2–3 cells deep) surrounding the entire vascular cylinder. Strikingly, there is no expression in xylem or petals, nor after wounding of leaves. Combined with the lack of similarity between gPAL-2 and gPAL-3 promoters, these data are compelling proof that different bean PAL promoters use idiosyncratic cis-elements in response to different stimuli. As well, it suggests that their respective enzyme products may serve slightly different functions.

C. PAL cis-Elements

These comparisons beg the question of whether sets of cis-elements, mutually exclusive between differentially expressed gene family members, are shared among similarly activated promoters of genes encoding different phenylpropanoid enzymes. Alternatively, all genes whose enzymatic activities are required in a certain context could receive their inductive signals independently. Experiments to address these questions are just beginning for PAL, 4CL, and CHS genes. Analysis of gPAL-2 promoter deletions, and in vivo footprinting (Church and Gibert, 1984; Saluz and Jost, 1989) of the parsley PAL-1 promoter reveal some tantalizing clues as to the nature and organization of *cis*-element networks among phenylpropanoid genes. As well, recent isolation and analysis of *Arabidopsis* PAL genes has opened this system up for further experiments (Ohl et al., 1990).

Unidirectional and internal deletion analysis of bean gPAL-2 promoter GUS fusions in transgenic tobacco show that *cis*-elements governing root apex expression are separable from those conferring xylem specificity (M. Bevan, pers. comm.). Sequences between -150 bp and -280 bp are both necessary and sufficient for root apex and xylem expression, and would responsiveness. Internal deletions of approximately 100 bp between -150 bp and -250 bp abolish xylem expression. Since the internal deletions contain a further 1 kb of gPAL-2 upstream sequence, it appears that *cis*-elements determining xylem expression are not present redundantly. This is mentioned only to contrast findings from light induction of both ribulose-biphosphate decarboxylase (SSU) and parsley CHS promoters (Kuhlemeier et al., 1988; Schulze-Lefert et al., 1989a, b; see also Sect. IV). Repositioning of gPAL-2 promoter sequences from -70 bp to -370 bp via reversal with respect to a minimal CaMV 35S promoter abrogates root apex expression (M. Bevan, pers. comm.). With these constructs, gPAL-2 promoter activity in xylem and root cortex is unaffected. The simplest interpretation is that *cis*-elements dictating root apex expression are TATA proximal and orientation and/or distance dependent.

Several candidates for the respective functional *cis*-elements reside between -70 bp and -260 bp of the gPAL-2 promoter (M. Bevan, pers. comm.). These sequences are highly related to putative *cis*-elements defined by in vivo footprinting of the parsley PAL-1 promoter (Lois et al., 1989). These footprints appear after application of either fungal elicitor or UV-containing white light to suspension cultured parsley cells. Two footprints seem to appear after both treatments, and another appears specifically upon elicitor treatment. Sequences similar to the generally induced footprints can be found in other phenylpropanoid genes from several species (9/12 consensus in one case, 8/12 in the other). Oddly, sequences defined by the elicitor specific footprint on the parsley PAL promoter are also found in promoters known to be elicitor non-responsive. The functional role of these putative *cis*-elements in these cases awaits further analysis. As well, although strong similarity to the footprinted parsley PAL sequences can be found within functionally relevant regions of both the gPAL-2 and bean CHS-15 promoters (see below) no direct test of their role in either tissue specificity or stress response exists to date.

III. 4-Coumarate:CoA Ligase (4CL)

Analysis of 4CL gene structure and regulation is limited to parsley and potato. In the following section, it will become apparent that control of 4CL transcription may offer a regulatory paradigm different from that outlined above for (at least) bean PAL. In contrast to the distinct bean PAL isoforms derived from differentially responding promoters, 4CL in parsley and potato is encoded by two highly homologous genes. Tissue specific and environmentally induced 4CL expression is nevertheless quite complex. Thus, each 4CL gene must contain all necessary *cis*-sequences required to ensure correct expression in response to many diverse signals.

A. Tissue Specific and Stress-induced Expression

Both within and between species, 4CL genes are very homologous (Douglas et al., 1987; Lozoya et al., 1988; Becker-André, 1989). This homology also extends to the respective promoter sequences, where the two parsley genes are nearly identical for several hundred base pairs and where conservation between parsley and potato is also obvious over the TATA proximal 200 bp (Douglas et al., 1987; Becker-André, 1989).

Transcription of 4CL increases drastically (above a detectable background) is suspension cultured parsley cells after elicitor or UV-light treatment (Chappell and Hahlbrock, 1984; Douglas et al., 1987; Lois et al., 1989). Douglas *et al.* (1987) exploited differences in intron sequence to demonstrate that both parsley 4CL genes respond to either exogenous stimulus. Potato 4CL mRNA from both genes accumulates in suspension cultured cells after treatment with elicitor or mercury chloride, but, oddly, not after UV-light treatment (Becker-André 1989).

In primary parsley leaves, 4CL mRNA was detected in situ in cell types also expressing PAL (see above, and Schmelzer et al., 1989). Cell-type specific expression of both PAL and 4CL is localized to epidermal cells, developing xylem tissue and oil duct epithelial cells. Also, as mentioned previously with respect to PAL expression, cell-type specificity is overcome by massive cell type independent transcription in response to *Phytophthora megasperma* sporulation. Schmelzer et al. (1989) demonstrated that both PAL and 4CL mRNA are first detectable in sharply localized regions of leaf cells surrounding sites of hypersensitive cell death 4 h post inoculation. These authors also show that wounding induces 4CL mRNA accumulation, where expression spreads in a gradient from the site of damage throughout the leaf by 6 h.

Tissue- and organ-specific expression of the two potato 4CL genes was analyzed via a novel polymerase chain reaction strategy (PATTA; Becker-André and Hahlbrock, 1989). Highest expression, on a per cell basis, was observed in stems, tubers, roots, sepals, and young leaves. As leaves aged, less 4CL mRNA per cell was found (Becker-André, 1989). These data further corroborate that 4CL expression is highest in young, developing tissue (leaves and roots) and in highly lignified tissue (stems and tubers). Both potato 4CL genes are expressed equivalently in all tested organs, except anthers where 4CL-2 expression is about 2 fold higher than 4CL-1 (Becker-André, 1989). Thus, as appears to be the case for parsley, differential expression of the two potato 4CL genes is unlikely in nearly all cases. It should be noted, however, that in other species, 4CL isoforms exist which have either differential substrate specificities or developmental expression profiles at the enzymes level (see Lozoya et al., 1988 and references therein).

RNA blot analysis also shows that potato 4CL mRNA accumulates after wounding, and fungal infection of either cultivar compatible or cultivar incompatible *Phytophthora infestans* races (Fritzemeier et al., 1987; Becker-André, 1989). In infections with both fungal races, an initial induction of mRNA amount is followed by a steep decline 5 h post-inoculation, which is, in turn, followed by a longer, steady accumulation of 4CL mRNA (Becker-André, 1989). This analysis of mRNA extracted from infected leaves is somewhat reminiscent of in situ data in parsley leaves discussed above, where early 4CL transcription activation is localized to the site of fungal penetration. The functional significance of the second, spreading 4CL mRNA accumulation is unclear, although it could be a protective response to further sporulation on the same leaf. Finally the magnitude of 4CL mRNA accumulation is higher in the compatible interaction, although this probably reflects only a larger number of cells being contacted by growing fungal hyphae.

B. 4CL cis-Elements

The identification of *cis*-controlling elements from 4CL genes is thus far limited to analysis of the parsley 4CL-1 promoter. Data described below support the conclusion that the promoter contains sufficient information to mediate tissue specific regulation, but that sequences within the coding region are additionally required for maximal activation by fungal elicitor or UV-light (Douglas et al., 1991; Hanffe et al., 1991). Three different approaches were used in this analysis. First, in vivo footprinting of DNA from suspension cultured cells revealed several constitutive sites of protein-DNA interaction between −78 bp and −195 pb. Only slight modulations in the intensity of these footprints were observed after elicitor or UV-light treatment (Schulze-Lefert, 1989; Lois et al., 1989). Interestingly, several of the sequences thus identified are also found in the potato 4CL-1 promoters, and a subset also shows strong constitutive footprints (Becker-André, 1989). Parsley 4CL-1 promoter sequences containing these footprints, and up to 1.5 kb further upstream, were fused to the gene encoding GUS. These constructs were analyzed in both transiently expressing parsley protoplasts and transgenic parsley calli. The parsley protoplast system is known to respond faithfully to elicitor and UV-treatment, and has been used to rapidly identify stress responsive *cis*-elements (Dangl et al., 1987) (see Sect. IV). A 4CL-1-GUS fusion containing 597 bp of promoter is expressed to significant levels in parsley protoplasts. To our surprise, however, no 4CL-GUS construct was activated in either transiently expressing or stably transformed parsley cells. Nonetheless, we used deletion analysis to reveal quantitative *cis*-elements associated with the in vitro footprints. A 36 bp deletion, from −210 bp to −174 bp, resulted in a 10-fold drop in GUS activity. Notably, this deletion removes the promoter distal footprints, suggesting a role for these sequences in, at least, determination of 4CL mRNA amount (Hanffe et al., 1991).

Although these promoter sequences are insufficient to mediate elicitor or UV-light response, they are sufficient to direct tissue specific expression in transgenic tobacco (Douglas et al., 1991). The same 597 bp 4CL-1 promoter GUS fusion, which is well expressed in parsley cells, directed high levels of expression in primary tobacco xylem tissue of axillary and developing leaf veins. This expression was highest in immature tracheary elements, and lower in mature xylem. As discussed above for expression of the bean gPAL-2 promoter, expression of the 4CL promoter parallels deposition of lignin in xylem trachaery elements. In old stem sections, GUS expression is limited to files of ray parenchyma cells lying between highly lignified tracheary elements. 4CL-GUS expression was also evident in vascular tissue, nectaries, and pollen in developing flowers (Hanffe et al., 1991). Importantly, GUS activity was observed in epidermal cells of pigmented petal tissue, but not in epidermal cells of young leaves. Since flavonoid biosynthesis is known to be light regulated in leaves, we concluded that expression from the 597 bp promoter in petals is light independent. This 597 bp 4CL promoter is also unresponsive to elicitor and UV-light treatment in transgenic tobacco. The same pattern of expression was observed with a 210 bp promoter fragment. As a control in these experiments, the endogenous tobacco 4CL gene is responsive to these treatments. Another 4CL-GUS fusion, containing 120 bp of promoter, is unable to generate tissue specific expression in transgenic tobacco, although GUS activity is detectable in whole leaf extracts, using the sensitive fluorescence assay for GUS activity. Thus, sequences between −210 bp and −120 bp are required for all normal developmental activity of the 4CL promoter except its light-triggered expression in epidermal tissues.

That sequences in the 4CL coding region are, in fact, required for optimal stress responsiveness was shown by experiments in stably transformed parsley cells and transgenic tobacco plants (Douglas et al., 1991). In the former, the 597 bp promoter fragment, in its natural context in combination with the entire 4CL-1 genomic clone, mediated elicitor and UV-light response as measured at the steady state mRNA level. Furthermore, this same promoter fragment, in conjunction with the 4CL cDNA, was also stress inducible. In transgenic tobacco, a parsley 4CL genomic clone containing 1.2 kb of promoter sequence is also responsive to wounding, elicitor and UV-light. We concluded therefore, that some sequence(s) in the coding region are necessary for high level 4CL mRNA accumulation after stress induction (Douglas et al., 1991). These *cis*-linked sequences may function as binding sites for transcription factors, or they may function to stabilize nascent 4CL transcripts during synthesis and transport. Further experiments are required to assess whether the 4CL coding region is sufficient to render a heterologous promoter stress responsive, as reported for light control of a pea ferredoxin gene (Elliot et al., 1989).

Regulation of the parsley 4CL-1 gene, then, illustrates the complexity with which phenylpropanoid genes may respond to divergent signals. Developmental cues, of primary import, impinge only on promoter sequences. Stress signals are either distinct from, or act in concert with, those transduced through *cis*-elements in the promoter. This separation of regulatory functions is fundamentally different from that of the bean PAL promoter described above. Thus, either parsley and bean have evolved very divergent control mechanisms, or separate perception of inductive signals exists between seemingly co-regulated phenylpropanoid pathway genes. Further comparisons and contrasts between the two systems, including analysis of bean 4CL structure and regulation can only enhance our limited knowledge.

IV. Chalcone Synthase (CHS)

CHS is the key step in the conversion of 4 courmaroyl-CoA esters into flavonoids. Different modes of CHS regulation are illustrative examples of how different plant species have evolved divergent regulatory mechanisms to meet specific demands for biosynthetic products. Flavonoids are ubiquitous as pigments in fruits and flowers, while isoflavonoids serve the specialized role of phytoalexins in legumes. As well, flavonoid derivatives are known to function in the rhizosphere as attractants for *Rhizobium* and *Agrobacterium*. It is therefore no surprise that structure and regulation of CHS genes have received a great deal of attention. I will concentrate here on recent advances in the understanding of CHS *cis*-elements. The focus will remain on selected systems, as it has been throughout. A recent review discusses detailed aspects of CHS gene organization and regulation (Dangl et al., 1989) and several others include helpful discussions of CHS expression in more general terms (Lamb et al., 1989; Hahlbrock and Scheel, 1989; Dixon and Harrison, 1990).

A. Developmental and Light Regulation in Parsley

Flavonoids function in the epidermal cell layer of many species as potential UV protectants. As such, CHS expression is highly regulated in epidermal cells of developing tissue. Cell-type specific regulation was first defined by in situ analyses. These experiments proved that the entire series of inductive events, culminating in vacuolar deposition of flavonoids, was restricted to epidermal cells of emerging seedlings or illuminated etiolated leaves (Jahnen and Hahlbrock, 1988; Schmelzer et al., 1988). A refined analysis of CHS promoter activity in transgenic tobacco has also been recently reported (Schmid et al., 1990). These events are mirrored by events in parsley suspension cultured cells, where transcriptional control of the single copy CHS gene was first demonstrated (Chappell and Hahlbrock, 1984; Herrmann et al., 1988). Maximal response at the transcriptional level is dependent on UV, but modified in a complex way by other wavelengths, notably blue (Ohl et al., 1989; R. Feinbaum and F. Ansubel, pers. comm.).

This general scheme for developmental and light regulation of CHS expression holds also for *Antirrhinum majus, Petunia hybrida*, and *Arabidopsis thaliana* (Kaulen et al., 1986; Lipphardt et al., 1988; van Tunen et al., 1988; Feinbaum and Ausubel, 1988). It must be noted, however, that other regulatory signals surely impact on CHS expression (Knogge et al., 1986) (see Sect. IV.D).

B. CHS cis-Elements Defined in a Simplified System

As mentioned above, transcriptional activation of the CHS gene by UV containing white light and subsequent vacuolar deposition of flavonoids, is observed in parsley cell cultures. Moreover, parsley protoplasts retain the ability to specifically respond to either light or fungal elicitor (Dangl et al., 1987). We utilized this simplified system, in combination with in vivo footprinting, to define light regulated *cis*-elements on the parsley CHS promoter (Schulze-Lefert et al., 1989a, b; Block et al., 1990). Four light dependant footprints appear on the CHS promoter after light treatment. They appear co-ordinately and in parallel with the onset of maximal CHS transcription. As well, each is induced by either UV, blue or mixed wavelengths (T. Merkle et al., submitted).

The functional significance and interplay between these *cis*-elements was determined in two systems: via transient expression of CHS–GUS fusions in parsley protoplasts, and by analysis in transgenic plants (Schulze-Lefert et al., 1989a; Donald and Cashmore, 1990). Two separable light responsive "units" were identified using the parsley system. Each contains two footprinted *cis*-elements. In each unit, one *cis*-elements is identical or related to a sequence highly conserved in a number of light and stress regulated plant gene promoters (the "G-box"; see Giuliano et al., 1988; Schulze-Lefert et al., 1989a). The other footprinted *cis*-elements in each unit are unrelated to each other, and not easily identifiable in other sequences. Although they are functionally redundant in this assay, the TATA proximal light-responsive unit is inherently stronger than the TATA distal unit. Their combination is synergistic, and each can be quantitatively enhanced by the presence of a non-footprinted sequence found further upstream (Schulze-Lefert et al., 1989a, b).

The necessity of these *cis*-elements for light responsiveness was directly proven by site-directed mutagenesis of the two footprinted sequences in the context of the TATA proximal light responsive unit. Substitution of 10 bp within either footprinted sequence abrogated light response, showing that their combinatorial presence is required (Schulze-Lefert et al., 1989a). Fine structure analysis has refined our understanding of this *cis*-element mutualism by defining the functional borders of each *cis*-element and by single base pair substitutions throughout the highly conserved "G-box" sequence (Block et al., 1990). Two critical points emerge. First, although the "G-box" is a palindrome, functionally relevant bases are asymmetrically distributed through it. Second, the slightly degenerate "G-box" copy found in the TATA distal light responsive unit can substitute for the first, but in an orientation dependent manner. Functional roles for sequences containing "G-box" homologies from other light responsive CHS genes have also been demonstrated using the parsley protoplast system (Lipphardt et al., 1988; R. Wingender and J. Schell, pers. comm.) (see Sect. IV. D).

The "G-box" sequence also binds specifically protein(s) from a number of species and tissues (Giuliano et al., 1988; Staiger et al., 1989). Several clones encoding protein factors binding this sequence have recently been cloned, and are members of the bZIP family of transcriptional activators (Weishaar et at., 1991; Guiltinan et al., 1990; Schindler et al., 1991). The presence of this factor(s) in dark adapted tissue suggests that modification may be required prior to acquisition of transcriptional activity (see Ditta et al., 1989, for an example). Alternatively, this factor's general presence may imply a need for its combinatorial association with one or more induction specific factors. This presumed combination could occur either before or during binding to the template. The functional architecture of the parsley CHS promoter, where binding to two *cis*-elements spaced 10 pb apart is necessary for function, could argue for such a model.

Definition of light-responsive *cis*-elements on the parsley CHS promoter suffers from two limitations. The first is that there is no direct evidence to prove that *cis*-elements defined in a protoplast system are required in planta. The second is the absence of a reliable in vitro transcription system for RNA polymerase II in plants. This now severly limits further purification and functional analysis of interactions between transcription factors on this promoter.

C. Cis-Elements Controlling Cell-Type Specificity
of CHS and CHI Expression

Petunia hybrida has proven useful for analysis of flower specific expression of CHS and CHI necessary for production of flavonoid derived pigments (see van Tunen and Mol, 1991, for review). Levels of CHS mRNA in immature *Petunia* flowers are highest in anthers and lower in the colored corolla (van Tunen et al., 1988). As floral development progresses, this distribution is reversed. In the lightly pigmented flower tubes, reduced CHS mRNA amounts were found. Although there are many full and defective copies of CHS genes in the *Petunia hybrida* genome, only two (CHS-A and CHS-J) are appreciably expressed (Koes et al., 1986, 1987, 1989). CHS-A accounts for 90% of CHS mRNA in floral tissues and CHS-J for 10%. This ratio is maintained after light induction of dark adapted seedlings, where red wavelengths trigger CHS mRNA accumulation (Koes et al., 1989). Oddly, light induction of suspension cultured cells reverses this ratio. Thus, there is no obvious differential regulation of these two genes during floral development (Koes et al., 1989).

Preliminary delineation of CHS-A *cis*-elements utilized transgenic analysis of fusions to the CAT or GUS reporter genes (van der Meer et al., 1990; Koes et al., 1990). An 800 bp fragment was sufficient to confer both tissue specific and light responsiveness on the reporter gene. CAT activity was restricted to pigmented floral organs and flower stems. Truncation to a 220 bp promoter did not alter this expression pattern, although CAT levels were diminished. Finally, further truncation to a 67 bp promoter resulted in very low level tissue-specific expression, but light inducibility was lost. The organ-specific transgenic expression was predicted based on transient analysis of CHS-A-*cat* fusions in protoplasts derived from either purple or white callus. This ingenious pre-screen highlights the many genetic tools available in the *Petunia* system (van der Meer et al., 1990).

Preliminary analysis of protein binding to CHS-A derived promoter sequences showed that the region from −142 to +81 is capable of binding specifically to a protein(s) from flower nuclear extracts (van der Meer et al., 1990). This fragment contains a copy of the "G-box" sequence discussed above, and two copies of the sequence 5′-TACPYAT-3′. This latter sequence was elegantly defined by analysis of transposon induced deletions on the *A. majus* promoter (Sommer et al., 1988). There, it is also re-iterated, and disruption on the TATA distal copy caused a 65% drop in flower specific CHS mRNA accumulation. Small additions or deletions on either side of this element had no effect, allowing its precise definition.

The 5′-TACPYAT-3′ sequence is strongly implicated as a critical determinant of flower specificity on the *Petunia* CHS-A promoter, as it is retained on the 67 pb fragment. In contrast, the "G-box" homology, and two other longer stretches of sequence related to the *A. majus* promoter are suggested to be involved in light response, since they are retained on the 220 bp light responsive fragment, but absent from the 67 bp promoter (van der Meer et al., 1990). This is also consistent with functional analysis of the *A. majus* CHS promoter, where a fragment from −197 to −39, containing these homologies, confers light regulation (Lipphardt et al., 1988). It should be noted that the 5′-TACPYAT sequence is absent from the parsley CHS promoter. This may be circumstantial evidence for the element's involvement in flower specific CHS expression, since parsley flowers are unpigmented.

The critical isomerization of the CHS product, naringenin chalcone, requires CHI activity (Dixon et al., 1988). As outlined below, it is apparent that CHS and CHI are tightly co-related in many cell types (van Tunen et al.,

1988, 1989, 1990; Mehdy and Lamb, 1987). Therefore, comparison of CHS and CHI control within a well characterized system should test whether or not shared *cis*-elements mediate the observed co-ordinate developmental regulation. In *Petunia*, present data suggests that this simple hypothesis is incorrect. CHI mRNA accumulation is largely coordinate with that of CHS in floral tissue and UV-light treated seedlings (van Tunen et al., 1988, 1989). In contrast to the lack of CHS differential expression in *Petunia*, however, the two CHI genes exhibit unexpected modes of control. Temporal and spatial expression of the CHI-A gene is determined by differential promoter usage, a situation thus far unique for phenylpropanoid genes. The coding region proximal start point yields a transcript found (coordinately with CHS) in floral tissues and tubes, which is also UV-light induced (CHI-A$_1$). Transcript from the distal start point at -437 is expressed in mature anthers and pollen grains. Expression of this second transcript (CHI-A$_2$) is neither UV-light responsive nor coordinated with CHS expression (van Tunen et al., 1989). Transcripts from the CHI-B gene are expressed only in immature anthers, yet no CHI activity is found (van Tunen et al., 1990). The authors suggest that this gene may encode a highly related (77% amino acid homology) but enzymatically different product. This points to a possible complication in the use of polyclonal antisera in screening λgt 11 expression libraries (van Tunen et al., 1990).

Tissue specific CHI expression has been confirmed and extended by analysis of GUS fusions in transgenic plants (van Tunen et al., 1990). Several salient points emerge. Most interestingly, the two CHI-A promoters function independently to mediate their respective specificities. A 437 bp CHI-A$_1$ promoter drives GUS expression in floral limbs, tube inner epidermis and parenchyma, and sepals. Activity from this promoter is also observed in the seed coat, seed endoderm and embryo, and in leaf and stem epidermis. In contrast, 440 bp of sequence upstream from the distal CHI-A$_2$ promoter drives GUS activity exclusively in male gametophyte tissue. The above distributions of CHI-A promoter activity appear to exactly parallel GUS expression driven by either the CHS-A or CHS-J promoters (Koes et al., 1990). The CHI-B promoter (750 bp), however, dictated GUS expression exclusively in tapetal and pollen cells (van Tunen et al., 1990).

Can the divergent regulatory strategies observed for *Petunia* CHS and CHI genes be reconciled with models predicting that transcription control is governed by shared *cis*-elements? Sequence comparison of CHS and CHI promoters suggested that this may be the case, although preliminary functional analyses show that it is not necessarily so (Koes et al., 1989; van Tunen et al., 1989, 1990; van der Meer et al., 1990). Most relevant is the observation that the *Petunia* line "Red Star", where CHS mRNA is absent in white petal tissue due to *trans*-acting mutations, CHI-A mRNA levels remain normal (van Tunen et al., 1988). Therefore, petal specific expression of CHS and CHI are separable. Promoter sequences highly conserved among several flavonoid genes expressed in immature anther tissue (CHS-A, CHS-J, CHI-B, and dihydro-flavonol reductase) were found, and postulated to be *cis*-elements involved in tissue specific expression (van Tunen et al., 1989). The first level of direct analysis, however, showed that these sequences were dispensible for anther specific CHS-A expression (van der Meer et al., 1990). Their functional role remains enigmatic.

D. Differential CHS Expression in Legumes

As mentioned in the introduction, CHS is the key enzyme in the biosynthesis of isoflavonoid phytoalexins in legume species (Ebel, 1986). There are also now many reports of flavonoid derived compounds playing critical roles in plant–microbe interactions of either pathogenic or symbiotic natures (Stachel et al., 1985; Peters et al., 1986; Redmond et al., 1986; Firmin et al., 1986; see Long, 1989, for review). Thus, legume phenylpropanoid metabolism in general, and flavonoid specific branches in particular, may have novel regulatory requirements. This has been discussed above with respect to PAL expression in French bean. In both French bean and soybean (*Glycine max*) CHS is encoded by multi-gene families (Ryder et al., 1987; Wingender et al., 1989).

Recent reviews deal extensively with CHS gene structure and expression in these two systems (Dangl et al., 1989; Lamb et al., 1989; Hahlbrock and Scheel 1989; Dixon and Harrison, 1990). Here, the emphasis will be on how differential expression CHS genes in legumes contrasts the modes of CHS control discussed above for parsley and *Petunia*.

In both systems, there has been extensive analysis of CHS expression induced by virulent and avirulent fungal pathogens (and elicitors derived from them). In both elicitor treated suspension cultured cells or infected cotyledons and roots, transcriptional regulation is responsible for the appearance of multiple CHS isoforms (Ebel et al., 1984; Schmelzer et al., 1984; Ryder et al., 1984; Grab et al., 1985; Cramer et al., 1985a; Lawton and Lamb, 1987). There are also differences in the timing and spatial distribution of both PAL and CHS accumulation after infection with either compatible or incompatible fungal races in either system (Bell et al., 1984, 1986; Hahn et al., 1985; Bonhoff et al., 1986; Habereder et al., 1989). Strong evidence also exists for the induction of legume CHS genes by wounding, UV-containing white light, and infection with either *Agrobacterium* or *Brady-rhizobium* (Ryder et al., 1987; Wingender et al., 1989).

Differential transcriptional activation of specific members of the CHS gene families in French bean was also inferred by the isolation of five polymorphic cDNA clones (Ryder et al., 1987). Appropriate 3' untranslated probes were chosen for both hybrid select translation and S1 analysis after various inductive treatments. One probe, detecting three closely related mRNA species, hybridizes to 40% of elicitor-induced CHS mRNA. This "class" of mRNAs is also highly represented after wounding of bean hypocotyls, and one member of this group gives rise to most of the CHS mRNA found after UV-light treatment. Strangely, none of the seven different mRNAs detected in this analysis is strongly induced after fungal infection of hypocotyls (Ryder et al., 1987). Gene specific regulation was also observed in soybean, where the CHS-1 gene encoded nearly all elicitor and uv-light induced CHS mRNA (Wingender et al., 1989).

For both systems, then, gene specific regulation is the norm. What is somewhat puzzling, however, is that this differential expression probably does not generate CHS enzymes of different catalytic ability; since the cDNAs sequenced from French bean encode nearly identical CHS enzymes (Ryder et al., 1987). Either these minor differences encode subtle functional changes, or this example of genome plasticity is driven only by the need for regulatory divergence. The fact that several French bean CHS genes still respond, for example, to elicitor treatment suggests that signal transduction to those templates is mediated by shared mechanisms.

Identification of French bean and soybean CHS cis-elements is only beginning. A 336 bp French bean CHS-15 promoter drives elicitor-induced CAT expression in electroporated soybean protoplasts (Dron et al., 1988), and is quoted as mediating tissue-specific and wound-or elicitor-responsiveness in transgenic tobacco (cited in Dixon and Harrison, 1990). Data from 5' CHS promoter deletions transiently expressed in soybean protoplasts suggests the presence of negative regulatory elements between -326 and -173 and elicitor-responsive sequences TATA proximal of -130 (Dron et al., 1988). These studies are hampered by the overall low and variable response of soybean protoplasts to fungal elicitor. The significance of these data is enhanced by the finding of three DNase I hypersensitive sites on this bean CHS promoter (Lawton et al., 1990). Two are elicitor inducible, and are mapped to positions TATA proximal to -130, while the third is constitutive and located in the putative silencer region. As well, several homologies with the bean gPAL-2 promoter are located within these CHS-15 promoter fragment (see Dixon and Harrison, 1990).

Analysis of the soybean CHS-1 promoter in both soybean and parsley protoplasts broadens the above findings (R. Wingender and J. Schell, pers. comm.). There is evidence for possibly two far upstream silencers on this promoter. Yet sequences mediating elicitor response in either homologous or heterologous protoplasts are located between -175 und -134. Interestingly, the same region is necessary for uv-light activation in parsley protoplasts; although sequences downstream of -75 are also necessary. No uv-light activation of these constructs was seen in soybean protoplasts. Those results are intriguing for two reasons. First, the same template requirements were defined in systems known to be responsive to very different molecules (a protein in parsley and a β-glucan in soybean, see Parker et al., 1989; Ebel and Scheel, 1992). This clearly implies a convergence of signals for transcriptional activation well before the binding of template by transcription machinery. Secondly, these results suggest, as has been discussed above, the presence of shared cis-elements between different genes responding to the same stimulus. The soybean CHS-1 promoter (and the French bean CHS-15 and gPAL-2 promoters) in fact all contain a sequence related to one of the in vivo footprinted sequences from the parsley PAL-1 gene described above (Lois et al., 1989). It is present on a soybean CHS-1 promoter construct which has lost both light and elicitor inducibility; it is thus insufficient to mediate three responses. There is also a "G-box" homology on the soybean CHS-1 promoter. Partial deletion of this sequence abrogates both light and elicitor response. These data again suggest that combinatorial usage of general and stress-specific transcription factors may generate the regulatory complexity governing phenylpropanoid and flavonoid gene activation.

V. Perspectives

This review was meant to detail current understanding of promoter elements mediating the overlapping developmental and environmental control of general phenylpropanoid and flavonoid gene expression. By concentrating on the most advanced systems, it is not an exhaustive review. What should be apparent is that the analyses required to define, in a critical manner, necessary cis-elements are proceeding apace. Sequence comparisons have proven useful as a guide for more detailed study, but one must remain cautious about the functional relevance of "homologies" until conclusive data exist. In this sense, in vivo footprinting has proven an extremely powerful tool to discretely define targets for subsequent in vivo and in vitro analyses. Clearly, however, this technique is limited thus far to analysis of cell cultures, where all cells should respond, more or less, identically. Functional analyses in transient expression systems require that protoplasts maintain specific responsiveness, a sometimes difficult demand. Even when rigorous conditions are established, it is still unwise to assume that data thus divined are applicable in planta. Therefore, the tedious, and often nearly impossible task of generating sufficient numbers of independent transgenic plants for each construct to be tested is to date the most reliable method to assess promoter function. The power of cell type analysis afforded by GUS histochemical techniques cannot be overstated. Here, also, caution is advised since it is apparent that this technology can only provide qualitative and, at best, approximate quantitative answers.

The complexity of putative transcription factors binding to phenylpropanoid gene promoters awaits unravelling. Several promoters contain "shared" sequences whose functional relevance remains to be tested. The paucity of firm data regarding trans-factors would be greatly alleviated by the development of reliable in vitro transcription systems. The availability of mutants, and eventual cloning of genes, affecting trans regulation, especially in Petunia and A. majus will also be helpful.

Our naive knowledge suggests that sets of cis-elements, binding combinations of both general and specific transcription factors, are responsible for the exquisite cell-type specific and stress-induced expression of PAL, 4CL, CHS, and CHI genes. Whether these responses are mediated by completely-, partially-, or non-overlapping sets of cis-elements present on each promoter responding to a given stimulus remains largely unknown, although recent evidence suggests a great deal of interplay between the effects of various stimuli (Lozoya et al., 1991). I have tried to highlight the several examples where simple co-ordinate regulation impinging on shared *cis*-elements appears not to be the case, hopefully not at the expense of the alternative.

Further comparison and contrast between these systems, and the development of others, will continue to make the control of phenylpropanoid gene regulation an enticing and rewarding field of study for those with the patience to withstand its many vagaries.

Acknowledgements

As with most reviews, this one relies heavily on the beneficence of many colleagues who shared the unpublished data cited throughout. Contributions from Mike Bevan (IPSR Cambridge Laboratories, John Innes Institute, Norwich, U.K.), Jos Mol and Arjen van Tunen (Department of Genetics, Free University Amsterdam, The Netherlands), Ruth Wingender and Jeff Schell (Max Planck Institut für Züchtungsforschung, Köln, Federal Republic of Germany), Rhonda Feinbaum and Fred Ansubel (Massachusetts General Hospital, Boston, U.S.A.), and Rich Dixon and Maria Harrison (Noble Foundation Plant Biology Labs, Ardmore, Oklahoma, U.S.A.) made this chapter not only more interesting, but more fun to write as well. Finally, I wish to thank Prof. Klaus Hahlbrock (Köln) for his past and continued support, and for a thorough critique of the manuscript.

VI. References

Bailey JA (1987) Phytoalexins: a genetic view of their significance. In: Day PR, Jellis GJ (eds) Genetics and plant pathogenesis. Blackwell, Oxford, pp 233–244

Becker-André M (1989) Untersuchungen zur Struktur, Expression und Regulation der 4-Coumarat:CoA Ligase Gene in Kartoffel (*Solanum tuberosum* L. cv. Datura). Thesis, Universitat zu Köln, Cologne, Federal Republic of Germany

Becker-André M, Hahlbrock K (1989) Absolute mRNA quantification using the polymerase chain reaction (PCR). A novel approach by a PCR aided transcript titration assay (PATTA). Nucleic Acids Res 17: 9439–9448

Bell JN, Dixon RA, Bailey JA, Rowell PM, Lamb CJ (1984) Differential induction of chalcone synthase mRNA acitivity at the onset of phytoalexin accumulation in compatible and incompatible plant:pathogen interactions. Proc Natl Acad Sci USA 81: 3384–3388

Bell JN, Ryder TB, Wingate VPM, Bailey JA, Lamb CJ (1986) Differential accumulation of plant defense gene transcripts in a compatible and incompatible plant:pathogen interaction. Mol Cell Biol 6: 1615–1623

Bevan MJ, Shufflebottom D, Edwards K, Jefferson R, Schuch W (1989) Tissue- and cell-specific activity of a phenylalanine ammonia lyase promoter in transgenic plants. EMBO J 8: 1899–1906

Binns AN, Chen RH, Wood HN, Lynn DG (1987) Cell division promoting activity of naturally occurring dehydrodiconiferyl glucosides: do cell wall components control cell division? Proc Natl Acad Sci USA 84: 980–984

Block A, Dangl JL, Hahlbrock K, Schulze-Lefert P (1990) Functional borders, genetic fine structure and distance requirements of *cis*-elements mediating light responsiveness of the parsley chalcone synthase promoter. Proc Natl Acad Sci USA 87: 5387–5391

Bolwell GP, Bell JN, Cramer CL, Schuch W, Lamb CJ, Dixon RA (1985) L-Phenylalanine ammonia-lyase from *Phaseolus vulgaris*: characterization and differential induction of multiple forms from elicitor-treated cell suspension cultures. Eur J Biochem 149: 411–419

Bonhoff A, Loyal R, Ebel J, Grisebach H (1986) Race:cultivar-specific induction of enzymes related to phytoalexin biosynthesis in soybean roots following infection with *Phytophthora megasperma* f. sp. *glycinea*. Arch Biochem Biophys 246: 149–154

Chappell J, Hahlbrock K (1984) Transcription of plant defense genes in response to UV light or fungal elicitor. Nature 311: 76–78

Church G, Gilbert W (1984) Genomic sequencing. Proc Natl Acad Sci USA 81: 1991–1995

Cramer CL, Bell JN, Ryder T, Bailey JA, Schuch W, Bolwell GP, Robbins MP, Dixon RA, Lamb CJ (1985a) Co-ordinated synthesis of phytoalexin biosynthetic enzymes in biologically-stresed cells of bean (*Phaseolus vulgaris* L.). EMBO J 4: 285–289

Cramer CL, Ryder TB, Bell JN, Lamb CJ (1985b) Rapid switching of plant gene expression by fungal elicitor. Science 227: 1240–1243

Cramer CL, Edwards K, Dron M, Liang X, Dildine SL, Bolwell GP, Dixon RA, Lamb CJ, Schuch W (1989) Phenylalanime ammonia-lyase gene organization and structure. Plant Mol Biol 12: 367–383

Dangl JL, Hauffe KD, Lipphardt S, Hahlbrock K, Scheel D (1987) Parsley protoplasts retain differential responsiveness to U.V. light and fungal elicitor. EMBO J 6: 2551–2556

Dangl JL, Hahlbrock K, Schell J (1989) Regulation and structure of chalcone synthase genes. In: Vasil IK, Schell J (eds) Cell culture and somatic cell genetics of plants, vol 4. Academic Press. New York, pp 155–173

Dixon RA, Harrison MJ (1990) Activation, structure, and organization of genes involved in microbial defense in plants. Adv Genet 28: 165–234

Dixon RA, Blyden ER, Robbins MP, van Tunen AJ, Mol JNM (1988) Comparative biochemistry of chalcone isomerases. Phytochemistry 27: 2801–2808

Donald RGK, Cashmore AR (1990) Mutation of either G-box or I-box sequences profoundly effects expression from the *Arabidopsis* rbc S-1A promoter. EMBO J 9: 1717–1726

Douglas C, Hoffmann H, Schulz W, Hahlbrock K (1987) Structure and elicitor or U.V.-light-stimulated expression of two 4-coumarate:CoA ligase genes in parsley. EMBO J 6: 1189–1195

Douglas CJ, Hanffe KD, Ites-Morales ME, Ellard M, Paszkowski U, Hahlbrock K, Dangl JL (1991) Exonic sequences are required for elicitor and light activation of a plant defense gene, but promoter sequences are sufficient for tissue specific expression EMBO J 10: 1767–1775

Dron M, Clouse SD, Lawton MA, Dixon RA, Lamb CJ (1988) Glutathione and fungal elicitor regulation of a plant defense gene promoter in electroporated protoplasts. Proc Natl Acad Sci USA 85: 6738–6742

Ebel J (1986) Phytoalexin synthesis: the biochemical analysis of the induction process. Annu Rev Phytopathol 24: 235–264

Ebel J, Scheel D (1992) Elicitor recognition and signal transduction. In: Boller T, Meins F (eds) Genes involved in plant defense. Springer, Wien New York, pp 183–205 [Dennis ES et al (eds) Plant gene research. Basic knowledge and application]

Ebel J, Schmidt WE, Loyal R (1984) Phytoalexin synthesis insoybean cells: elicitor induction of phenylalanine ammonia-lyase and chalcone synthase mRNAs and correlation with phytoalexin accumulation. Arch Biochem Biophys 232: 240–248

Edwards K, Cramer CL, Bolwell GP, Dixon RA, Schuch W, Lamb CJ (1985) Rapid transient induction of phenylalanine ammonia-lyase mRNA in elicitor-treated bean cells. Proc Natl Acad Sci USA 82: 6731–6735

Elliot RC, Dickey LF, White MJ, Thompson WF (1989) cis-acting elements for light regulation of Pea ferrodoxin I gene expression are located within transcribed sequences. Plant Cell 1: 691–698

Feinbaum RL, Ausubel FM (1988) Transcriptional regulation of the Arabidopsis thaliana chalcone synthase gene. Mol Cell Biol 8: 1985–1992

Firmin JL, Wilson KE, Rossen L, Johnston AWB (1986) Flavonoid activation of nodulation genes in Rhizobium reversed by other compounds. Nature 324: 90–92

Fritzemeier K-H, Cretin C, Kombrink E, Rohwer F, Taylor J, Scheel D, Hahlbrock K (1987) Transient induction of phenylalanine ammonia-lyase and 4-coumarate:CoA ligase mRNAs in potato leaves infected with virulent or avirulent races of Phytophthora infestans. Plant Physiol 85: 34–41

Giuliano G, Pichersky E, Makik VS, Timko MP, Scolnik PA, Cashmore AR (1988) An evolutionarily conserved protein binding sequence upstream of a plant light-regulated gene. Proc Natl Acad Sci USA 85: 7089–7093

Grab D, Loyal R, Ebel J (1985) Elicitor-induced phytoalexin synthesis in soybean cells: changes in the activity of chalcone synthase mRNA and the total population of translatable mRNA. Arch Biochem Biophys 243: 423–529

Guiltinan MJ, Marcotte WR, Quatrano RS (1990) A plant leucine zipper protein that recognizes an abscistic acid response element. Science 250: 267–271

Habereder H, Schroder G, Ebel J (1989) Rapid induction of phenylalanine ammonia-lyase and chalcone synthase mRNAs during fungus infection of soybean (Glycine max L.) roots or elicitor treatment of soybean cell cultures at the onset of phytoalexin synthesis. Planta 177: 58–65

Hahlbrock K, Scheel D (1989) Physiology and molecular biology of phenylpropanoid metabolism. Annu Rev Plant Physiol Plant Mol Biol 40: 347–369

Hanffe KD, Paszkowski U, Schulze-Lefert P, Hahlbrock K, Dangl JL (1991) A parsley 4CL-1 promoter fragment specifies complex expression patterns in transgenic tobacco. Plant Cell 3: 435–443

Herrmann A, Schulz W, Hahlbrock K (1988) Two alleles of the single-copy chalcone synthase gene in parsley differ by a transposon-like element. Mol Gen Genet 212: 93–98

Jefferson RA (1987) Assaying chimeric genes in plants: the GUS gene fusion system. Plant Mol Biol Rep 5: 387–405

Jahnen W, Hahlbrock K (1988) Differential regulation and tissue-specific distribution of enzymes of phenylpropanoid pathways in developing parsley seedlings. Planta 173: 453–458

Kaulen H, Schell J, Kreuzaler F (1986) Light induced expression of chimaeric chalcone synthase-NPT11 gene in tobacco cells. EMBO J 5: 1–8

Knogge W, Schmelzer E, Weissenböck G (1986) The role of chalcone synthase in the regulation of flavonoid biosynthesis in developing oat primary leaves. Arch Biochem Biophys 250: 364–372

Koes RE, Spelt CE, Reif HJ, van den Elzen PJM, Veltkamp E, Mol JMN (1986) Floral tissue of Petunia hybrida (V30) expresses only one member of the chalcone synthase multigene family. Nucleic Acids Res 14: 5229–5239

Koes RE, Spelt CE, Mol JNM, Gerats AGM (1987) The chalcone synthase multigene family of Petunia hybrida: sequence homology, chromosomal localization and evolutionary aspects. Plant Mol Biol 10: 159–169

Koes RE, Spelt CE, Mol JNM (1989) The chalcone synthase multigene familiy of Petunia

324   Jeffery L. Dangl

hybrida (V30): differential, light-regulated expression during flower development and UV-light induction. Plant Mol Biol 12: 213–225

Koes RE, van Blokland RJ, Qattrochio F, van Tunen AJ, Mol JNM (1990) Chalcone synthase promoters in petunia are active in pigmented and unpigmented cell types. Plant Cell 2: 379–392

Kuhlemeier C, Green PS, Chua N-H (1987) Regulation of gene expression in higher plants. Annu Rev Plant Physiol 38: 221–257

Lamb CJ, Lawton MA, Dron M, Dixon M, Dixon RA (1989) Signals and transduction mechanisms for activation of plant defenses against microbial attack. Cell 56: 215–224

Lawton MA, Lamb CJ (1987) Transcriptional activation of plant defense genes by fungal elicitor, wounding and infection. Mol Cell Biol 7: 335–341

Lawton MA, Clouse SD, Lamb CJ (1990) Glutathiane-elicited changes inchromatin structure within the promoter of the plant defense gene chalcone synthase. Plant Cell Rep 8: 561–564

Liang X, Dron M, Cramer CL, Dixon RA, Lamb CJ (1989a) Differential regulation of phenylalanine ammonia-lyase genes during plant development and by environmental cues. J Biol Chem 264: 14486–14492

Liang X, Dron M, Schmid J, Dixon RA, Lamb CI (1989b) Development and environmental regulation of a phenylalanine ammonia-lyase-β glucuronidase gene fusion in transgenic tobacco plants. Proc Natl Acad Sci USA 86: 9284–9288

Lipphardt S, Brettschneider R, Kreuzaler F, Schell J, Dangl JL (1988) UV-inducible transient expression in parsley protoplasts identifies regulatory *cis*-elements of a chimeric *Antirrhinum majus* chalcone synthase gene. EMBO J 7: 4027–4033

Lois R, Dietrich A, Hahlbrock K (1989) A phenylalanine ammonia-lyase gene from parsley: structure, regulation and identification of elicitor and light responsive *cis*-acting elements. EMBO J 8: 1641–1648

Long SR (1989) *Rhizobium*–legume nodulation: life together in the underground. Cell 56: 203–214

Lozoya E, Hoffmann H, Douglas C, Schulz W, Scheel D, Hahlbrock K (1988) Primary structures and catalytic properties of isozymes encoded by the two 4-coumarate : CoA ligase genes in parsley. Eur J Biochem 176: 661–667

Lozoya E, Block A, Lois R, Hahlbrock K, Scheel D (1991) Transcriptional repression of light-induced flavanoid synthesis by elicitor treatment of cultured parsley cells. Plant J 1: 227–234

Mehdy M, Lamb CJ (1987) Chalcone isomerase cDNA cloning and mRNA induction by fungal elicitor, wounding and infection. EMBO J 6: 1527–1533

Mol JNM, Stuitje AR, Gerats AMG, Koes RE (1988) Cloned genes of plant phenylpropanoid metabolism. Plant Mol Biol Rep 6: 274–279

Ohl S, Hahlbrock K, Schaefer E (1989) A stable blue light derived signal modulates UV light chalcone synthase gene activation in cultured parsley cells. Planta 177: 288–236

Ohl S, Hedrick SA, Chory J, Lamb CJ (1990) Functional properties of a phenylalanine ammonia lyase promoter from *Arabidopsis*. Plant Cell 2: 837–848

Peters NK, Frost JW, Long SR (1986) A plant flavone, luteolin, induces expression of *Rhizobium meliloti* nodulation genes. Science 233: 977–980

Redmond JW, Batley M, Djoerdjevic MA, Ines RW, Krempel PL, Rolfe BG (1986) Flavones induce expression of the nodulation genes in *Rhizobium*. Nature 323: 632–635

Ryder TB, Cramer CL, Bell JN, Robbins MP, Dixon RA, Lamb CJ (1984) Elicitor rapidly induces chalcone synthase mRNA in *Phaseolus vulgaris* cells a the onset of the phytoalexin response. Proc Natl Acad Sci USA 81: 5724–5728

Ryder TB, Hedrick SA, Bell JN, Liang X, Clouse SD, Lamb CJ (1987) Organization and differential activation of a gene family encoding the plant defense enzyme chalcone synthase in *Phaseolus vulgaris*. Mol Gen Genet 210: 219–233

Saluz HP, Jost JP (1989a) Genomic footprinting with Taq polymerase. Nature 338: 277

Schindler U, Ecker JR, Cashmore AR (1991) An *Arabidopsis thaliana* G-box binding protein similar to the wheat leucine zipper protein identified as HBP-1. In: Jenkins GI, Schuch W (eds) Molecular biology of plant development. The Company of Biologists, Cambridge, pp 211–218

Schmelzer E, Börner H, Grisebach H, Ebel J, Hahlbrock K (1984) Phytoalexin synthesis in soybean (*Glycine max*). Similar time courses of mRNA induction in hypocotyls infected with a fungal pathogen and in cell cultures treated with fungal elicitor. FEBS Lett 172: 59–63

Schmelzer E, Jahnen W, Hahlbrock K (1988) In situ localization of light-induced chalcone synthase mRNA, chalcone synthase, and flavonoid end products in epidermal cells of parsley leaves. Proc Natl Acad Sci USA 85: 2989–2993

Schmelzer E, Krüger-Lebus S, Hahlbrock K (1989) Temporal and spatial patterns of gene expression around sites of attempted fungal infection in parsley leaves. Plant Cell 1: 993–1001

Schmid J, Doerner PW, Clouse SD, Dixon RA, Lamb CJ (1990) Developmental and environmental regulation of a bean chalcone synthase promoter in transgenic tobacco. Plant Cell 2: 619–631

Schulze-Lefert P (1989) Identifizierung und Charakterisierung *cis*-aktiver Kontrollelemente des lichtregulierten Chalkonsynthase-Gens in Petersilie. Dissertation, Universität zu Köln, Cologne, Federal Republic of Germany

Schulze-Lefert P, Dangl JL, Becker-André M, Hahlbrock K, Schulz W (1989a) Inducible in vivo DNA footprints define sequences necessary for UV-light activation of the parsley chalcone synthase gene. EMBO J 8: 651–657

Schulze-Lefert P, Becker-André M, Schulz W, Hahlbrock K, Dangl JL (1989b) Functional architecture of the light-responsive chalcone synthase promoter from parsley. Plant Cell 1: 707–714

Sommer H, Bonas U, Saedler H (1988) Transposon-induced alterations in the promoter region affect transcription of the chalcone synthase gene of *Antirrhinum majus*. Mol Gen Genet 211: 49–55

Stachel SE, Messens E, van Montagu M, Zambryski P (1985) Identification of the signal molecules produced by wounded plant cells that activate T-DNA transfer in *Agrobacterium tumefaciens*. Nature 318: 624–629

Staiger D, Kaulen H, Schell J (1989) A CACGTG motif of the *Antirrhinum majus* chalcone synthase promoter is recognized by an evolutionarily conserved nuclear protein. Proc Natl Acad Sci USA 86: 6930–6934

Tepper CS, Albert FG, Anderson AJ (1989) Differential mRNA accumulation in three cultivars of bean in response to elicitors from *Colletotrichum lindemuthianum*. Physiol Mol Plant Pathol 34: 85–98

van der Meer I, Spelt CE, Mol JNM, Stuitje AR (1990) Promoter analysis of the chalcone synthase (*chs A*) gene of *Petunia hybrida*: a 67 bp promoter region directs flower-specific expression. Plant Mol Biol 15: 95–109

van Tunen AJ, Mol JNM (1990) Control of flavanoid synthesis and manipulation of flower color. In: Grierson D (ed) Developmental regulation of plant gene expression. Blackie and Son, Glasgow, pp 94–130 [Grierson D (ed) Plant-biotechnology Series, vol 2]

van Tunen AJ, Koes RE, Spelt CE, van der Krol AR, Stuitje AR, Mol JNM (1988) Cloning of the two chalcone flavanone isomerase genes from *Petunia hybrida*: coordinate, light-regulated and differential expression of flavonoid genes. EMBO J 7: 1257–1263

van Tunen AJ, Hartman SA, Mur LA, Mol JNM (1989) Regulation of chalcone flavanone
    isomerase (CHI) gene expression in *Petunia hybrida*: the use of alternative promoters in
    corolla, anthers and pollen. Plant Mol Biol 12: 539–551

van Tunen AJ, Mur LA, Brouns GA, Rienstraj JD, Koes RE, Mol JNM (1990) Pollen- and
    anther-specific *chi* promoters from *Petunia hybrida*: tandem promoter regulation of the
    chi A gene. Plant Cell 2: 393–401

Weishaar B, Armstrong GA, Block A, da Costa e Silva O, Hahlbrock K (1991) Light-
    inducible and constitutively expressed DNA-binding proteins recognizing a plant pro-
    moter element with functional relevance in light responsiveness. EMBO J 7: 1777–1786

Wingender-Drissen R, Röhrig H, Höricke C, Wing S, Schell J (1989) Differential regulation
    of soybean chalcone synthase genes in plant defense, symbiosis and upon environmental
    stimuli. Mol Gen Genet 218: 315–322

Chapter 13

Regulation of Lignification in Defense

Michael H. Walter

Institut für Pflanzenphysiologie, Universität Hohenheim, D-W-7000 Stuttgart 70,Federal Republic of Germany

With 1 Figure

Contents

I. Introduction

This chapter deals with some aspects of structural barriers impermeable to attempted invasions of pathogens as opposed to more direct pathogen controls by way of phytoalexin antibiotics, hydrolases or fungitoxic thionins. There is growing evidence that such barriers are reinforced or even newly erected in plant defense responses, and that they contribute to disease resistance (Aist, 1983; Hargreaves and Keon, 1986; Delmer and Stone,

1988). Major structural barriers are the cellulosic cell walls, in particular epidermal cell walls, impregnated with lignin polymers and other wall-bound phenolics and the cuticle with cutin and soluble waxes as major constituents (Vance et al., 1980; Kolattukudy, 1980, 1985; Kolattukudy and Soliday, 1985).

Lignin is one of the major organic materials in the biosphere and second only to cellulose in abundance. During development lignin is deposited in secondary walls of lignifying tissues, especially xylem, thus providing rigidity and structural support to the otherwise elastic polysaccharide cell walls and enabling solute conductance in the vascular system (Sarkanen and Hergert, 1971; Lewis and Yamamoto, 1990). Lignin is one of the prerequisites for terrestrial plant life. Lignified cell walls are considered to be very effective in limiting the progression of microorganisms, as most obviously demonstrated by the slow biodegradation of highly lignified tissues and the negative correlation between lignin content and digestibility of forages by ruminants.

Lignification has also been implicated as a mechanism of inducible disease resistance (Vance et al., 1980). In active defense responses to wounding or microbial attack induced lignification is but one of many biochemical events resulting in an ultrastructurally different reinforced cell wall. For recent reviews on cell wall structure and biosynthesis and its modification in plant defense the reader is referred to Aist (1983); Hargreaves and Keon (1986); Bolwell (1988); Delmer and Stone (1988), and Stone (1989). Depositions on or into the cell wall of suberin, hydroxyproline-rich glycoproteins, callose and a variety of monomeric or polymeric phenolic material collectively called wall-bound phenolics (including hydroxycinnamic acids, cinnamoyl esters, cinnamoyl amides and benzoic acid derivatives) are all thought to contribute to barrier formation. Suberin, a mixed polymer with lignin-like aromatic and hydroxyfatty acid aliphatic domains (Kolattukudy, 1980) and condensed phenolics (Matern and Kneusel, 1988) with frequent crosslinks to other cell wall material (Fry, 1986) are difficult to distinguish from lignin due to their common biosynthetic origin (e.g., Ampomah and Friend, 1988; Bruce and West, 1989). As in other mechanisms of disease resistance gene activation seems to be involved in several (Roberts et al., 1988; Lamb et al., 1989), but not all of these responses (Kauss, 1987; Matern and Kneusel, 1988).

The discussion in this chapter must for the most part be confined to lignification and its regulation in plant defense. Recent reports on local or systemic increases in lignin content following an encounter with a pathogen or with pathogen-derived elicitors will be reviewed. Another focus will be on the regulation of lignin biosynthetic enzymes in situations of biological stress. The evidence for a role of lignification in plant disease resistance has been summarized by Friend (1976), Vance et al. (1980), and Ride (1983) and is supplemented by new data showing a loss of hypersensitive resistance of wheat to stem rust by specifically inhibiting lignin precursor biosynthesis (Moerschbacher et al., 1990).

II. Lignin: Matrix Polymer of a Structural Barrier
Against Pathogen Ingress

Lignin may be viewed as a matrix polymer enclosing cellulosic and other cell wall materials, which renders the cellulosic fibrils inaccessible by microbial enzymes. Induced lignin deposited in wound- or infection boundary zones may act similarly providing a protective shield. The metabolic inertness of lignin is based on its structural complexity in a network of non-repetitive units linked by many different types of stable C–C and C–O bonds.

A. Biosynthesis

Lignin is a heteropolymer built from phenylpropanoid units. Phenylpropanoids are common structural elements of a number of other natural compounds including flavonoids, isoflavonoids, coumarins, stilbenes, phenolic esters and suberin (Hahlbrock and Scheel, 1987, 1989). In view of the common biosynthetic origin of these compounds the branch pathway committed to lignin biosynthesis is placed in the context of phenylpropanoid metabolism and its various end products, but is specified therein by structural formula of specific products (Fig. 1).

Following the deamination of phenylalanine to cinnamic acid by phenylalanine ammonia-lyase (PAL), four different classes of enzymes are involved in a successive modification of cinnamic acid: hydroxylases, O-methyltransferases, CoA ligases for activation and reductases/dehydrogenases (Hahlbrock and Grisebach, 1979; Grisebach, 1981; Gross, 1985; Higuchi, 1985). Particular characteristics and mechanistic details of these enzymes will be discussed below.

Specific for the lignin pathway are two consecutive reductive steps leading from the three hydroxy- and methoxylated cinnamoyl-CoA intermediates to the corresponding cinnamyl-alcohols, the direct monomeric precursors of lignin. These two steps are catalyzed by cinnamoyl-CoA reductase (CCR) and cinnamyl-alcohol dehydrogenase (CAD) via a cinnamaldehyde intermediate. The resulting cinnamyl-alcohols are polymerized in a random fashion from radicals created by wall-bound peroxidases. The non-enzymatic polymerization of mesomeric radical monomers leads to various types of linkages between the three building blocks 4-coumaryl-alcohol, coniferyl-alcohol and sinapyl-alcohol. The final structure of the polymer has been approached by several model and synthetic lignins, which seem to still not fully represent the three-dimensional structure of true lignin (Lewis et al., 1987). During polymerization the establishment of covalent bonds to both carbohydrates and proteins in the cell wall is likely.

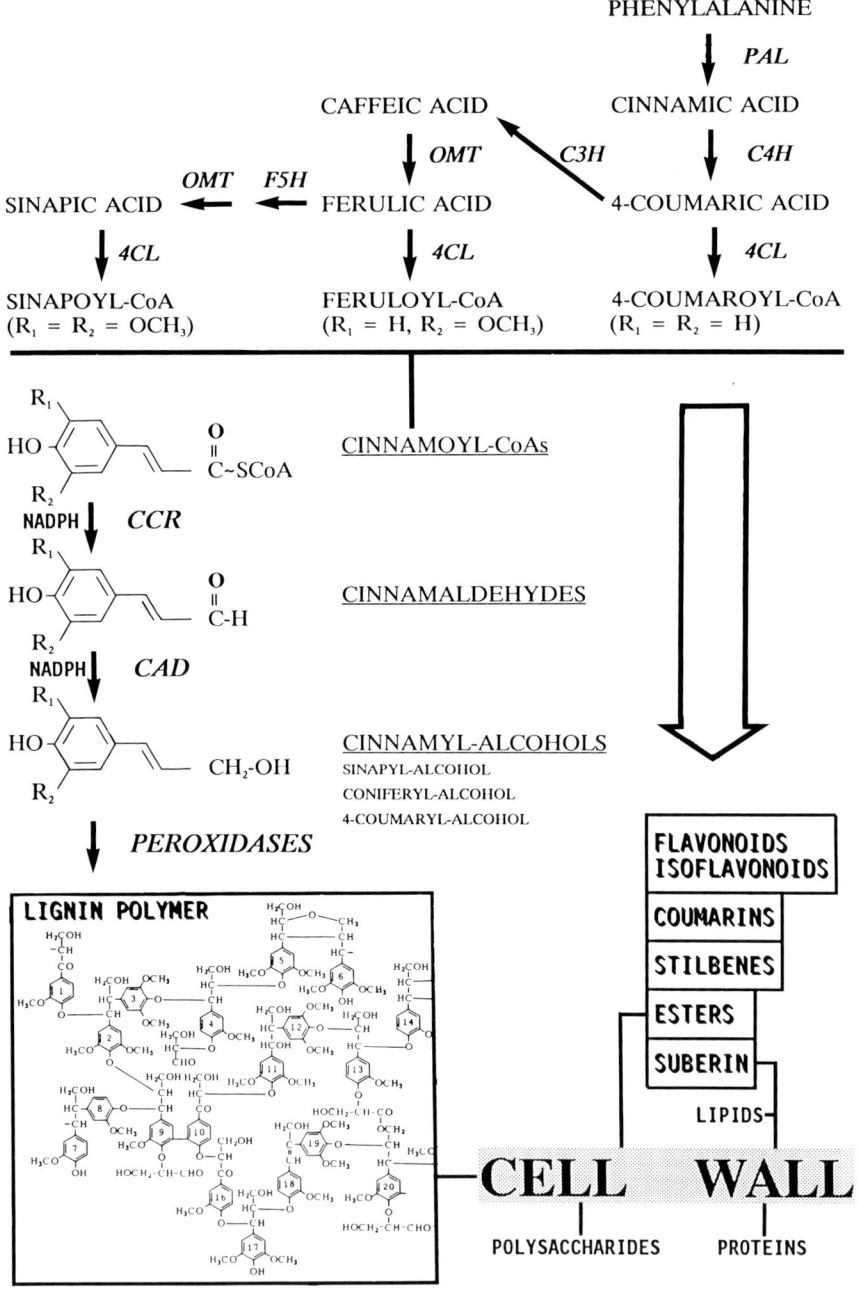

PHENYLALANINE

↓ *PAL*

CAFFEIC ACID   CINNAMIC ACID

↓ *OMT*   *C3H*   ↓ *C4H*

OMT *F5H*

SINAPIC ACID ◄── FERULIC ACID   4-COUMARIC ACID

↓ *4CL*   ↓ *4CL*   ↓ *4CL*

SINAPOYL-CoA   FERULOYL-CoA   4-COUMAROYL-CoA
($R_1 = R_2 = OCH_3$)   ($R_1 = H, R_2 = OCH_3$)   ($R_1 = R_2 = H$)

CINNAMOYL-CoAs

NADPH ↓ *CCR*

CINNAMALDEHYDES

NADPH ↓ *CAD*

CINNAMYL-ALCOHOLS
SINAPYL-ALCOHOL
CONIFERYL-ALCOHOL
4-COUMARYL-ALCOHOL

↓ *PEROXIDASES*

LIGNIN POLYMER

FLAVONOIDS
ISOFLAVONOIDS

COUMARINS

STILBENES

ESTERS

SUBERIN

LIPIDS

CELL WALL

POLYSACCHARIDES   PROTEINS

B. Lignin Determination

Owing to its heterogeneity and chemical complexity as well as to its association with other compounds in the cell wall, lignin is difficult to measure. The evidence for and against the presence of a true lignin polymer in particular plant species based on results from numerous chemical, histochemical and physical techniques of analysis has recently been re-evaluated (Lewis and Yamamoto, 1990). The verification of true lignin is difficult in developmental processes and this applies also to local lignin deposition in defense responses.

A wide range of methods for lignin determination mostly designed by wood scientists can be found in recent reviews (Vance et al., 1980; Kirk and Obst, 1988; Lewis and Yamamoto, 1990). Qualitatively, lignin can be determined by many color stains. The two most widely used are the Wiesner method, staining with phloroglucinol-HCl for cinnamaldehyde groups and the Mäule test specific for syringyl groups. An additional fairly selective color reaction is the chlorine-sulfite test. A stabilized diazonium salt (*p*-nitrobenzene-diazonium-tetrafluoroborate) stain for lignin has recently been developed (Barber and Ride, 1988). All these staining methods suffer from their inability to clearly distinguish between lignin and other related phenolic material. If combined and complemented by specific stains for associated material the histochemical methods are thought to provide sufficient proof for the presence of lignin or at least lignin-like material.

Chemical degradation or modification procedures are used to assay for lignin more specifically or more quantitatively. Rather harsh chemical treatments are usually required to derivatize and solubilize lignin components from insoluble polymers (Kirk and Obst, 1988). Nitrobenzene oxidation in an alkaline solution produces vanillin from guaiacyl lignin, vanillin plus syringaldehyde from guaiacyl/syringyl lignin and usually small amounts of 4-hydroxybenzaldehyde, which can then be quantified by gas chromatography. Certain flavonoids, tannins or suberins can interfere with the assay. The traditional thioglycolic acid method has been employed successfully by several authors to demonstrate lignin deposition in defense responses (Hammerschmidt, 1984; Doster and Bostok, 1988; Bruce and West, 1989). The procedure involves acid catalyzed derivatization with

---

Fig. 1. Schematic presentation of lignin biosynthesis with conversion steps committed to formation of the lignin polymer shown by structural formula. The enzymes involved are: phenylalanine ammonia-lyase (PAL), cinnamic acid 4-hydroxylase (C4H), hydroxycinnamic acid 3-hydroxylase (C3H), ferulic acid 5-hydroxylase (F5H), *O*-methyltransferase (OMT), 4-coumarate:CoA ligase (4CL), cinnamoyl-CoA reductase (CCR), cinnamyl-alcohol dehydrogenase (CAD) and peroxidase(s). Cinnamoyl-CoA intermediates are precursors of various other defense-related compounds (boxed products), some of which can be cross-linked and can, associated with lignin, enter the cell wall structural barrier. The polymeric structure of lignin is taken from Nimz (1974)

thioglycolic acid to produce base-soluble, acid insoluble lignin thioglycolate. Interference by non-lignin components like polyphenols is apparently minimal. After additional analyses to ensure purity the precipitate can be weighed or resolubilized and measured quantitatively by its UV absorption. Newly formed lignin has also been determined by quantifying radioactive nitrobenzene oxidation products from cell walls after feeding the plant with [^{14}C]phenylalanine or [^{14}C]cinnamic acid (Dean and Kuć, 1987). More detailed information concerning also the monomeric composition of the induced material could be obtained from cupric oxide oxidation of cell walls followed by gas chromatography and mass spectrometry analysis of oxidation products (Robertsen and Svalheim, 1990) (see below). A very promising physical technique is ^{13}C NMR, which can now also be applied in solid-state measurements (Lewis et al., 1987).

C. Deposition of Lignin or Lignin-like Material in Defense Responses

The monomeric composition of lignin deposited in cell walls during plant development differs between plant species. Gymnosperm lignin is made up mainly of coniferyl-alcohol, whereas in angiosperms lignin consists of both guaiacyl and syringyl units in approximately equal amounts. Grasses contain all three alcohol units. The composition may also vary with the age of the plant. Particular substrate specificities of CAD from either angiosperm or gymnosperm sources have been implicated in bringing about these structural differences (Lüderitz and Grisebach, 1981; Kutsuki et al., 1982a).

There have been reports in the literature that the composition of lignin induced in defense responses is different from the polymer in healthy tissue. Asada and Matsumoto (1972) determined a high content of syringyl units in vessel walls, but predominantly guaiacyl lignin in diseased parenchyma cells of japanese radish roots. Ride (1975) found a much higher proportion of syringyl- and 4-coumaryl units in induced lignin from infected wheat leaves than in untreated control tissue. A 4-coumaryl-rich, guaiacyl-poor material unlike normal angiosperm lignin was isolated from epidermal cell walls of several cucurbits lignified as a response to infection (Hammerschmidt et al., 1985). Phloroglucinol-positive material induced in oligogalacturonide-treated cucumber hypocotyls has similarly been classified as an atypical lignin mainly derived from 4-coumaryl-alcohol (Robertsen and Svalheim, 1990). However, the majority of the investigators cited in the following have not distinguished between particular compositions of vascular "developmental" lignin and pathogen-induced "defense" lignin and the material has not clearly been identified as lignin in all cases.

Wheat leaves challenged by fungi have been used extensively to characterize induced lignification. Ride (1975) has identified induced lignin by histochemical stains and ionization difference spectra in wound margins, yet only following infection of wounds with non-host pathogens. The rate of lignification was slower in response to the pathogenic *Septoria nodorum* and *Septoria tritici*. Improved methods using fluorescence microscopy and autoradiography have localized fungus-induced lignification in necrotic cells generated in the hypersensitive response (Ride and Pearce, 1979; Beardmore et al., 1983). A quantitative assay for "defense" lignin has recently been developed in this system (Barber and Ride, 1988). Wounded wheat leaves treated with fungal material or elicitors (see below) were stained with a diazonium salt and subsequently scanned with a densitometer to measure lignification in situ. The stain is fairly unspecific, since it will couple with phenolic groups. To overcome this difficulty preextractions were done to remove interfering phenolic material without disintegrating the tissue. The results compared fairly well to the established thioglycolic acid method of analysis, which cannot be applied in situ.

Rapid deposition of lignin at the surface of potato tuber discs following inoculation with non-pathogens was demonstrated by a combination of histochemical staining procedures with the thioglycolate assay (Hammerschmidt, 1984). More recent reports on newly formed "defense" lignin employing the latter method came from a cucumber/*Colletotrichum lagenarium* interaction (Dean and Kuć, 1987) and from almond bark tissue challenged by *Phytophthora* fungi (Doster and Bostok, 1988). Less well-defined lignin-like material was identified in carnations infected by *Fusarium* (Baayen, 1988) and in the interaction of one particular fungus (*Phytophthora cinnamomi*) with a large group of plant species ranging from fully susceptible to fully resistant (Cahill et al., 1989). "True" lignin has now also been detected in a cell suspension culture from castor bean elicited by pectic fragments (Bruce and West, 1989).

D. Elicitors of Lignification

Following the use of elicitors in the characterization of phytoalexin accumulation (Ebel and Scheel, 1992), compounds of fungal or chemical origin with an ability to elicit lignification have been sought for. Initially, elicitation of lignin deposition seemed more specific than that of phytoalexin accumulation, the former being induced almost exclusively by fungi except for mercuric chloride as an abiotic elicitor (Pearce and Ride, 1980). More recent investigations identified pectin, polygalacturonic acid and chitosan as elicitors of lignification in cucumber hypocotyls (Robertsen, 1986). An oligomer of 11 galacturonosyl units was the most potent inducer. Similarly, Bruce and West (1989) found galacturonide oligomers between 6 and 20 residues to elicit lignification in castor bean cell suspension cultures. Chitin, a component of fungal cell walls, and again chitosan as well as some related compounds were elicitor-active when applied to wounded wheat leaves (Pearce and Ride, 1982). Chitin oligomers, initially found to be non-reactive, were shown to possess significant elicitor activity after the inclusion of a prewounding step (Barber et al., 1989). Tetramers through hexamers were active, monomers and dimers were inactive. These chitin oligomers could easily be created from fungal chitin by the action of plant chitinases known to be rapidly induced in plant defense responses to fungal infection and in turn act as elicitors of lignification as proposed on the basis of work with cultured carrot cells (Kurosaki et al., 1986). Using the diazonium salt stain described above a "Onozuka R10" cellulase and a β-glucan from *Phytophthora megasperma* f. sp. *glycinea* capable of eliciting phytoalexin synthesis were identified in a screen for additional elicitors (Barber and Ride, 1988). However, the hepta-β-glucoside specific in phytoalexin formation in soybean was totally inactive. Three elicitors of glycoprotein nature have been described. The first one has been isolated from the wheat rust *Puccinia graminis* (Moerschbacher et al., 1986a). The carbohydrate portion of the molecule consisting mainly of mannose and galactose contains the elicitor-active part (Kogel et al., 1988). Preparations of this elicitor from two races of the pathogen containing different genes for avirulence did not exhibit specificity on near isogenic wheat lines differing in their genes for resistance (Moerschbacher et al., 1989). A phytotoxic glycoprotein from the barley pathogen *Rhynchosporium secale* has been reported to induce the accumulation of lignin-like material in cell walls of barley, bean and sunflower (Mazars et al., 1990). Asada and Matsumoto (1987) isolated a so-called lignification-inducing factor (LIF) from Japanese radish roots infected with *Peronospora*. LIF was tentatively identified as a glycoprotein. LIF seems to originate from the plant and not from the pathogen, since LIF activity could be readily extracted from homogenized uninfected and from heat treated plant tissue. It was postulated that LIF is localized in plant cell walls and liberated in an active form by stress factors. The LIF preparation from radish roots did elicit lignification in other plants, but only in those which have the basic capability to form wound lignin. Spraying LIF on cucumber leaves gave protection not only in the treated leaf, but resulted in a systemic protection of the plant against infection by *Colletotrichum lagenarium*.

III. Regulation of Enzymes in Lignin Biosynthesis

In this section the enzymes of lignin biosynthesis are discussed with respect to their possible involvement in the regulation of lignification with particular emphasis on their defense-related expression. Within a considerable number of enzymatic steps, all of which are now known, only the two reducing steps catalyzed by CCR and CAD are specific to lignification (Fig. 1). The question thus arises as to whether there are mechanisms for channelling phenylpropanoids into different branch pathways at one of the earlier enzymatic steps, e.g, by isoenzymes committed to a particular branch by substrate specificity or by end-product specific multienzyme complexes. It is also conceivable that different isoenzymes serve for developmental and defense needs.

A. Phenylalanine Ammonia-Lyase

Phenylalanine ammonia-lyase (PAL, E.C. 4.3.1.5) has been implicated as a control enzyme in the flux of phenylpropanoids (Camm and Towers, 1973). Local PAL activity correlated with lignin accumulation in infected wheat tissue (Maule and Ride, 1976). Today with a plethora of information available on PAL regulation (Jones, 1984; Dangl, this volume) as well as on number and structure of PAL genes in different systems (Cramer et al., 1989; Lois et al., 1989) it seems unlikely that PAL alone could exert effective control over the lignin pathway. Enzyme levels of PAL are extraordinarily sensitive to the physiological state of a plant irrespective of being brought about by environmental or developmental stimuli. Microbe-related defenses require a partitioning of cinnamate into a number of different pathways (Hahlbrock and Scheel, 1987, 1989).

On the other hand, if PAL is blocked by the specific chemical inhibitor L-α-aminooxy-β-phenylpropionic acid (AOPP), a substrate analogue of the enzyme, this clearly has dramatic consequences for the formation of the lignin polymer. Severe impairment of the vascular system was observed in AOPP-treated plants unless lignin precursors could be recruited from storage pools of sinapin or coniferin (Amrhein et al., 1983; Smart and Amrhein, 1985). AOPP was found to be a potent inhibitor of the accumulation of insoluble lignin-like polymers in tobacco reacting hypersensitively to viral infection (Massala et al., 1987) and in cucumber hypocotyls treated with the oligogalacturonide elicitor (Robertsen and Svalheim, 1990).

B. Hydroxylases

Cinnamic acid 4-hydroxylase (C4H, E.C. 1.14.13.11) is a cytochrome P450 dependent mixed function oxygenase introducing the p-hydroxyl group in 4-coumaric acid. Enzymes linked to cytochrome P450 have been difficult to purify (Bolwell, 1988). In elicitor-treated bean cell suspension cultures C4H activity increases concomitantly with PAL (Bolwell et al., 1985). Local increases of C4H and 4-coumarate:CoA ligase were found in infected wheat leaves, which do form lignin (Maule and Ride, 1983).

The subsequent enzymic introduction of a 3-hydroxyl group to form caffeic acid, which has earlier been thought to be performed by a phenolase of low specificity, has recently received new attention. A novel 3-hydroxylase activity was identified in extracts of cultured parsley cells, which requires Zn^{2+} and ascorbate (Kneusel et al., 1989). The substrate is 4-coumaroyl-CoA, not 4-coumaric acid. A possible requirement of a modification of 4-coumaric acid prior to 3-hydroxylation had already been suggested, yet at the opposite end of the molecule, with 5-O-(4-coumaroyl) shikimate as substrate for an elicitor-inducible 3-hydroxylase (Heller and Kühnl, 1985). Boniwell and Butt (1988) reported on 3-hydroxylation of 4-coumaric acid to caffeic acid with an requirement of NADH and FAD by a particulate preparation from potato tubers showing some characteristics of a phenolase. Ascorbate could not substitute for the other electron donors.

A novel mechanism has been proposed for the recruitment of the Zn^{2+} dependent enzyme described above in a defense situation: A shift in cytoplasmic pH observed in elicitor-treated parsley cells may rapidly activate this enzyme from an inactive form obviating the need for de novo enzyme synthesis (Kneusel et al., 1989).

Only a single study is available for the 5-hydroxylation of ferulic acid to the direct precursor of sinapic acid. This reaction is performed by a cytochrome P450 mixed function oxygenase (Grand, 1984). Comparison of regulatory properties and tissue distribution with C4H in xylem tissue of poplar stems suggested that these hydroxylase activities depend on two distinct cytochrome P450 systems.

C. O-Methyltransferases

S-adenosyl-L-methionine (S-AdoMet) dependent o-diphenol-O-methyl-transferases (OMTs, E.C. 2.1.1.6) catalyze the O-methylation of substrates like caffeic acid or 5-hydroxyferulic acid. Three such enzymes, which are serologically related, have been purified from tobacco leaves (Hermann et al., 1987). Rapid purification techniques allowed a direct comparison of increases in OMT activity and enzyme protein amounts. Parallel changes indicated a de novo-synthesis of all three different OMT enzymes during the hypersensitive response to tobacco mosaic virus (Dumas et al., 1988). OMT cDNAs are being isolated and should provide new tools for a molecular analysis of OMT gene expression (B. Fritig, pers. comm.). One of the tobacco OMTs was earlier shown to be the major enzyme in healthy tissue and to exhibit specificity towards phenylpropanoids, possibly channelling them into vascular lignin. The two other enzymes, which are induced in the hypersensitive response, accept a much wider array of o-diphenols. This may indicate that they are less committed to lignin like end-products and produce a complex mixture of methylated phenolic material (Legrand et al., 1978; Collendavello et al., 1981). Carrot cell cultures contain an elicitor-inducible S-AdoMet dependent caffeoyl-CoA 3-O-methyltransferase (Kühnl et al., 1989). The enzyme has a pronounced substrate specificity for the caffeoyl-CoA ester; less than 2% of the activity was found with free caffeic acid. An enzyme of similar, but not identical specificity has been isolated from cultured parsley cells (Pakusch et al., 1989). It is not yet clear, if modification of o-diphenols prior to O-methylation is a more general feature or confined to these particular systems. In infected wheat leaves both a caffeic acid OMT and a 5-hydoxyferulic acid OMT had elevated levels of extractable enzyme activities, but there was no deviation from the ratio of reaction products found with the enzymes from healthy tissue (Maule and Ride, 1976). It appears that the different monomeric composition of pathogen-induced "defense" lignin and vascular "developmental" lignin reported from this system (Ride, 1975) does not arise from a control in this step. On the other hand, distinct OMTs typical for either gymnosperm or angiosperm vascular lignin seem to exist (Kuroda et al., 1981).

D. CoA Ligases

Prior to directing the substituted 4-coumaric acids into the different branch pathways, an activation by way of a CoA-ester intermediate is required. 4-Coumarate:CoA ligase (4CL, E.C. 6.2.1.12) activities in xylem tissue of gymnosperms and angiosperms were readily found with 4-coumaric acid and ferulic acid as substrates. Surprisingly, quite a number of angiosperms did not convert sinapic acid—a reaction thought to be indispensable for syringyl lignin (Kutsuki et al., 1982b). In soybean cell suspension cultures two types of 4CL isoenzymes have been isolated, each with characteristic properties (Knobloch and Hahlbrock, 1975). Isoenzyme I displayed preference for 4-coumaric acid, ferulic acid and sinapic acid suggesting an essential function in lignin formation. A second isoenzyme had a high affinity towards 4-coumaric acid and caffeic acid possibly channelling the flux of phenylpropanoids into the flavonoid, isoflavonoid phytoalexin or caffeic ester branches. In cultured parsley cells, which do not produce isoflavonoid phytoalexins, the studies on 4CL have now progressed to the gene level. Two enzyme isoforms exist each of which is encoded by a single copy gene. The two genes have a very similar structure. They are both activated to a similar extent by UV irradiation and fungal elicitor. A differential role in phenylpropanoid branch pathways is unlikely in this system, since expression of the genes in *Escherichia coli* yielded isoenzymes with virtually identical catalytic properties (Lozoya et al., 1988). Similarly detailed molecular studies in other systems, where isoenzymes with different specificities have been found [soybean (Hahlbrock and Knobloch, 1975), petunia (Ranjeva et al., 1976), pea (Wallis and Rhodes, 1977) and poplar (Grand et al., 1983)] are needed on this important branch point enzyme of phenylpropanoid metabolism.

E. Reductases/Dehydrogenases

The two lignin-specific enzymes catalyzing the reduction of cinnamoyl-CoA esters to cinnamyl-alcohols have received surprisingly little attention as targets for gene isolation and characterization, and it is only recently that studies using the methods of molecular biology were initiated. Early evidence for a reduction of ferulic acid to coniferyl-alcohol via a conifer-aldehyde intermediate came from tracer experiments. A direct reduction of the acid has, for thermodynamic reasons, always been considered unlikely but, unexpectedly, a carboxylic acid reductase requiring tungsten as a cofactor has now been described from a bacterial source (White et al., 1989).

In plants it is still generally accepted that the reduction requires a carboxyl-activated intermediate, a CoA ester, and proceeds via a cinnamaldehyde. Three distinct enzymatic activities, each dependent on specific cofactors, were demonstrated in *Forsythia*, namely ferulic acid: CoA ligase, feruloyl-CoA reductase and coniferyl-alcohol oxidoreductase (dehydrogenase) (Gross et al., 1973).

Cinnamoyl-CoA reductases (CCR, E.C. 1.2.1.44) have been purified from three different sources. The enzyme from soybean cell suspension cultures exhibits high affinity to feruloyl-CoA and sinapoyl-CoA (Wengenmayer et al., 1976). No reaction was observed with acetyl CoA. An improved purification procedure was developed, and apparently homogeneous enzymes from soybean and spruce could be directly compared (Lüderitz and Grisebach 1981). Pronounced differences in substrate specificities were found, which probably are reflected in the different monomeric compositions of lignin from a dicotyledoneous angiosperm and a gymnosperm. Sinapoyl-CoA was a substrate only for the soybean, but not for the spruce enzyme. The enzymes from both sources converted 4-coumaroyl-CoA only poorly (Lüderitz and Grisebach, 1981). Characteristics similar to the soybean CCR were found for a CCR from poplar, which is a woody angiosperm (Sarni et al., 1984). In all cases studied the CCR had an exclusive or almost exclusive requirement for NADPH as a cofactor. Highest extractable enzyme activities were obtained from tissue undergoing xylogenesis (stems, cambial sap). An induction of this enzyme in defense responses has not been shown. Stimulation of CCR activity in neoformed bean calli after infection with *Agrobacterium rhizogenes* is considered not to be a defense response, since it was dependent on a bacterial strain harboring phytohormone genes. Elevated levels of phytohormones in the bean calli provided by the bacterial genes seemed responsible for an onset of differentiation and xylogenesis associated with stimulation of lignin biosynthetic enzymes (Grima-Pettenati et al., 1989).

Cinnamyl-alcohol dehydrogenase (CAD, E.C. 1.1.1.195) catalyzing the subsequent reduction step has often been studied in conjunction with the CCR. It is possible that in vivo the two enzymes are associated in a multienzyme complex, although on affinity columns such a reductase/dehydrogenase interaction could not be detected (Lüderitz and Grisebach, 1981). Moreover, there have been no particular problems reported in the separation of the enzymes during purification. Since additional information is not available for CCR, but only for CAD, this second of the two specific enzymes in lignin biosynthesis will be discussed in detail separately in Sect. IV.

F. Peroxidases

The polymerization of lignin monomers within the cell wall is initiated by oxidation of cinnamyl-alcohols to the mesomeric phenoxy-radicals, which then combine randomly to a highly heterogeneous mixture of different C–C and C–O bonds. These free radical species are created by wall-bound peroxidases in the presence of hydrogen peroxide. Peroxidases (E.C. 1.11.1.7) are a large group of isoenzymes with an extreme range of isoelectric points, serving a multitude of functions (van Huystee, 1987). Their isoenzyme profile is different in particular physiological situations. At least 12 distinguishable isoenzymes have been characterized from tobacco, which fall into three subgroups: the anionic, the moderately anionic and the cationic isoenzymes. Each group is thought to have a different function in the cell. The function of cationic isoenzymes is as yet unclear. They might provide H_2O_2 for other peroxidases. H_2O_2 is produced rapidly in elicitor-treated cell cultures and might act as a second messenger for phytoalexin production (Lindner et al., 1988; Apostol et al., 1989). The moderately anionic peroxidases of tobacco are highly expressed in wounded stem tissue (Lagrimini and Rothstein, 1987). They have only a moderate activity towards lignin precursors and may be involved in wound healing and suberization.

Anionic peroxidases are the isoenzymes most likely involved in lignin formation. They are wall-associated and have a high affinity for cinnamyl-alcohols in vitro, but can probably also crosslink extensin monomers and feruloylated polysaccharides. One of the two tobacco anionic peroxidases attributed to lignin · polymerization has been cloned and sequenced (Lagrimini et al., 1987). Its mRNA was readily detectable in lignifying stem tissue, but wounding of the stem did not result in an appreciable increase in mRNA amounts. In contrast, mRNA encoding a highly anionic peroxidase from potato implicated in suberization was wound-inducible (Roberts et al., 1988). The same highly anionic peroxidase cDNA probe was used to show differential activation by fungal elicitor of a homologous gene in tomato cell lines resistant and susceptible to *Verticillium albo-atrum* (Mohan and Kolattukudy, 1990).

Biochemical studies in the wheat/stem rust pathosystem revealed a general increase in peroxidase activities with notable changes in particular isoenzymes (Flott et al., 1989). An attempt to find lignification-specific peroxidases by using substrates of different specificity (guaiacol, syringaldazin, and coniferyl-alcohol) failed. Apparently all isoperoxidases converted all substrates, at least in vitro (Moerschbacher et al., 1988). Recent evidence for differential accumulation of anionic and cationic peroxidases was obtained in infected barley leaves and in elicitor-treated castor bean cell cultures (Kerby and Somerville, 1989; Bruce and West, 1989). Induction of one anionic and four cationic isoperoxidases correlated with the deposition of lignin in the castor bean system. Apart from such correlative evidence a specific assignment for one or several plant isoperoxidases to lignification has yet to be demonstrated.

Despite the immense amount of information accumulated in recent years on lignin biosynthetic enzymes an overall picture on enzyme regulation in lignin biosynthesis is still lacking. Most of these enzymes have now been shown to be elicitor- or pathogen-inducible, ranging from very minor to considerable increases in extractable activities. Unfortunately, these studies have been done in many different experimental systems and the information on specific control points is still too fragmentary to define "lignin-specific" isoenzymes, which are suggested in several cases by narrow substrate specificities or isoenzymes committed to either developmental or defense tasks. From the biochemical data available a characterization of soybean 4CL genes and tobacco OMT genes should provide valuable further insights into regulatory mechanisms in lignin biosynthesis.

IV. Cinnamyl-Alcohol Dehydrogenase—a Committed Enzyme

A. Alcohol Dehydrogenases

The conversion of aldehydes to alcohols or the opposite reaction is a frequently observed reaction in metabolism and not limited to lignification. The alcohol dehydrogenases (ADHs) reversibly catalyzing this general reaction have been extensively characterized from many sources. While insects form a separate class of highly different ADHs, the zinc-containing long-chain ADHs in yeast, fungi, mammals and plants are of common ancestry. A large number of DNA-deduced ADH protein sequences are available and have been compared (Jörnvall et al., 1987). From the extensive variations evolutionary relationships can be discerned and the general importance of particular amino acids can be recognized. Strictly conserved residues include two of several ligands to the active site zinc atom, a Cys-46 and a His-67. Furthermore, many glycines are invariable; among them Gly-199, Gly-201 and Gly-204 have been ascribed special roles in the NAD coenzyme binding fold. Other residues may influence the size of the substrate binding pocket and thereby substrate specificity. A correlation of amino acid substitutions with changes in substrate specificity of different human ADH isoenzymes has been attempted (Eklund et al., 1987). The dimeric mammalian and the tetrameric yeast ADHs are capable of oxidizing a wide range of aliphatic and aromatic alcohols. Some human class I ADH isoenzymes even have a preference for bulkier substrates and accept short chain alcohols only poorly (Wagner et al., 1983). Yeast ADH readily converts aromatic aldehydes to the corresponding alcohols (Long et al., 1989).

In plants ADH is known to be involved in the survival during flooding as part of the anaerobic response (Sachs and Freeling, 1978). Up to three genes code for a series of isoenzymes not all of which are expressed in stress responses (Tihanyi et al., 1989). ADH gene transcripts are among those mRNAs, which are induced in potato by anaerobiosis, but also by pathogen-related fatty acid elicitors and salicylic acid (Matton et al., 1990). The metabolic role of elicitor-induced ADH is unknown.

Plant ADHs have rarely been analysed for substrate specificity. In the cases where substrates other than ethanol have been tested, several plant

ADHs seem to be fairly specific for short-chain aliphatic alcohols. They are not active with aromatic alcohols (Felder et al., 1973; Tihanyi et al., 1989).

B. Cinnamyl-Alcohol Dehydrogenases

In 1973 Davies et al. isolated three different ADHs from potato tubers, one active with NAD (the preferred coenzyme for the yeast, mammalian and plant ADHs described above) and aliphatic alcohols, one active with NADP and terpene alcohols and one active with NADP and various aromatic alcohols including cinnamyl-alcohols. This latter enzyme was suggested to be involved in the metabolic route from cinnamic acid to lignin. Mansell et al. (1976) have performed an extended comparative screening of some 90 plant species for ADHs with preference for coniferyl-alcohol. Multiple forms of ADHs were detectable on starch gels with ethanol and NAD substrates. In contrast, only one form of ADH was commonly found with coniferyl-alcohol and NADP. These specific cinnamyl-alcohol dehydrogenase (CAD) bands were distinct from ADH bands and found only with NADP as coenzyme. Subsequently, CADs with a narrow specificity for cinnamyl-alcohols were purified from soybean cell suspension cultures (Wyrambik and Grisebach, 1975, 1979), spruce (Lüderitz and Grisebach, 1981) and poplar (Sarni et al., 1984). Two isoenzymes were isolated from soybean and were shown to convert specifically coniferyl-alcohol (isoenzyme 1) or cinnamyl-alcohols and a number of substituted cinnamyl-alcohols (isoenzyme 2). Neither isoenzyme reacted with benzyl-alcohol, anisic alcohol or ethanol. Iso-enzyme 2, consisting of two subunits with apparent molecular weights of 40,000, as judged from sodium dodecylsulfate PAGE, bears resemblance to the zinc-containing long-chain ADHs from yeast and horse liver with respect to its subunit molecular weight, zinc content, steady state kinetics and stereospecificity with regard to coenzyme and substrate (Wyrambik and Grisebach, 1979). Similar characteristics were found for a poplar CAD (Sarni et al., 1984). There is, however, one important difference to the evolu-tionarily conserved ADH class, which was not reported by these authors. The CADs from all sources analyzed to date have an exclusive or almost exclusive requirement for NADP as a cofactor as opposed to the ADHs, which have a strong preference for NAD (Jörnvall et al., 1987). It is also noteworthy that 1 mol zinc/mol enzyme was found by Wyrambik and Grisebach (1979), whereas most ADHs contain an additional zinc atom outside the active site (Vallee and Auld, 1990).

The picture arising from these biochemical studies of CAD is that of an enzyme strictly committed to the ultimate step of forming the lignin monomers, as concluded from its narrow substrate specificity and its characteristic preference for the NADP coenzyme. Information on the pri-mary structure of CAD is needed to assess its relationship to the conserved ADH class introduced above.

This task has been approached in cell suspension cultures of bean challenged by a fungal elicitor from *Colletotrichum lindemuthianum*, the causal agent of the anthracnose disease. This system has been employed in molecular studies of a number of active plant defenses including phytoalexin accumulation and induction of cell wall proteins. (Lamb et al., 1989). Initial experiments showed a pronounced increase in extractable CAD activity in elicitor-treated bean cells (Grand et al., 1987). This was preceded by the induction of translatable mRNA for a polypeptide (size 65 kDa) thought to represent CAD, based on in vitro translation followed by immunoprecipitation with an antibody raised against a CAD purified to apparent homogeneity from young poplar stems (Grand et al., 1987). The same antibody was used for immunoscreening a λ gt11 cDNA library derived from elicitor-treated bean cells (Walter et al., 1988). Antibody-positive clones were further analysed by hybrid-selected translation and translation of partial in vitro transcripts. Both methods confirmed the antibody recognition of the polypeptide encoded by λCAD4, which was then subjected to a sequence analysis. The protein deduced from the single long open reading frame had a size of 65 kDa in accordance with the size of the in vitro translation product (Grand et al., 1987) and a CAD purified from bean. In a search for the two strictly conserved amino acid residues of the zinc-binding domain of ADHs, a cysteine and a histidine were indeed found in the deduced amino acid sequence at similar respective positions as in ADHs [Cys-46, His-67 in ADHs (Jörnvall et al., 1987) and Cys-51, His-73 in the putative CAD (Walter et al., 1988)]. Moreover, Gly residues are located in the deduced sequence at positions 193, 195, and 198, comparable to NADbinding glycines at positions 199, 201 and 204 in ADHs. Outside these regions the sequence had no significant homology to ADH consensus sequences. A putative NADP binding site could not be identified, since no consensus sequence was available from NADP dependent ADHs, which exist in bacteria (Lamed and Zeikus, 1981) and probably also in plants (Jaaska, 1984). However, recent sequence information from a NADP-dependent ADH of a thermophilic bacterium indicates that the Gly-X-Gly-X-X-Gly coenzyme binding motif is also found in NADP-linked ADHs (Peretz and Burstein, 1989). The similarities in important cofactor binding sites were taken as a confirmation that the λCAD4 clone encoded CAD. It was subsequently found that expression of the gene corresponding to λCAD4 is correlated with lignification in at least one system (Grima-Pettenati et al., 1989).

We have now isolated and sequenced the bean gene corresponding to the λCAD4 cDNA. The coding sequence is interrupted by an unusually high number of 18 small introns (M.H. Walter and D. Hess, unpubl.). The promoter region contains two boxes, which exhibit 75% homology to consensus sequences for elicitor-responsive *cis*-elements (M.H. Walter and D. Hess, unpubl.; Lois et al., 1989) suggesting a defense function for this gene consistent with its rapid activation in elicitor treated bean cell cultures (Walter et al., 1988). The defense-related expression may be complemented by hitherto unknown expression signals for developmental activation.

However, we have now found a 73% sequence homology of the protein deduced from the λCAD4 clone to a recently published amino acid sequence of maize malic enzyme (Rothermel and Nelson, 1989), which raises doubts about the identity of the λCAD4 cDNA (Walter et al., 1990). Studies on expression of the λCAD4 sequence in *Escherichia coli*, followed by assays of enzyme activities, have been initiated and are expected to soon clarify the true identity of the λCAD4 cDNA. Furthermore, two other cDNAs encoding products recognized by the CAD antibody are being investigated.

If clone λCAD4 should turn out to encode a bean malic enzyme, it might still be a valuable tool in future studies, even in relation to lignin biosynthesis. NADP-dependent malic enzyme (E.C. 1.1.1.40) catalyzes the oxidative decarboxylation of malate to pyruvate generating CO_2 and reducing equivalents (Wedding, 1989). If the antibody raised against puri-fied CAD is directed to a NADP binding site epitope, it may well recognize other NADP-dependent enzymes including malic enzyme. Malate is a mobile storage form of CO_2 and reducing equivalents (Lance and Rustin, 1984; McKelvey and Fioravanti, 1984), the latter of which are needed for the reduction of cinnamoyl-CoAs to cinnamyl-alcohols. Sources of NADPH for lignin synthesis have long been sought for (Pryke and ap Rees, 1977). The generally large malate pools in higher plants might provide such reducing power both for defense and development.

While there arises doubt now about the identity of the λCAD4 clone, other conclusions drawn on CAD remain valid. With respect to elevated levels of CAD enzyme activity in defense situations the available information is limited. Apart from the elicitor-challenged bean cell culture system (Grand et al., 1987) an increase in CAD activity was observed in wheat leaves injected with a glycoprotein elicitor from a wheat rust fungus (Moerschbacher et al., 1986b, 1989). However, elevated levels of CAD activity in wheat leaves sprayed with live fungal uredospores might have been brought about by the fungus, since axenically grown vegetative mycelium of the stem rust did exhibit CAD activity (Moerschbacher et al., 1988).

The question recently raised (Walter et al., 1988), as to whether there are separate CAD genes for "developmental" and "defense" lignin is still un-answered. The different compositions of vascular and pathogen-induced lignin might be controlled in this step. A "defense" CAD gene might also be one of a number of genes from phenylpropanoid metabolism concomitantly activated in defense. It now also seems possible that the delivery of reducing equivalents is induced in defense responses and may play a role in the regulation of the two last steps in the biosynthesis of lignin monomers. For any specific manipulation of the lignin pathway the CAD gene(s) remain an interesting target due to the strict commitment of CAD to the lignin branch pathway.

V. Possible Roles of Lignin Monomers and Polymers in the Structural Barrier

Cell walls contribute to plant defenses in a variety of ways. It has been discussed here that they act as natural barriers, which can be reinforced upon pathogen attack or in other stress situations. They also play an important role in recognition and communication events involved in the initiation of both specific and non-specific plant responses to pathogens including changes in wall composition. Elicitors of lignification of galacturonide, glucan or glycoprotein nature are believed to be released from plant cell walls by the pathogen or, possibly via the action of rapidly induced plant hydrolases like chitinases, from the chitin-containing cell walls of fungi. Many other exchanges of signal molecules are likely to occur. Coniferyl-alcohol is a potent inducer of the *vir-* region of *Agrobacterium tumefaciens*, approaching the activity of the well-known inducer acetosyringone (Spencer and Towers, 1988). It has been suggested that *Agrobacterium* may be capable of sensing cells which are undergoing cell wall rigidification by the induction of lignification and target those defending and possibly damaged cells for transformation. Dehydro-diconiferyl glucosides have been shown to exert regulatory effects in plants exhibiting cytokinin-like activities (Binns et al., 1987). Such regulatory properties of lignin monomers or derivatives thereof may provide an alternative explanation for the rapid stimulation of CAD activity in elicitor-treated bean cell cultures (Grand et al., 1987). Protein components of the cell wall (hydroxyproline-rich glycoproteins) are induced with a delay in this system (Lamb et al., 1989).

The rapidity of CAD induction might have other reasons as well. Both coniferaldehyde and coniferyl-alcohol have antifungal properties when tested in vitro (Keen and Littlefield, 1979; Hammerschmidt and Kuć, 1982). It is possible that the two compounds can act as defense components in their own right comparable to phytoalexins. The compounds seem to attain significant steady-state levels in infected flax leaves but not in wounded or healthy tissue, supporting a role as phytoalexins in the flax system. However, such a phytoalexin potential of lignin monomers has only been reported in two cases, and it is possible that the monomers have been caught in the process of polymerization. Whatever the extent of polymerization is, fungitoxic effects of monomers during the construction of the barrier can add to a successful pathogen control.

The true polymeric nature of a major part of the phenolic material deposited in defense responses is clearly suggested by positive results with lignin derivatization and modification procedures and by its insolubility. The histochemical stains are unable to distinguish polymers from monomers. What is lacking is an assessment of the chemical nature and the frequency of intermonomer linkages in pathogen-induced lignin. Such intermonomer bonds as well as crosslinks to associated material may profoundly affect the effectiveness of the structural barrier. In vascular lignin this question has been successfully approached by the use of ^{13}C-NMR spectroscopy (Nimz, 1974; Lewis et al., 1987).

The newly formed structural polymer may also consist of components other than lignin monomers. "Lignosuberized" boundary zones are common in wound periderms. The distinction between lignin and suberin is frequently a matter of controversy. Consistent with the structural model proposed by Kolattukudy (1980) phenolic domains very lignin-like in nature are suggested for suberin by a positive reaction of suberized tissue in the lignin-specific thioglycolic acid procedure (Hammerschmidt, 1985). Moreover, nitrobenzene oxidation of suberin-containing potato tissue yielded typical lignin degradation products (Cottle and Kolattukudy, 1982). The almost complete lack of syringaldehyde generation in this procedure indicated that the material, although lignin-like, was different from the typical angiosperm lignin found in vascular tissue, but it might be similar to the pathogen-induced "defense-type" lignin of particular monomeric composition found in wheat leaves, japanese radish and cucumber hypocotyls (Ride, 1975; Asada and Matsumoto, 1972; Robertsen and Svalheim, 1990). Perhaps the best distinction between the two polymers can be accomplished by complementing the thiogycolic acid procedure with specific stains for the aliphatic (fatty acid) domains of suberin. By a positive reaction with thioglycolic acid but negative results with the lipid-specific stains Sudan black B and Nile blue, Bruce and West (1989) identified induced lignin, but not suberin, in castor bean cell cultures. Other polymers of phenolic nature may be present in defense barriers (Matern and Kneusel, 1988) and may be linked to cell wall carbohydrates or cell wall proteins. The diversity of possible crosslinks between wall polymers has been discussed by Fry (1986).

VI. Perspectives

The evidence for a key role of structural barriers in disease resistance is still largely correlative. A more precise understanding of the biochemistry of wall rigidification is needed, e.g., a clearer distinction of phenolic wall polymers. This involves also the distinction between the structures of vascular and pathogen-induced lignin and the extent of polymerization.

Creation and use of DNA probes will be important to evaluate the contribution of particular defense reactions to disease resistance. Anti-sense RNA or ribozyme strategies may be tools to create artificial mutants. Suitable targets for such a specific block in defense gene product accumulation are genes committed to a particular branch pathway such as the CCR and the CAD for the lignin pathway. Alternatively, particular isogenes within the large gene family of peroxidases can be targeted by gene specific probes once assignments of particular functions to single isoperoxidases can be made. It may also be feasible to constitutively express lignin biosynthetic genes, which are normally selectively activated in defense or development and investigate any alterations in the phenotype of the transgenic plants. One such experiment introducing a bean PAL gene into tobacco under the control of the CaMV 35S promoter resulted in severe detrimental effects on the recipient plant (Y. Elkind and C.J. Lamb, pers. comm.). The CaMV 35S promoter has also been used to drive high expression levels of an anionic peroxidase cDNA implicated in lignin formation (Lagrimini et al., 1987) in transgenic tobacco. The transformed plants had the unique phenotype of chronic severe wilting, which was initiated at the time of flowering (Lagrimini et al., 1990). Both results indicate that it may be necessary to limit the expression of putative defense genes to certain plant tissues, developmental stages or defense situations, in particular if the enzymes encoded also serve important functions in development.

Note added in proof. Wolfgang Schuch and coworkers have recently isolated authentic CAD cDNA clones from tobacco unrelated to the λ CAD4 clone. The deduced CAD protein exhibits approximately 30% sequence homology to yeast ADH. Putative zinc binding amino acids and NAD(P)H binding domain are found at the expected positions as predicted in this review (Knight et al. (1992) Plant Mol Biol, submitted). Using a CAD cDNA as a probe I have now isolated a CAD gene and its promoter from a tobacco library.

Acknowledgements

The author's unpublished work reported herein was supported in part by the Vater und Sohn Eiselen Stiftung, Ulm, Federal Republic of Germany, within a grant awarded to D. Hess and by the Commission of the European Communities (ECLAIR-AGRE 0021). During this work the author was recipient of a research position donated by the Vater und Sohn Eiselen Stiftung.

VII. References

Aist JR (1983) Structural responses as resistance mechanisms. In: Bailey JA, Deverall BJ (eds) The dynamics of host defense. Academic Press, Sidney, pp 33–70

Amrhein N, Frank G, Lemm G, Luhmann HB (1983) Inhibition of lignin formation by L-α-aminooxy-β-phenylpropionic acid, an inhibitor of phenylalanine ammonia-lyase. Eur J Cell Biol 29: 139–144

Ampomah YA, Friend J (1988) Insoluble phenolic compounds and resistance of potato tuber discs to *Phytophthora* and *Phoma*. Phytochemistry 27: 2533–2541

Apostol I, Heinstein PF, Low PS (1989) Rapid stimulation of an oxidative burst during elicitation of cultured plant cells. Plant Physiol 90: 109–116

Asada Y, Matsumoto I (1972) The nature of lignin obtained from downy mildew-infected japanese radish root. Phytopathol Z 73: 208–214

Asada Y., Matsumoto I (1987) Induction of disease resistance in plants by a lignification-inducing factor. In: Nishimura S, Vance CP, Doke N (eds) Molecular determinants of plant diseases. Japan Scientific Society Press, Tokyo, Springer, Berlin Heidelberg New York Tokyo, pp 223–231

Baayen RP (1988) Responses related to lignification and intravascular periderm formation in carnations resistant to *Fusarium* wilt. Can J Bot 66: 784–792

Barber MS, Ride JP (1988) A quantitative assay for induced lignification in wounded wheat leaves and its use to survey potential elicitors of the response. Physiol Mol Plant Pathol 32: 185–197

Barber MS, Bertram RE, Ride JP (1989) Chitin oligomers elicit lignification in wounded wheat leaves. Physiol Mol Plant Pathol 34: 3–12

Beardmore J, Ride JP, Granger JW (1983) Cellular lignification as a factor in the hypersensitive resistance of wheat to stem rust. Physiol Plant Pathol 22: 209–220

Binns AN, Chen RH, Wood HN, Lynn DG (1987) Cell division promoting activity of naturally occuring dehydrodiconiferyl glucosides: do cell wall components control cell division? Proc Natl Acad Sci USA 84: 980–984

Bolwell GP (1988) Synthesis of cell wall components: aspects of control. Phytochemistry 27: 1235–1253

Bolwell GP, Robbins MP, Dixon RA (1985) Metabolic changes in elicitor-treated bean cells. Enzymic responses associated with rapid changes in cell wall composition. Eur J Biochem 148: 571–578

Boniwell JM, Butt VS (1988) Flavin nucleotide-dependent 3-hydroxylation of 4-hydroxyphenylpropanoid carboxylic acids by particulate preparations from potato tubers. Z Naturforsch 41c: 56–60

Bruce RJ, West CA (1989) Elicitation of lignin biosynthesis and isoperoxidase activity by pectic fragments in suspension cultures of castor bean. Plant Physiol 91: 889–897

Cahill D, Legge, N, Grant B, Weste G (1989) Cellular and histological changes induced by *Phytophthora cinnamoi* in a group of plant species ranging from fully susceptible to fully resistant. Phytopathology 79: 417–424

Camm EL, Towers GHN (1973) Phenylalanine ammonia-lyase. Phytochemistry 12: 961–973

Collendavelloo J, Legrand M, Geoffroy P, Barthelemy J, Fritig B (1981) Purification and properties of the three *o*-diphenol-*O*-methyltransferases of tobacco leaves. Phytochemistry 20: 611–616

Cottle W, Kolattukudy PE (1982) Biosynthesis, deposition, and partial characterization of potato suberin phenolics. Plant Physiol 69: 394–399

Cramer CL, Edwards K, Dron M, Liang X, Dildine S, Bolwell GP, Dixon RA, Lamb CJ, Schuch W (1989) Phenylalanine ammonia-lyase gene organization and structure. Plant Mol Biol 12: 367–383

Davies DD, Ugochukwu EN, Patil KD, Towers GHN (1973) Aromatic alcohol dehydrogenases from potato tubers. Phytochemistry 12: 531–536

Dean RA, Kuć J (1987) Rapid lignification in response to wounding and infection as a mechanism for induced systemic protection in cucumber. Physiol Mol Plant Pathol 31: 69–81

Delmer DP, Stone BA (1988) Biosynthesis of plant cell walls. In: Preiss J (ed) The biochemistry of plants, vol 14, carbohydrates. Academic Press, San Diego, pp 373–420

Doster MA, Bostok RM (1988) Quantification of lignin formation in almond bark in response to wounding and infection by *Phytophthora* species. Phytopathology 78: 473–477

Dumas B, Legrand M, Geoffroy P, Fritig B (1988) Purification of tobacco *O*-methyltransferases by affinity chromatography and estimation of the rate of synthesis of the enzymes during the hypersensitive reaction to virus infection. Planta 176: 36–41

Ebel J, Scheel D (1992) Elicitor recognition and signal transduction. In: Boller T, Meins F (eds) Genes involved in plant defense. Springer, Wien New York, pp 183–205 [Dennis ES et al (eds) Plant gene research. Basic knowledge and application]

Eklund H, Horjales E, Vallee BL, Jörnvall H (1987) Computer-graphics interpretations of residue exchanges between α, β and γ subunits of human-liver alcohol dehydrogenase class I isozymes. Eur J Biochem 167: 185–193.

Felder MR, Scandalios JG, Liu, EH (1973) Purification and partial characterization of two genetically defined alcohol dehydrogenase isozymes in maize. Biochim Biophys Acta 318: 149–159

Flott BE, Moerschbacher BM, Reisener H-J (1989) Peroxidase isoenzyme patterns of resistant and susceptible wheat leaves following stem rust infection. New Phytol 111: 413–421

Friend J (1976) Lignification in infected tissue: In: Friend J, Threlfall DR (eds) Biochemical aspects of plant-parasite relationships. Academic Press, New York, pp 291–303

Fry SC (1986) Crosslinking of matrix polymers in the growing cell wall of angiosperms. Annu Rev Plant Physiol 37: 165–186

Grand C (1984) Ferulic acid 5-hydroxylase: a new cytochrome P-450-dependent enzyme from higher plant microsomes involved in lignin synthesis. FEBS Lett 169: 7–11

Grand C, Boudet A, Boudet AM (1983) Isoenzymes of hydroxycinnamate:CoA ligase from poplar stems: properties and distribution. Planta 158: 225–229

Grand C, Sarni F, Lamb CJ (1987) Rapid induction by fungal elicitor of the synthesis of cinnamyl-alcohol dehydrogenase, a specific enzyme of lignin synthesis. Eur J Biochem 169: 73–77

Grima-Pettenati J, Chriqui D, Sarni-Manchado P, Prinsen E (1989) Stimulation of lignification in neoformed calli induced by *Agrobacterium rhizogenes* on bean hypocotyls. Plant Sci 61: 179–188

Grisebach H (1981) Lignins. In: Conn EE (ed) The biochemistry of plants, vol 7, secondary plant products. Academic Press, New York, pp 457–478

Gross GG (1985) Biosynthesis and metabolism of phenolic acids and monolignols. In: Higuchi T (ed) Biosynthesis and biodegradation of wood components. Academic Press, Orlando, pp 229–272

Gross GG, Stöckigt J, Mansell RL, Zenk MH (1973) Three novel enzymes involved in the reduction of ferulic acid to coniferyl-alcohol in higher plants: ferulate:CoA ligase, feruloyl-CoA reductase and coniferyl-alcohol oxidoreductase. FEBS Lett 31: 283–286

Hahlbrock K, Grisebach H (1979) Enzymic controls in the biosynthesis of lignin and flavonoids. Annu Rev Plant Physiol 30: 105–130

Hahlbrock K, Scheel D (1987) Biochemical responses of plants to pathogens. In: Chet I (ed) Innovative approaches to plant disease control. Wiley, New York, pp 229–254

Hahlbrock K, Scheel D (1989) Physiology and molecular biology of phenylpropanoid metabolism. Annu Rev Plant Physiol Plant Mol Biol 40: 347–369

Hammerschmidt R (1984) Rapid deposition of lignin in potato tuber tissue as a response to fungi non-pathogenic on potato. Physiol Plant Pathol 24: 33–42

Hammerschmidt R (1985) Determination of natural and wound-induced potato tuber suberin phenolics by thioglycolic acid derivatization and cupric oxide oxidation. Potato Res 28: 123–127

Hammerschmidt R, Kuć J (1982) Lignification as a mechanism for induced systemic resistance in cucumber. Physiol Plant Pathol 20: 61–71

Hammerschmidt R, Bonnen AM, Bergstrom GC, Baker K (1985) Association of epidermal lignification with nonhost resistance of cucurbits to fungi. Can J Bot 63: 2393–2398

Hargreaves JA, Keon JPR (1986) Cell wall modifications associated with resistance of cereals to fungal pathogens. In: Bailey JA (ed) Biology and molecular biology of plant–pathogen interactions. Springer, Berlin Heidelberg New York Tokyo, pp 133–140

Heller W, Kühnl T (1985) Elicitor induction of a microsomal 5-*O*-(4-coumaroyl) shikimate 3′-hydroxylase in parsley cell suspension cultures. Arch Biochem Biophys 241: 453–460

Hermann C, Legrand M, Geoffroy P, Fritig B (1987) Enzymatic synthesis of lignin: purification to homogeneity of the three *O*-methyltransferases of tobacco and production of specific antibodies. Arch Biochem Biophys 253: 367–376

Higuchi T (1985) Biosynthesis of lignin. In: Higuchi T (ed) Biosynthesis and biodegradation of wood components. Academic Press, Orlando, pp 141–161

Jaaska V (1984) NAD-dependent aromatic alcohol dehydrogenases in wheats (*Triticum* L.) and goatgrasses (*Aegilops* L.): evolutionary genetics. Theor. Appl Genet 67: 535–540

Jones DH (1984) Phenylalanine ammonia-lyase: regulation of its induction and its role in plant development. Phytochemistry 23: 1349–1359

Jörnvall H, Persson B, Jeffery J (1987) Characteristics of alcohol/polyol dehydrogenases. The zinc-containing long-chain alcohol dehydrogenases. Eur J Biochem 167: 195–201

Kauss H (1987) Some aspects of calcium-dependent regulation in plant metabolism. Annu Rev Plant Physiol 38: 47–72

Keen NT, Littlefield LJ (1979) The possible association of phytoalexins with resistance gene expression in flax to *Melampsora lini*. Physiol Plant Pathol 14: 265–280

Kerby K, Somerville S (1989) Enhancement of specific intercellular peroxidases following inoculation of barley with *Erysiphe graminis* f. sp. *hordei*. Physiol Mol Plant Pathol 35: 323–337

Kirk TK, Obst JR (1988) Lignin determination. In: Wood WA, Kellogg ST (eds) Biomass. Academic Press, San Diego, pp 87–100 (Methods in enzymology, vol 161B)

Kneusel RE, Matern U, Nicolay K (1989) Formation of *trans*-caffeoyl CoA from *trans*-4-coumaroyl-CoA by Zn^{2+}-dependent enzymes in cultured plant cells and its activation by an elicitor-induced pH shift. Arch Biochem Biophys 269: 455–462

Knobloch K-H, Hahlbrock K (1975) Isoenzymes of p-coumarate:CoA ligase from cell suspension cultures of *Glycine max*. Eur J Biochem 52: 311–320

Kogel G, Beissmann B, Reisener H-J, Kogel K-H (1988) A single glycoprotein from *Puccinia graminis* f. sp. *tritici* cell walls elicits the hypersensitive lignification response in wheat. Physiol Mol Plant Pathol 33: 173–185

Kolattukudy PE (1980) Biopolyester membranes of plants: cutin and suberin. Science 208: 990–1000

Kolattukudy PE (1985) Biosynthesis of cutin, suberin, and associated waxes. In: Higuchi T (ed) Biosynthesis and biodegradation of wood components. Academic Press, Orlando, pp 162–208

Kolattukudy PE, Soliday CL (1985) Effect of stress on the defensive barrier of plants. In: Key JL, Kosuge T (eds) Cellular and molecular biology of plant stress. AR Liss, New York, pp 381–400

Kühnl T, Koch U, Heller W, Wellmann E (1989) Elicitor-induced S-adenosyl-L-methionine: caffeoyl-CoA 3-*O*-methyltransferase from carrot cell suspension cultures. Plant Sci 60: 21–25

Kuroda H, Shimada M, Higuchi T (1981) Characterization of a lignin-specific *O*-methyltransferase in aspen wood. Phytochemistry 20: 2635–2639

Kurosaki F, Amin M, Nishi A (1986) Induction of phytoalexin production and accumulation of phenolic compounds in cultured carrot cells. Physiol Mol Plant Pathol 28: 359–370

Kutsuki H, Shimada M, Higuchi T (1982a) Regulatory role of cinnamyl-alcohol dehydrogenase in the formation of guaiacyl and syringyl lignins. Phytochemistry 21: 19–23

Kutsuki H, Shimada M, Higuchi T (1982b) Distribution and roles of p-hydroxycinnamate:CoA ligase in lignin biosynthesis. Phytochemistry 21: 267–271

Lagrimini L M, Rothstein S (1987) Tissue specificity of tobacco peroxidase isoenzymes and their induction by wounding and tobacco mosaic virus infection. Plant Physiol 84: 438–442

Lagrimini LM, Burkhardt W, Moyer M, Rothstein S (1987) Molecular cloning of complementary DNA encoding the lignin-forming peroxidase from tobacco: molecular analysis and tissue-specific expression. Proc Natl Acad Sci USA 84: 7542–7546

Lagrimini LM, Bradford S, Rothstein S (1990) Peroxidase-induced wilting in transgenic tobacco plants. Plant Cell 2: 7–18

Lamb CJ, Lawton MA, Dron M, Dixon RA (1989) Signal and transduction mechanisms for activation of plant defenses against microbial attack. Cell 56: 215–224

Lamed RJ, Zeikus JG (1981) Novel NADP-linked alcohol-aldehyde/ketone oxidoreductase in thermophilic ethanologenic bacteria. Biochem J 195: 183–190

Lance C, Rustin P (1984) The central role of malate in plant metabolism. Physiol Vég 22: 625–641

Legrand M, Fritig B, Hirth L (1978) O-diphenol O-methyltransferases of healthy and tobacco-mosaic-virus-infected hypersensitive tobacco. Planta 144: 101–108

Lewis NG, Yamamoto E (1990) Lignin: occurrence, biogenesis and biodegradation. Annu Rev Plant Physiol Plant Mol Biol 41: 455–496

Lewis NG, Yamamoto E, Wooten JB, Just G, Ohashi H, Towers GHN (1987) Monitoring biosynthesis of wheat cell-wall phenylpropanoids in situ. Science 237: 1344–1346

Lindner WA, Hoffmann C, Grisebach H (1988) Rapid elicitor-induced chemiluminiscence in soybean cell suspension cultures. Phytochemistry 27: 2501–2503

Lois R, Dietrich A, Hahlbrock K, Schulz W (1989) A phenylalanine ammonia-lyase gene from parsley: structure, regulation and identification of elicitor and light responsive cis-acting elements. EMBO J 8: 1641–1648

Long A, James P, Ward OP (1989) Aromatic aldehydes as substrates for yeast and yeast alcohol dehydrogenase. Biotechnol Bioengineer 33: 657–660

Lozoya E, Hoffmann H, Douglas C, Schulz W, Scheel D, Hahlbrock K (1988) Primary structures and catalytic properties of isoenzymes encoded by the two 4-coumarate:CoA ligase genes in parsley. Eur J Biochem 176: 661–667

Lüderitz T, Grisebach H (1981) Enzymic synthesis of lignin precursors. Comparison of cinnamoyl-CoA reductase and cinnamyl-alcohol: NADP + dehydrogenase from spruce (Picea abies L.) and soybean (Glycine max L.). Eur J Biochem 119: 115–124

McKelvey JR, Fioravanti CF (1984) Coupling of malic enzyme and NADPH:NAD trans-hydrogenase in the energetics of Hymenolepis diminuta (Cestoda). Comp Biochem Physiol 77B: 737–742

Mansell RL, Babbel GR, Zenk MH (1976) Multiple forms and specificity of coniferyl alcohol dehydrogenase from cambial regions of higher plants. Phytochemistry 15: 1849–1853

Massala R, Legrand M, Fritig B (1987) Comparative effects of two competitive inhibitors of phenylalanine ammonia-lyase on the hypersensitive resistance of tobacco to tobacco mosaic virus. Plant Physiol Biochem 25: 217–225

Matern U, Kneusel RE (1988) Phenolic compounds in plant disease resistance. Phytoparasitica 16: 153–170

Matton DP, Constabel P, Brisson N (1990) Alcohol dehydrogenase gene expression in potato following elicitor and stress treatment. Plant Mol Biol 14: 775–783

Maule AJ, Ride JP (1976) Ammonia-lyase and O-methyltransferase activities related to lignification in wheat leaves infected with Botrytis. Phytochemistry 15: 1661–1664

Maule AJ, Ride JP (1983) Cinnamate 4-hydroxylase and hydroxycinnamate:CoA ligase in wheat leaves infected with Botrytis cinerea. Phytochemistry 22: 1113–1116

Mazars C, Lafitte C, Marquet PY, Rossignol M, Auriol P (1990) Elicitor-like activity of the toxic glycoprotein isolated from Rhynchosporium secalis (Oud.) Davis culture filtrates. Plant Sci 69: 11–17

Moerschbacher B, Kogel K-H, Noll U, Reisener HJ (1986a) An elicitor of the hypersensitive lignification response in wheat leaves isolated from the rust fungus Puccinia graminis f. sp. tritici. I. Partial purification and characterization. Z Naturforsch 41c: 830–838

Moerschbacher B, Heck B, Kogel K-H, Obst O, Reisener HJ (1986b) An elicitor of the hypersensitive lignification response in wheat leaves isolated from the rust fungus *Puccinia gramini* f. sp. *tritici*. II. Induction of enzymes correlated with the biosynthesis of lignin. Z Naturforsch 41c: 839–844

Moerschbacher BM, Noll UM, Flott BE, Reisener H-J (1988) Lignin biosynthetic enzymes in stem rust infected resistant and susceptible near-isogenic wheat lines. Physiol Mol Plant Pathol 33: 33–46

Moerschbacher BM, Flott BE, Noll U, Reisener H-J (1989) On the specificity of an elicitor preparation from stem rust which induces lignification in wheat leaves. Plant Physiol Biochem 27: 305–314

Moerschbacher BM, Noll U, Gorrichon L, Reisener H-J (1990) Specific inhibition of lignification breaks hypersensitive resistance of wheat to stem rust. Plant Physiol 93: 465–470

Mohan R, Kolattukudy PE (1990) Differential activation of expression of a suberization-associated anionic peroxidase gene in near-isogenic resistant and susceptible tomato lines by elicitors of *Verticillium albo-atratum*. Plant Physiol 92: 276–280

Nimz H (1974) Das Lignin der Buche—Entwurf eines Konstitutionsschemas. Angew Chem 86: 336–344

Pakusch A-E, Kneusel RE, Matern U (1989) S-adenosyl-L-methionine: *trans*-caffeoyl-coenzyme A 3-*O*-methyltransferase from elicitor-treated parsley cell suspension cultures. Arch Biochem Biophys 271: 488–494

Pearce RB, Ride JP (1980) Specificity of induction of the lignification response in wounded wheat leaves. Physiol Plant Pathol 16: 197–204

Pearce RB, Ride JP (1982) Chitin and related compounds as elicitors of the lignification response in wounded wheat leaves. Physiol Plant Pathol 20: 119–123

Peretz M, Burstein Y (1989) Amino acid sequence of an alcohol dehydrogenase from the thermophilic bacterium *Thermoanaerobium brockii*. Biochemistry 28: 6549–6555

Pryke JA, ap Rees T (1977) The pentose phosphate pathway as a source of NADPH for lignin synthesis. Phytochemistry 16: 557–560

Ranjeva R, Boudet AM, Faggion R (1976) Phenolic metabolism in petunia tissues. IV. Properties of p-coumarate:coenzyme A ligase isoenzymes. Biochimie (Paris) 58: 1255–1262

Ride JP (1975) Lignification in wounded wheat leaves in response to fungi and its possible role in resistance. Physiol Plant Pathol 5: 125–134

Ride JP (1983) Cell walls and other structural barriers in defense. In: Callow JA (ed) Biochemical plant pathology. Wiley, New York, pp 215–236

Ride JP, Pearce RB (1979) Lignification and papilla formation at sites of attempted penetration of wheat leaves by non-pathogenic fungi. Physiol Plant Pathol 15: 79–92

Roberts E, Kutchan T, Kolattukudy PE (1988) Cloning and sequencing of cDNA for a highly anionic peroxidase from potato and the induction of its mRNA in suberizing potato tubers and tomato fruits. Plant Mol Biol 11: 15–26

Robertsen B (1986) Elicitors of the production of lignin-like compounds in cucumber hypocotyls. Physiol Mol Plant Pathol 28: 137–148

Robertsen B, Svalheim O (1990) The nature of lignin-like compounds in cucumber hypo-cotyls induced by α-1,4-linked oligogalacturonides. Physiol Plant 79: 512–518

Rothermel BA, Nelson T (1989) Primary structure of the maize NADP-dependent malic enzyme. J Biol Chem 264: 19587–19592

Sachs MM, Freeling M (1978) Selective synthesis of alcohol dehydrogenase during anaerobic treatment of maize. Mol Gen Genet 161: 111–115

Sarkanen KV, Hergert HL (1971) Lignins in the plant kingdom. Classification and distribution. In: Sarkanen KV, Ludwig S (eds) Lignins: occurrence, formation, structure and reactions. Wiley, New York, pp 43–94

Sarni F, Grand C, Boudet AM (1984) Purification and properties of cinnamoyl-CoA reductase and cinnamyl alcohol dehydrogenase from poplar stems (*Populus* × *euramericana*). Eur J Biochem 139: 259–265

Smart C, Amrhein N (1985) The influence of lignification on the development of vascular tissue in *Vigna radiata* L. Protoplasma 124: 87–95

Spencer PA, Towers GHN (1988) Specificity of signal compounds detected by *Agrobacterium tumefaciens*. Phytochemistry 27: 2781–2785

Stone BA (1989) Cell walls in plant-microorganism associations. Aust J Plant Physiol 16: 5–17

Tihanyi K, Talbot B, Brzezinski R, Thirion J-P (1989) Purification and characterization of alcohol dehydrogenase from soybean. Phytochemistry 28: 1335–1338

Vallee BL, Auld DS (1990) Zinc coordination, function, and structure of zinc enzymes and other proteins. Biochemistry 29: 5647–5659

Vance CP, Kirk TK, Sherwood RT (1980) Lignification as a mechanism of disease resistance. Annu Rev Phytopathol 18: 259–288

van Huystee RB (1987) Some molecular aspects of plant peroxidases biosynthetic studies. Annu Rev Plant Physiol 38: 205–219

Wagner FW, Burger AR, Vallee BE, (1983) Kinetic properties of human liver alcohol dehydrogenase: oxidation of alcohols by class I isozymes. Biochemistry 22: 1857–1863

Wallis PJ, Rhodes MJC (1977) Multiple forms of hydroxycinnamate:CoA ligase in etiolated pea seedlings. Phytochemistry 16: 1891–1894

Walter MH, Grima-Pettenati J, Grand C, Boudet AM, Lamb CJ (1988) Cinnamyl-alcohol dehydrogenase, a molecular marker specific for lignin synthesis: cDNA cloning and mRNA induction by fungal elicitor. Proc Natl Acad Sci USA 85: 5546–5550

Walter MH, Grima-Pettenati J, Grand C, Boudet AM, Lamb CJ (1990) Extensive sequence similarity of the bean "cinnamyl-alcohol dehydrogenase" to a maize malic enzyme. Plant Mol Biol 15: 525–526

Wedding RT (1989) Malic enzymes of higher plants. Characteristics, regulation and physiological function. Plant Physiol 90: 367–371

Wengenmayer H, Ebel J, Grisebach H (1976) Enzymic synthesis of lignin precursors. Purification and properties of a cinnamoyl-CoA:NADPH reductase from cell suspension cultures of soybean (*Glycine max*). Eur J Biochem 65: 529–536

White H, Strobel G, Feicht R, Simon H (1989) Carboxylic acid reductase: a new tungsten enzyme catalyses the reduction of non-activated carboxylic acids to aldehydes. Eur J Biochem 184: 89–96

Wyrambik D, Grisebach H (1975) Purification and properties of isoenzymes of cinnamyl-alcohol dehydrogenase from soybean cell suspension cultures. Eur J Biochem 59: 9–15

Wyrambik D, Grisebach H (1979) Enzymic synthesis of lignin precursors. Further studies on cinnamyl-alcohol dehydrogenase from soybean cell suspension cultures. Eur J Biochem 97: 503–509

Subject Index

O.W. Barnett (ed.)

Potyvirus Taxonomy

Archives of Virology / Supplementum 5

1992. 56 figs. Approx. 400 pages.
Soft cover DM 290,-, öS 2030,-
Reduced price for subscribers
to "Archives of Virology":
Soft cover DM 261,-, öS 1827,-
ISBN 3-211-82353-0

Prices are subject to change without notice

A number of economically important diseases are caused by potyviruses, the largest group of plant viruses. Many of these diseases are distributed world-wide. The development of effective control strategies against viruses is dependent on the availability of reliable methods of identification and detection.

To date this has not seemed possible for the potyvirus group, because of its size, complexity, and immense variation.

This book brings together the collaborative efforts of experts in the field. It summarizes characteristics of potyviruses which relate to their taxonomy and points to areas which require consideration before an international consensus can be reached.

Main topics dealt with in detail are: serological relationships, nucleic acid sequence information, biological properties, and specific problems with several virus subgroups or pairs of viruses.

Springer-Verlag Wien New York

Plant Gene Research

Basic Knowledge and Application

Editors: E.S. Dennis, B. Hohn, T. Hohn, P.J. King, F. Meins, J. Schell, D.P.S. Verma

The first volume
D.P.S. Verma, T. Hohn (eds.)
Genes Involved in Microbe-Plant Interactions

1984. 54 figs. XIV, 393 pages.
Cloth DM 169,-, öS 1180,-. ISBN 3-211-81789-1

Knowledge of gene transfer occurring in nature opens new perspectives for its future utilization in plant breeding. The first volume of the series *Plant Gene Research* provides an overview of the important aspects of plant-microbe interactions and the various methods of research.

The second volume
B. Hohn, E.S. Dennis (eds.)
Genetic Flux in Plants

1985. 40 figs. XII, 253 pages.
Cloth DM 109,-, öS 760,-. ISBN 3-211-81809-X

This volume gathers together for the first time the most recent information on plant genome instability. The plant genome can no longer be looked upon as a stable entity. Many examples of change and disorder in the genetic material have been reported recently. Chloroplast DNA sequences have been found in nuclei and mitochondria. Mitochondrial DNA molecules can switch between various forms by recombination processes. Stress on plants or on cells in culture can cause changes in chromosome organization. DNA can be inserted into the plant genome by transformation with the Ti plasmid of *Agrobacterium tumefaciens*, and transposable elements produce insertions and deletions.

The third volume
A.D. Blonstein, P. J. King (eds.)
A Genetic Approach to Plant Biochemistry

1986. 30 figs. XI, 291 pages.
Cloth DM 128,-, öS 896,-. ISBN 3-211-81912-6

This volume brings together for the first time some interesting examples of the contributions being made by genetics to the study of plant biochemistry, including some biochemical aspects of plant development. A wide range of topics is reviewed including plant hormones, photosynthesis, nitrogen metabolism, protein synthesis, and resistance to pathogens. Two chapters deal with new methods for isolating mutants at the plant level and in protoplast culture.

The fourth volume
T. Hohn, J. Schell (eds.)
Plant DNA Infectious Agents

1987. 76 figs. XIV, 348 pages.
Cloth DM 198,-, öS 1380,-. ISBN 3-211-81995-9

In the past few years rapid progress has been made transforming plant tissue by introducing foreign DNA. Methods make use either of viruses an soilbacteria, or involve technical manipulations of single cells followed by plant regeneration. It is now possible to spread certain genes systemically in a plant by rubbing hybrid virus nucleic acid onto a leaf and to transform its germline by infecting it with manipulated agrobacteria, thus bringing us closer to the prospect of developing new seed stocks with favourable properties such as past resistance and high nutritional value. This volume gives an account of these technologies, in addition providing basic knowledge on the strategies of natural cell invadors.

The fifth volume
D.P.S. Verma, R.B. Goldberg (eds.)
Temporal and Spatial Regulation of Plant Genes

1988. 55 figs. XIII, 344 pages.
Cloth DM 213,-, öS 1490,-. ISBN 3-211-82046-9

Genes expressed in different plant organs and processes are discussed with the emphasis on identifying various regulatory circuits controlling plant gene expression in a temporal and spatial manner. Regulation of foreign genes in transgenic plants is addressed with respect to a number of agriculturally important traits. This book illustrates the complexity of gene expression in plants and outlines strategies towards manipulating specific genes of interest.

The sixth volume
E.S. Dennis, D.J. Llewellyn (eds.)
Molecular Approaches to Crop Improvement

1991. 34 figs. IX, 166 pages.
Cloth DM 112,- , öS 784,-. ISBN 3-211-82230-5

Recent advances in the application of molecular biology techniques to crop improvement are reviewed. A range of agricultural and horticultural problems are approached including the genetic engineering of insect resistance, herbicide resistance and suppression of gene expression using anti-sense RNA. Major advances in cereal transformation, particularly rice are considered, and some of the uses of molecular biology in potato and other crop improvement are documented. The molecular biology of seed protein genes and self-incompatibility genes are examined and the potential benefits of this data in the engineering of protein quality or male sterility, respectively, are considered.

The seventh volume
R. Herrmann (ed.)
Cell Organelles

1992. 49 figs. Approx. 470 pages.
Cloth DM 229,-, öS 1600,-. ISBN 3-211-82264-X

The book presents a comprehensive overview of developments in the biology of plastids, mitochondria, glyoxisomes and peroxisomes. Written by the field's most accomplished authors, the topics of its 12 articles range from organelle biogenesis and heredity to evolution, with coverage of both molecular and classical aspects. It includes a modern treatment of genetic compartmentalization in the plant cell, which is one of the fundamental characteristics of eurkaryozes. The treatise serves as an introduction of pertinent literature, is an invaluable resource for research professionals and advanced graduate students in plant molecular and cell biology, and will prove useful to all botanists and teachers who demand a conceptual framework.

Prices are subject to change withouth notice

Springer-Verlag Wien New York